BRYOZOAN STUDIES 2022

Bryozoan Studies 2022 contains nineteen papers presented at the 19th International Conference of the International Bryozoology Association held at Trinity College Dublin in August 2022.

Bryozoans are complex and fascinating colonial organisms that range from Cambrian to the present day and which are found in marine and freshwater environments from pole to pole and subtidal to abyssal.

Recent tomographic techniques have revolutionised the study of modern and fossil taxa where internal structures are revealed through non-destructive methodologies. Here the internal structure of some Ordovician and Eocene taxa is illustrated through these methods. Phylogenetic studies of bryozoans question the classic classification of the group; here the phylogeny of species from California and Japan is described. Other topics covered are assessments of Recent faunas from Haiti and fossil assemblages from Mexico and the United States. Distributional patterns from the Arctic and New Zealand and the dispersal patterns of bryozoans on rafted pumice and on slipper lobsters are characterised. Further contributions provide descriptions of a rare Permian taxon, clarification of species assigned to the Upper Palaeozoic genus *Stenopora*, avicularia in *Wilbertopora*, enigmatic structures in a fenestrate bryozoan, repair structures in trepostomes, and an assessment of skeletonisation in the families of Palaeozoic orders.

PROCEEDINGS OF THE NINETEENTH INTERNATIONAL BRYOZOOLOGY ASSOCIATION CONFERENCE (DUBLIN, IRELAND, 22–26 AUGUST 2022)

Bryozoan Studies 2022

Edited by

Marcus M. Key, Jr.
Department of Geosciences, Dickinson College, Carlisle, USA

Joanne S. Porter
Heriot Watt University, Orkney Campus, Stromness, Orkney, Scotland

Patrick N. Wyse Jackson
Department of Geology, Trinity College, Dublin, Ireland

CRC Press
Taylor & Francis Group
Boca Raton London New York Leiden

CRC Press is an imprint of the
Taylor & Francis Group, an **informa** business
A BALKEMA BOOK

Cover illustration: Oeciophylloporina sp. from the early Silurian of Iran. Top, external 3D image; bottom, internal 2D slice producing a digital thin section through the colony (See Buttler *et al.*).

First published 2023
by CRC Press/Balkema
4 Park Square, Milton Park, Abingdon, Oxon, OX14 4RN

and by CRC Press/Balkema
2385 NW Executive Center Drive, Suite 320, Boca Raton FL 33431

CRC Press/Balkema is an imprint of the Taylor & Francis Group, an informa business

British Library Cataloguing-in-Publication Data
A catalogue record for this book is available from the British Library

Library of Congress Cataloging-in-Publication Data
A catalog record has been requested for this book

ISBN: 978-1-032-59487-3 (hbk)
ISBN: 978-1-032-59488-0 (pbk)
ISBN: 978-1-003-45491-5 (ebk)

DOI: 10.1201/9781003454915

Typeset in Times New Roman
by MPS Limited, Chennai, India

Bryozoan Studies 2022 – Key, Porter & Wyse Jackson (eds)
© 2023 The Editors, ISBN: 978-1-032-59487-3

Table of Contents

Dedication

Roger Cuffey in 2009, at one of his favourite places, Gettysburg, Pennsylvania, USA.
[Photograph by John H. Barnes]

This volume is dedicated with great affection and appreciation to Roger J. Cuffey (1939–2022). An assessment of Cretaceous bryozoans from Kansas by Paul Taylor and Roger is published in this volume.

Bryozoan Studies 2022 – Key, Porter & Wyse Jackson (eds)
© 2023 The Editors, ISBN: 978-1-032-59487-3

Foreword

This conference was scheduled to take place in Recife, Brazil thanks to a generous offer from Leandro Vieira. Unfortunately due to Covid restrictions, it was decided to move the conference to Houston, Texas, USA. Penny Morris stepped up and put together a proposal to host our conference there, but Covid restrictions were still too limiting. At the last minute, Patrick Wyse Jackson organized this conference in Dublin, Ireland thanks to the lifting of most Covid travel restrictions, and to the support provided by Trinity College Dublin and its School of Natural Sciences and Department of Geology.

The conference was held between 22nd and 26th August in the Museum Building of Trinity College Dublin, an iconic Victorian building in which many examples of Irish decorative stone are used as columns in the entrance and main hallways. Four days of presentations were held. On Monday afternoon, the delegates visited the Book of Kells and the Long Room in Trinity's Old Library. The manuscript is the most important illuminated gospel known. On Tuesday evening, delegates met in Patrick's house for beer, wine, quiche and salads. Wednesday was set aside as the traditional mid-conference free day. Rather than participate on a set touristic visit, delegates undertook a range of activities that suited themselves: these included visiting a small island with a monastic site in north County Dublin or walking a nineteenth century pier and enjoying ice cream at its seaward end. On Thursday, the conference dinner was held in an Indian restaurant in the city centre. During the meeting Abby Smith organised a sale of bryozoan-related arts and crafts which included some hats knitted by Caroline Buttler. Proceeds of this sale went to the IBA student award funds for future allocation.

In total 60 delegates drawn from 24 countries attended the meeting: 38 were in-person (63%) (Group photograph below) and 22 online (37%). Six accompanying members also attended. 54 presentations were delivered which comprised 46 talks (85%) and 8 posters (15%). 19 papers were submitted for publication in this volume.

14 students participated in the meeting which was just under a quarter of the attendance. IBA travel awards totalling close to NZ$6,500 were made to six students from five countries but unfortunately one was unable to travel at the last minute due to illness; his paper was presented by his co-author. This continued allocation of funds donated by IBA members to student travel awards is essential as it allowed these students to attend the meeting and to make highly valuable contributions and provided the opportunity for them to network with their peers and more experienced bryozoologists.

The conference logo was designed by Katie Wyse Jackson and featured an Irish harp in which the strings were represented by a lophophore.

The triennial IBA business meeting was presided over by Antonietta Rosso during which a number of deceased members were fondly remembered. Paul Taylor and Tim Wood were awarded Ellis Medals by the President, and Paul was reached by telephone in France during the meeting where the news was passed on to him. Later Tim emailed to express his gratitude. Student awards were made for PhD presentations to Lara Baptista, Julian Bibermair, Sebastian Decker and Sarah Leventhal and for MSc presentations to Sheena Stephens, Ismael Chowdhury, Jason Lopiccola and Tyler Feary.

Masato Hirose outlined details for the 20th triennial meeting which he proposed hosting in Japan, and his offer was enthusiastically accepted. Melissa Boonzaaier-Davids and Sebastian Decker on behalf of Thomas Schawa gave brief presentations on Capetown, South Africa and Roscoff, France respectively, potential venues for the 2028 meeting. Details will be distributed for members to consider.

Towards the end of the meeting a presentation was made to Patrick for hosting the meeting and Antonietta Rosso passed her presidential role to Caroline Buttler. The 19th triennial meeting then ended.

Without doubt the importance of this meeting was in bringing members of the IBA family together for both scientific discussion and socialising. Following Covid, this provided some normality to the IBA and set it up well for future activities.

We are most grateful to the following benefactors whose generosity allowed for the publication of this volume: Dickinson College's Research and Development Committee for a publication grant, and the Paleontological Society for a small conference support grant.

Patrick Wyse Jackson & Marcus M. Key, Jr.

Bryozoan Studies 2022 – Key, Porter & Wyse Jackson (eds)
© 2023 The Editors, ISBN: 978-1-032-59487-3

List of Participants

Delegates

Katerina Achilleos (Dunedin, New Zealand)
Lara Baptista (Azores, Portugal)
Julian Bibermair (Vienna, Austria)
Melissa Boonzaaier-Davids (Cape Town, South Africa)
Klaus Breitenback (Frankfurt, Germany)
Caroline Buttler (Cardiff, Wales, UK)
Ismael A. Chowdhury (Anaheim, CA, USA)
Sean Craig (Arcata, CA, USA)
Hans De Blauwe (Dudzele, Belgium)
Sebastian Decker (Vienna, Austria)
Nina Denisenko (St Petersburg, Russia)
Emanuela Di Martino (Oslo, Norway)
Tyler Feary (Dunedin, New Zealand)
Blanca Figuerola (Barcelona, Spain)
Paola Flórez (Santa Marta, Colombia)
Ernest Gilmour (Cheney, WA, USA)
Sergio González-Mora (Mexico City, Mexico)
Dennis Gordon (Wellington, New Zealand)
Eckart Håkansson (Crawley, WA, Australia)
Amy Joy Hess (Wallace, ID, USA)
Masato Hirose (Kanagawa, Japan)
Natasha Gray Hitchcock (Edgewater, MD, USA)
Mildred Johnson (Vienna, Austria)
Olga Kotenko (St Petersburg, Russia)
Marcus Key (Carlisle, PA, USA)
Chihiro Kubo (Kanagawa, Japan)
Hannah Lee (Elk Grove, CA, USA)
Sarah Leventhal (Boulder, CO, USA)
Lee Hsiang Liow (Oslo, Norway)
Jason Lopiccolo (Eureka, CA, USA)
Junye Ma (Nanjing, China)
Megan McCuller (Raleigh, NC, USA)
Hans Arne Nakrem (Oslo, Norway)
Andrew Ostrovsky (Vienna, Austria)

Anna Piwoni-Piórewicz (Gdynia, Poland)
Joanne Porter (Stromness, Orkney, Scotland, UK)
Arthur Porto (Ashland, OR, USA)
Mali Hamre Ramsfjell (Oslo, Norway)
Catherine Reid (Christchurch, New Zealand)
Antonietta Rosso (Catania, Italy)
Kate Runciman (Ontario, Canada)
John Ryland (Swansea, Wales, UK)
Ahmed Saadi (Vienna, Austria)
Maya Samuels-Fair (California, USA)
Carolann Schack (Oslo, Norway)
Thomas Schwaha (Vienna, Austria)
Natalia Shunatova (St Petersburg, Russia)
Abby Smith (Dunedin, New Zealand)
Mary Spencer Jones (London, England, UK)
Sheena Stephens (Eureka, CA, USA)
Juan Suárez Andres (Asturias, Spain)
Yuta Tamberg (Dunedin, New Zealand)
Paul Taylor (London, England, UK)
Leandro Vieira (Recife, Brazil)
Mark Wilson (Wooster, Ohio, USA)
Judith Winston (Hutchinson Island, FL, USA)
Emmy Woss (Vienna, Austria)
Patrick Wyse Jackson (Dublin, Ireland)
Vanessa Yepes Narváez (Santa Marta, Colombia)
Kamil Zagorsek (Liberec, Czech Republic)

Accompanying Members

Natalie Arroyo (Eureka, CA, USA)
Ann De Love (Dudzele, Belgium)
Venera Gilmour (Cheney, WA, USA)
Covadonga González (Asturias, Spain)
Veronika Licaverova (Liberec, Czech Republic)
Eva Lyngbakken (Oslo, Norway)

Bryozoan Studies 2022 – Key, Porter & Wyse Jackson (eds)

Applications for X-ray Tomography/Microscopy of Palaeozoic palaeostome bryozoans

Caroline J. Buttler*
Department of Natural Sciences, Amgueddfa Cymru – Museum Wales, Cardiff, Wales, UK
ORCID ID: 0000-0002-1577-926X

Ria L. Mitchell
Department of Materials Science and Engineering, The University of Sheffield, Sheffield, UK
ORCID ID: 0000-0002-6328-3998

Mark A. Wilson
Department of Earth Sciences, The College of Wooster, Wooster, Ohio, USA
ORCID ID: 0000-0002-4651-0589

Richard E. Johnston
Advanced Imaging of Materials (AIM) Core Facility, Faculty of Science and Engineering, Swansea University, Bay Campus, Swansea, UK
ORCID ID: 0000-0003-1977-6418

ABSTRACT: Non-destructive 3D imaging technology is used to determine if it can aid taxonomic identification of Palaeozoic palaeostome bryozoans. In previous studies it has not proved very successful with calcified specimens because the density contrast between the walls and zooecial infill is too weak to resolve the interior details. However, X-ray Micro Computed Tomography (X-ray μCT) and Microscopy (XRM) can be valuable in other ways and here we present two case studies. The interiors of extremely small, silicified Silurian fenestrate colonies have been visualised enabling us to create 'digital thin sections' at resolutions (voxel sizes) between 2.75 and 8.47 μm. The same technique can be used to gain a greater understanding of the borings common in Ordovician bryozoans, enabling construction of a 3D image of the structures produced.

1 INTRODUCTION

The development of non-destructive 3D X-ray imaging techniques has proved to be a highly effective method of examining the previously unknown interiors of palaeontological specimens (Rahman & Smith 2014; Raicicot 2017), and have been applied to both recent and fossil bryozoan material (Key & Wyse Jackson 2022). Here, two case studies are described to show different applications of X-ray Micro Computed Tomography (X-ray μCT) and Microscopy (XRM) of palaeostome bryozoans: the first using XRM showing how the identification of Silurian microbryozoans can be improved leading to greater knowledge about the preservation, and diversity and palaeoecology of Palaeozoic faunas, and the second using μCT examining borings in large hemispherical trepostome colonies to gain insights into their structure and morphology.

μCT and XRM vary in their set up and resolution and consequently in their applications to image fossil material. Standard laboratory-based μCT instruments typically have a relatively simple set up of an X-ray source and detector, where the sample stage moves in relation to both of these to achieve geometric magnification, affecting field of view, and voxel (3D pixel) size. Conversely, XRMs have a similar set up, but instead magnification is achieved through objective lenses mounted between the scintillator and the camera/detector, and the source and detector move in relation to the sample, affecting magnification, field of view, and voxel size (Merkle & Gelb 2013). In the case of the Zeiss Xradia Versa series XRMs used in this study the sample size typically needs to be small for geological samples, with the average μCT having larger detectors and higher X-ray penetrating

*Corresponding Author:
Caroline.Buttler@museumwales.ac.uk

DOI: 10.1201/9781003454915-1

capability allowing for larger samples to be scanned. This is why we have used a combination of both µCT and XRM in this study, as each instrument is suited better to each of the case studies.

2 HISTORY OF 3D X-RAY IMAGING OF FOSSIL BRYOZOANS

The examination of the exterior of palaeostome bryozoans does not typically reveal sufficient characters to identify specimens to below order and family level, and so the interiors must be studied (Wyse Jackson & Buttler 2015). However, the preservation state of the material, and how the fossil material is preserved (i.e. whether it is calcified, silicified, phosphatized, pyritized etc.), can affect the ability to establish taxonomic information. Identification of calcified palaeostome bryozoans to genus and species level is usually done using oriented thin sections of colony interiors. Since 1870, rock thin sections have been used routinely to identify the interior structure of bryozoans (Wyse Jackson & Buttler 2015). This technique is well established in the geological sciences and produces results that have longevity, but requires the colony to be sectioned destructively, often resulting in little left of the original specimen. This makes this method not ideally suited to scientifically valuable and rare museum specimens. Acetate peels (Boardman & Utgaard 1964; Nye *et al.* 1972; Wilson & Palmer 1989) have also been used which, although requiring sectioning of the specimen, retain more material. However, the resolution of acetate peels is not as good as thin sections, peels become brittle and degrade with time, the measurements taken from them are not as accurate, and the flexing and bending can affect morphology interpretation.

As an alternative to these destructive processes, 3D X-ray imaging techniques have more recently been used as a non-destructive method to examine the interiors of bryozoan colonies and to find them in rocks and artifacts (Key *et al.* 2023). This way, museum specimens can be scanned with no alteration of the specimen at all, and 'digital thin sections' can instead be used to investigate features inside the fossils from multiple angles. The 3D imaging data can also be kept forever as a digital representation of the museum specimen. Key & Wyse Jackson (2022) reviewed the history of 3D X-ray µCT imaging in bryozoans since it was first used in 2008. They noted the huge growth in its use, with 64% of the publications looking at fossil bryozoans. A number of these publications have described Palaeozoic palaeostome bryozoans where X-ray µCT has been used in conjunction with thin sections to visualize morphological features and aid identification e.g. through a multimodal workflow. Examples of this approach include Fedorov & Koromyslova (2019), Koromyslova & Fedorov (2021), Koromyslova *et al.* (2020) and Wyse Jackson & McKinney (2013).

However, there are some limitations with using µCT on palaeostome bryozoans. Ideally, there will be sufficient X-ray contrast in different parts of the bryozoan that will allow for digital segmentation and further study, showing up as different grayscales, or X-ray attenuations, in the collected data. This is usually from different elemental compositions, or from the presence of voids containing air. Palaeostome bryozoans commonly have both calcite colony walls and calcite infill in the zooecial chambers. This similar elemental material results in a lack of contrast between these different parts of the bryozoan colony (Buttler *et al.* 2012). This is problematic because it is not possible to identify the intrazooecial features such as diaphragms and cystiphragms, and structures within the zooecial walls including boundary type and acanthostyles, all essential for taxonomic identification. Despite this, X-ray imaging can produce an accurate reconstruction of fossil exteriors and provides information relating to how well preserved they are for potential further study.

3 CASE STUDY 1: SILICIFIED MICROBRYOZOANS

Taylor (2020), when discussing prospective future research in bryozoan taxonomy, indicated important gaps in our knowledge of fossil bryozoans. One particular area he highlighted were small Palaeozoic bryozoans which, in comparison to large colonies, have received little attention. The problem is the difficultly in examining the internal morphology. The size, often just a few millimetres in length, makes them difficult to thin section except at random angles. Silicified material produced from acid residues can have well preserved exteriors (Buttler 1994), but the specimens are very fragile and often lack internal features.

Fedorov & Koromyslova (2019) examined small colonies (around 2.5 mm in length) of *Revalotrypa* from the Ordovician of the Leningrad Oblast, Russia. They used combined X-ray CT and traditional thin sections. X-ray CT enabled visualisation of the whole colony showing some internal morphology, but where the zooecia were partly filled by calcite, thin sections had to be used due to the lack of contrast in the calcite walls and infill. Wyse Jackson & McKinney (2013) also used a combination of µCT and thin sections to investigate the internal morphology of the Mississippian fenestrate bryozoan *Polyfenestella.*

3.1 *Material*

In this study, we examined a fauna of silicified micro-bryozoans from the Early Silurian (Aeronian) of the Derenjal Mountains in central Iran which shows well preserved external features (Figure 1). The specimens were extracted by dissolving the

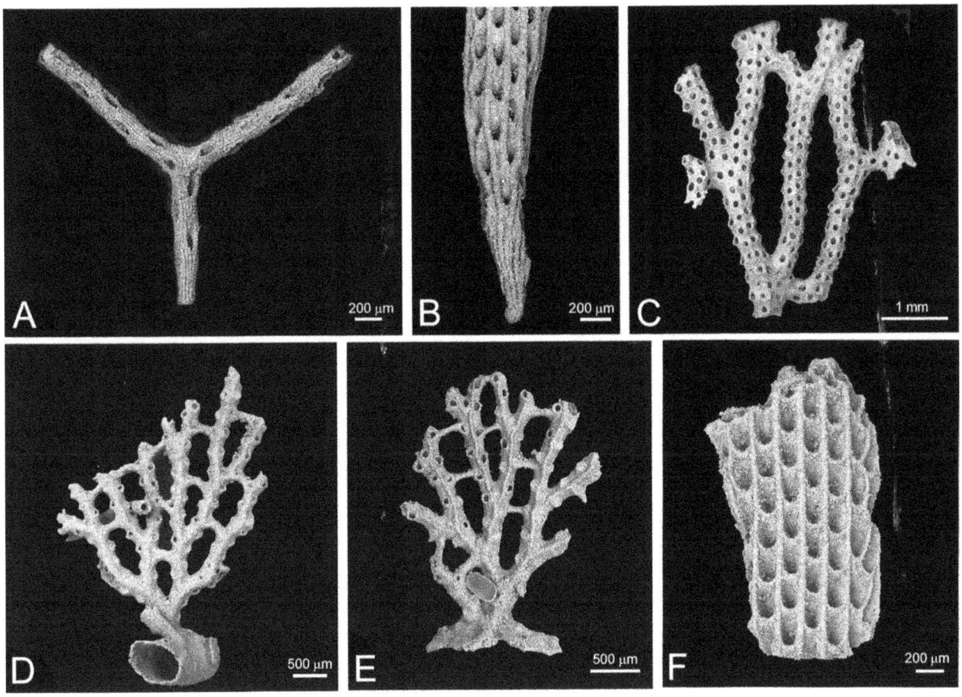

Figure 1. Silicified micro-bryozoans from the early Silurian (Aeronian) of central Iran. A, *Nematopora*, NMW 2022.7G.7. B, *Ulrichostylus*, NMW 2022.7G.28. C, *Oeciophylloporina*, NMW 2022.7G.24. D, *Fenestella* NMW 2022.7G.20. E. *Fenestella* NMW 2022.7G.4. F, *Ensipora* NMW 2022.7G.18. (NMW = Amgueddfa Cymru-National Museum Wales).

calcareous host rock in acid and sieving the residue which included the silicified colonies. The colonies comprised primarily of fenestrate and cryptostome bryozoans. The majority of the colonies measure under 5 mm in length and are extremely delicate so it was impossible to produce properly oriented thin sections without damaging the specimens. Identifications could not be made beyond generic level without internal structures. In our case, the specimens were examined with XRM to see if any internal morphology was preserved and, if so, if individual features could be digitally segmented to examine individual zooecial chamber shape.

3.2 *Method*

A ZEISS Xradia 520 Versa X-ray Microscope (XRM) was used to perform high resolution non-destructive 3D imagine of the silicified bryozoans. The XRM consists of a CCD detector system with scintillator-coupled visible light optics and a tungsten transmission target. Scans were collected at a tube voltage of 60 kV and power of 5 W, using both the 0.4x and 4x objectives, and 3201 projections were collected for each scan at exposure times between 2 and 8 seconds. Voxel (3D pixel) sizes ranged between and 2.75 and 8.47 µm, and scan times ranged between 2 and 10 hours. The tomograms were reconstructed from 2D projections using a ZEISS Microscopy commercial software package (XMReconstructor), and an automatically generated cone-beam reconstruction algorithm based on filtered back-projection. Scans were obtained within the Advanced Imaging of Materials (AIM) Core Facility at Swansea University, UK, and all correlation, segmentation and visualisation was achieved in ORS Dragonfly software.

3.3 *Results*

The results from the XRM revealed the preservation and structure of the intrazooidal walls of some the bryozoan colonies, and we were able to generate digital thin sections through all levels the colonies, which could aid identification (Figure 2).

In the study of fenestrate bryozoans, the shape of individual zooidal chambers are reconstructed from carefully oriented thin sections (e.g., Wyse Jackson *et al.* 2006). The scans obtained here allowed the digital segmentation of different sections of the colony, and it was possible to use segmentation to reveal the shape of the individual fenestrate zooidal chambers (Figure 3). This can provide an accurate reconstruction of the interior of the zooecial chambers. The individual interior

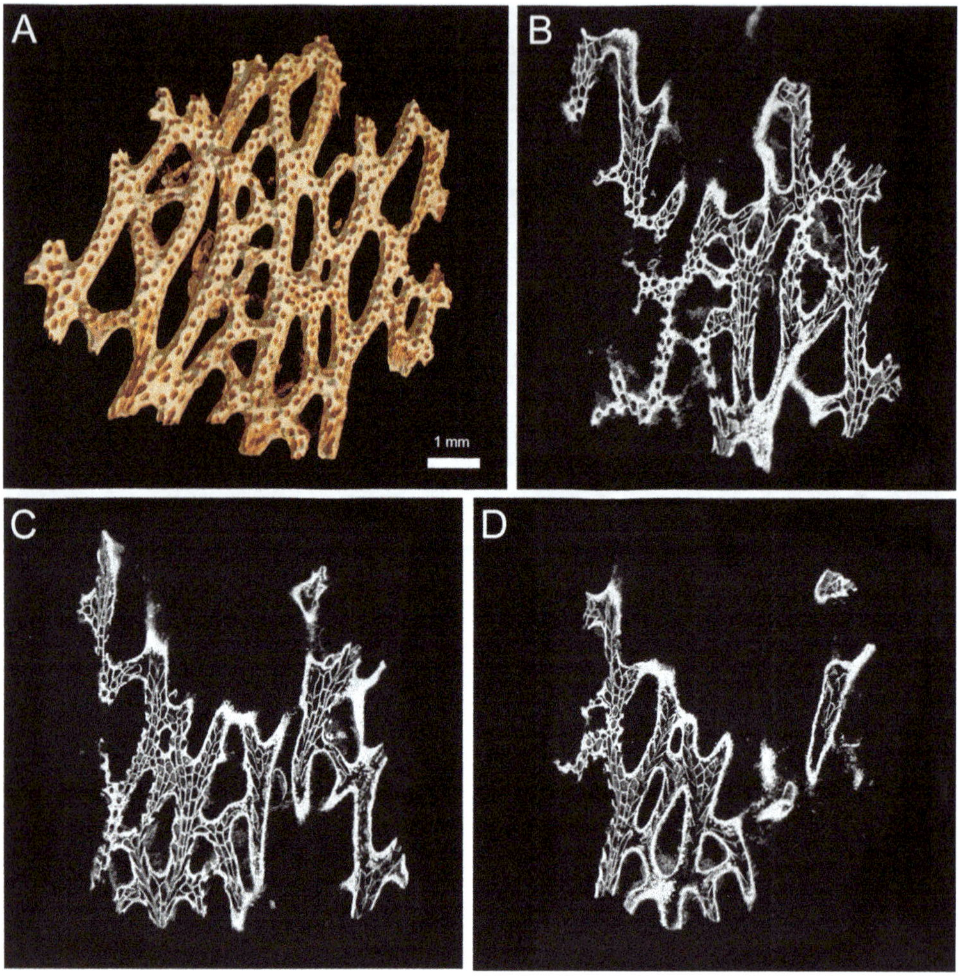

Figure 2. Fenestrate *Oeciophylloporina* bryozoan, NMW 2022.7G.1. A, External 3D image, B-D, Internal 2D slice producing digital thin section through colony. (NMW = Amgueddfa Cymru-National Museum Wales).

chambers of fenestrate (Figure 3A-C and cryptostome colonies (Figure 3D-F) could be imaged by digitally removing the colony walls.

Not all colonies had internal structures preserved due to poor preservation, and some had coarse silicification resulting in limited features being preserved internally (Figure 4).

One of the advantages of using XRM over conventional laboratory-based μCT is the straightforward ability to collect multi-scale volumes of regions of interest of the same sample. In this method, each targeted successive volume is at higher resolution compared to the original lower resolution scan, providing improvement in the amount of detail that can be seen. In our case, this was through using the Scout and Zoom method on the ZEISS Xradia Versa.

4 CASE STUDY 2: BORINGS IN ORDOVICIAN TREPOSTOME BRYOZOANS

4.1 *Material*

The thick calcitic skeletons of Ordovician trepostome bryozoans were prominent substrates for bioeroding organisms, especially worm-like excavators of tubular macroborings and bioeroding bivalves (Wilson 2007; Wyse Jackson & Key 2007; Vinn *et al.* 2019, 2022). Ordovician bryozoans, in fact, record a dramatic increase in boring intensity and diversity known as the Ordovician Bioerosion Revolution (Mángano *et al.* 2016; Wilson & Palmer 2006). Because Ordovician trepostomes are often large and hemispherical, they can have borings that

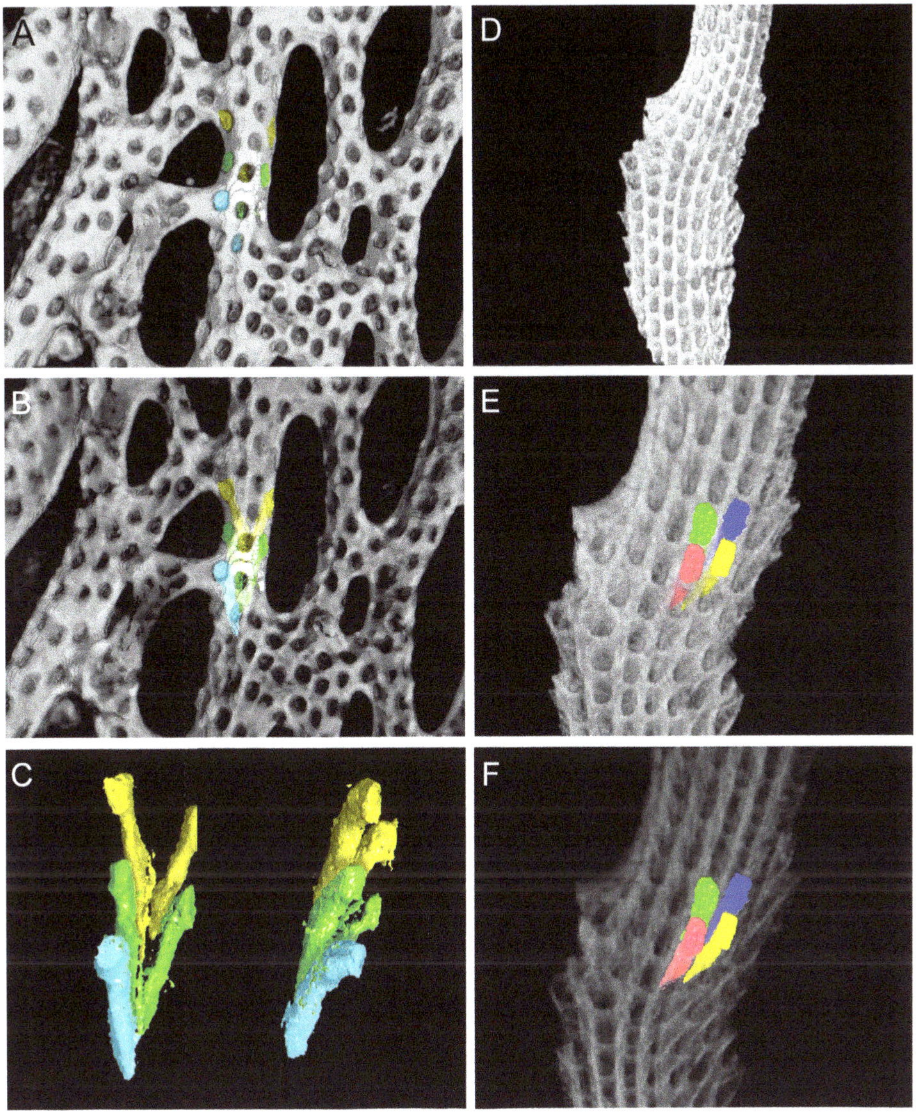

Figure 3. Segmentation used to reveal the shape of the individual zooidal chambers. A-C, through fenestrate *Oeciophylloporina* colony, NMW 2022.7G.1. D-F, cryptostome *Ptilodictya* colony, NMW 2022.7G.2. (NMW = Amgueddfa Cymru-National Museum Wales).

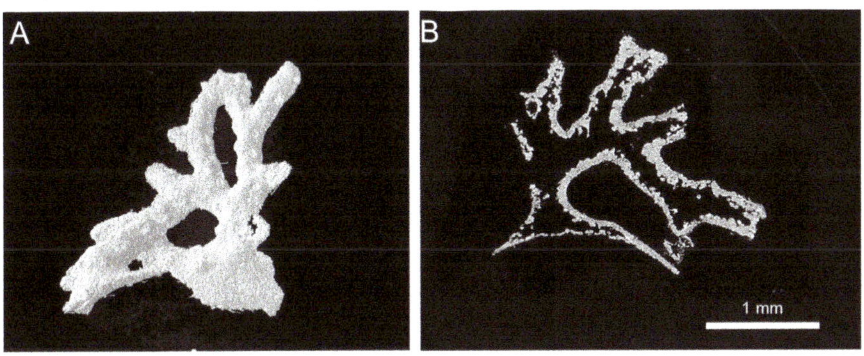

Figure 4. Silicified micro-bryozoan from the early silurian (Aeronian) of central Iran, showing coarse silicification, NMW 2022.7G.3 A., Exterior of colony with course silicification, B. Interior of colony with no internal features preserved. (NMW = Amgueddfa Cymru-National Museum Wales).

develop internally in three dimensions. Describing these three-dimensional borings is mostly done with 2D sections (Wyse Jackson & Key 2007), which provides only a limited view of the boring's morphology. μCT scanning produces much more realistic and useful renderings of boring morphology in trepostomes. For example, Heřmanová *et al.* (2020) discovered what are likely the earliest clionaid sponge borings in trepostomes from the Middle Ordovician (Darriwilian) of Russia. It would have been very difficult to recognize these borings with traditional sectioning. Here, we used μCT to scan a large hemispherical trepostome bryozoan from the Upper Ordovician (Sandbian) of Estonia. (Specimen C/W-231-1, Kohtla Quarry, location coordinates N59.35665°, E27.22343°; exposure now lost to mining activity.)

4.2 *Method*

The colonies were non-destructively scanned using a Nikon Metrology HMX ST 225 X-ray μCT scanner with a tungsten reflection target. This scanner offers the flexibility of having higher power and a larger flat panel detector, therefore being able to scan much larger samples compared with the Versa. Scan parameters include a tube voltage of 190 kV and a tube current of 190 μA, while a 0.5 mm tin filter was used to pre-harden the X-ray beam and remove unwanted lower energy X-rays. Average scan time was ~35 minutes, and voxel sizes ranged between 35.4 and 71.3 μm. Scans were reconstructed into 3D tomographic datasets as .tiff stacks using CT Pro Software (Nikon Metrology). Scans were obtained within the Advanced Imaging of Materials (AIM) Core Facility at Swansea University, UK, and all correlation, segmentation and visualisation was achieved in ORS Dragonfly software.

4.3 *Results*

The scans revealed extensive borings within the bryozoan colony. The borings appear on the surface of the bryozoan to be simple tubular *Trypanites* or *Palaeosabella*, but the scanning revealed that these borings are much larger and complex internally. The scans show that the borings expand internally and intersect, producing significant internal cavities and branching. These three-dimensional renderings challenge our traditional ichnological classification of these borings. Segmentation is employed to make the colony digitally transparent (Figure 5).

5 DISCUSSION

Non-destructive 3D X-ray imaging will not replace thin sections unless it becomes possible to detect a difference in the biogenic calcite in the zooecial walls and chamber infill.

However, there are valuable applications for X-ray imaging as shown here. The interiors of delicate silicified microbryozoans can retain structures which when identified can aid taxonomy. Using X-ray imaging to undertake digital segmentation enables the visualization of elements within colonies such as zooidal chamber shape. It

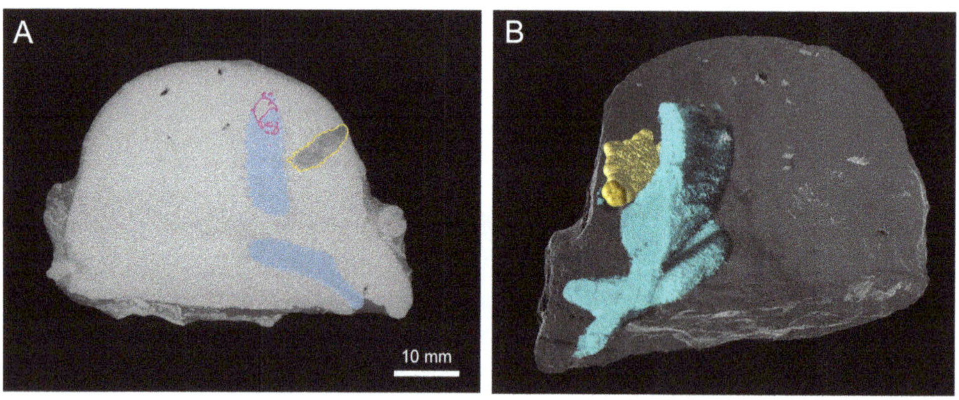

Figure 5. Hemispherical trepostome bryozoan from the Upper Ordovician (Sandbian) of Estonia, C/W-231-1, A, borings identified in transverse section, B, segmentation of borings with the rest of colony made digitally transparent. (C/W = College of Wooster).

also has great potential for resolving complex 3D boring structures within bryozoan colonies.

In terms of instrument set up for scanning, there can be variations that have an effect on the resulting data. For example, choosing the correct resolution and/or voxel size is important to make sure we are collecting as much useful information as possible. This is usually dictated by the size of the features of interest, but sometimes this is not known until an initial low-resolution scan is collected. Scanning at very high resolutions can result in observing much smaller fields of view than is useful, so a combined approach of multi-scale is often required. Further, as explained above, there is an issue with contrast. This could be remedied by undertaking approaches such as propagation phase contrast imaging, which show up subtle differences in greyscale and phases within the sample by moving the source and detector to specific ratios of distance to each other and the sample. Limitations are that phase contrast scanning generally takes longer than conventional absorption-based μCT, and the resulting data produces a bright fringe near the interface between phases – useful for identifying phase boundaries but means greyscale-based segmentation has to be used with caution. Another approach could be to use dual energy μCT. In this study we have used two different instruments for the collection of 3D data for the two case studies selecting based primarily on the size of specimen and contrast required.

ACKNOWLEDGEMENTS

The imaging work was supported by the Advanced Imaging of Materials (AIM) Core Facility (EPSRC Grant No. EP/M028267/1), the European Social Fund (ESF) through the European Union's Convergence programme administered by the Welsh Government (80708), and the Welsh Government Enhancing Competitiveness Grant (MA/KW/5554/19). Leonid Popov and Mansoureh Gobadi Pour collected the silicified micro-bryozoans. We would like to thank Marcus Key and Thomas Schwaha for helpful comments that improved the manuscript.

REFERENCES

Boardman, R.S. & Utgaard, J. 1964. Modifications of Study Methods for Paleozoic Bryozoa. *Journal of Paleontology* 38: 768–770.

Buttler, C.J. 1994. Ordovician (Ashgill) Microbryozoans from the Kildare Limestone, Ireland. In J. Ryland, P.J. Hayward & Taylor, P.D. (eds), *Biology and Palaeobiology of Bryozoans*: 37–40. Fredsenburg: Olsen & Olsen.

Buttler, C.J., Rahman, I.A. & Slater, B. 2012. Three-Dimensional Computer Reconstruction of a Lower Palaeozoic Trepostome Bryozoan. *Palaeontological Association Newsletter, Abstracts Annual Meeting* 81: 62.

Fedorov, P.V. & Koromyslova, A.V. 2019. New Findings of the Genus *Revalotrypa*, the Oldest Bryozoan Genus of Baltoscandia, in North-western Russia. *Carnets Geol* 19 (11): 199–209.

Heřmanová, Z., Bruthansová, J., Holcová, K., Mikuláš, R., Kočová Veselská, M., Kočí, T., Dudák, J. & Vohník, M. 2020. Benefits and Limits of X-ray Micro-computed Tomography for Visualization of Colonization and Bioerosion of Shelled Organisms. *Palaeontologica Electronica* 23 (2): a23

Key, M.M., Jr. & Wyse Jackson, P.N. 2022. History of Micro-computed Tomographic Three-dimensional Imaging in Bryozoology. In P.N. Wyse Jackson & M.E. Spencer Jones (eds), *Annals of Bryozoology 7: Aspects of the history of Research on Bryozoans*: 57–72. Dublin: International Bryozoology Association.

Key, M.M., Jr., Shaw, J. & Ward, I. 2023. Three-dimensional Imaging of Fossil Cheilostome Bryozoans in Eocene chert by Synchrotron Radiation Micro-Computed Tomography. In M.M. Key, Jr., J. S. Porter and P.N. Wyse Jackson (eds), *Bryozoan Studies 2022*: 55–63. Rotterdam: Balkema.

Koromyslova, A.V. & Fedorov, P.V. 2021. The Oldest Bifoliate Cystoporate and Two Other Bryozoan Taxa from the Dapingian (Middle Ordovician) of North-western Russia. *Journal of Paleontology* 95: 24–39.

Koromyslova, A.V., Fedorov, P.V. & Tolokonnikova, Z. A. 2020. Results of the Study of Ordovician and Permian Bryozoans by Using X-ray Micro-CT. *Proceedings of the Paleontological Society* 3: 23–31.

Mángano, G., Buatois, L., Wilson, M.A. & Droser, M. 2016, The Great Ordovician Biodiversification Event. In G. Mángano & L. Buatois (eds), *The Trace-Fossil Record of Major Evolutionary Events: Topics in Geobiology*: 127–156. Dordrecht: Springer.

Merkle, A. & Gelb, J. 2013. The Ascent of 3D X-ray Microscopy in the Laboratory. *Microscopy Today* 21: 10–15.

Nye, O.B., Jr., Dean, D.A. & Hinds, R.W. 1972. Improved Thin Section Techniques for Fossil and Recent Organisms. *Journal of Paleontology* 46: 271–275.

Rahman, I.A. & Smith, S.Y. 2014. Virtual Paleontology: Computer-aided Analysis of Fossil Form and Function. *Journal of Paleontology* 88: 633–635

Raicicot, R. 2017. Fossil Secrets Revealed: X-Ray CT Scanning and Applications in Paleontology. *The Paleontological Society Papers* 22: 21–38

Taylor, P.D. 2020. *Bryozoan Palaeobiology*. Hoboken, NJ: Wiley-Blackwell

Vinn, O., Wilson, M.A. & Toom, U. 2019. Earliest *Petroxestes* Borings from Sandbian (earliest late Ordovician) Bryozoans of Northern Estonia. *Palaios* 34: 453–457.

Vinn, O., Ernst, A., Wilson, M.A. & Toom, U. 2022. Borings and Bioclaustrations in Bryozoans from the Kunda Regional Stage (Darriwilian; Middle Ordovician) of northern Estonia and NW Russia.

Neues Jahrbuch für Geologie und Paläontologie 303: 219–225.
Wilson, M.A. 2007. Macroborings and the Evolution of Marine Bioerosion. In W. Miller (ed.), *Trace Fossils: Concepts, Problems, Prospects*: 356–367. Amsterdam: Elsevier.
Wilson, M.A. & Palmer, T.J. 1989. Preparation of Acetate Peels. In R.M. Feldmann, R.E. Chapman & J.T. Hannibal (eds), *Paleotechniques*. The Paleontological Society Special Publication 4: 142–145.
Wilson, M.A. & Palmer, T.J. 2006. Patterns and Processes in the Ordovician Bioerosion Revolution. *Ichnos*: 13, 109–112.
Wyse Jackson, P.N. & Buttler, C.J. 2015. Part G, Revised, Volume 2, Chapter 3: Preparation, Imaging, and Conservation of Paleozoic Bryozoans for Study. *Treatise Online* 63: 1–15.
Wyse Jackson, P.N. & Key, M.M., Jr. 2007. Borings in Trepostome Bryozoans from the Ordovician of Estonia: two Ichnogenera Produced by a Single Maker, a Case of Host Morphology Control. *Lethaia* 40: 237–252.
Wyse Jackson, P.N. & McKinney, F.K. 2013. A Micro-computed Tomography and Scanning Electron Microscopy Investigation of the Structure of *Polyfenestella* Bancroft, 1986 (Bryozoa, Fenestrata), from the Mississippian of Scotland: Revealing the Nature of its Heteromorphs. *Irish Journal of Earth Sciences* 31: 19–24.
Wyse Jackson, P.N., McKinney, F.K. & Bancroft, A. J. 2006. Fenestrate Bryozoan Genera Based on Species from Ireland Originally Described by Frederick M'Coy in 1844. *Palaeontology* 49: 741–767.

Bryozoan Studies 2022 – Key, Porter & Wyse Jackson (eds)

Species richness and diversity patterns of bryozoans of the Arctic region: Spatial variation of bryozoan diversity

Nina V. Denisenko*
Zoological Institute of the RussianAcademy of Sciences, Saint Petersburg, Russia
ORCID ID: 0000-0003-0284-5026

ABSTRACT: Using the data collected at 3133 stations in ten areas of the Arctic region during 1980 – 2000s, an examination of large-scale spatial distribution of bryozoan species richness was made. Detailed station grids allowed estimation of the relationship between bryozoan species richness and geographic factors: diversity always decreases with depth, but along latitude and longitude gradients, it changes ambiguously and is not always statistically significant.

1 INTRODUCTION

Study of patterns of spatial distribution of animals and plants in the World is one of the most significant aspects of biology. Knowledge of regularity of distribution allows a reconstruction of the formation of the modern fauna and flora and to predict possible future changes in case of terminal changes in environments. A decline in wildlife diversity from the tropics toward temperate latitudes has been generally accepted for the terrestrial fauna since the 1960s (McArthur & Wilson 1967; Pianka 1966). For the marine basins, the studies concerned with the assessment of variation in faunal diversity along the geographic gradient have also been conducted for over half a century (Sanders 1968; Stehli 1968; Thorson 1957), but, while the latitudinal decline in species richness for the tropical or temperate regions of the oceans has already been confirmed (Chaudhary *et al.* 2017; Clarke & Lidgard 2001; Mokievsky & Azovsky 2002; Rex *et al.* 1993; Roy *et al.* 2000), this pattern is not always evident in the Arctic region (Gray 1997, 2001; Kendal & Ashan 1993; Thorson 1957; Warwick & Rusvahuni 1987). Some authors stress that even if material was collected from the same expeditions, different taxonomic groups do not yield similar results that could be taken as evidence of a statistically significant relationship between species richness and latitude. In some groups, for instance, in isopods this relationship is weakly expressed, while in other taxa such as amphipods and fish it is statistically significant (Virnnstein *et al.* 1984). In bryozoans, studies of this kind have been conducted only for the fauna of the tropical and temperate latitudes of the Atlantic Ocean (Clarke & Lidgard 2001). The results of this analysis established a statistically significant relationship between the decline in bryozoan species richness and increasing latitude. Similar researches for the bryozoan fauna of the Arctic region only recently became possible, after the material collected during numerous surveys of the 1980–2010s had been processed. The registered check-list of bryozoans is present in Denisenko (2022a).

The aim of this study was to investigate spatial changes in bryozoan diversity in the Arctic subregions.

2 STUDY AREA AND BACKGROUND

The defined Arctic region boundaries were based on a scheme proposed by the Hydrography Service of Russia (Gorshkov 1980) and were coordinated with the scheme established by the Conservation of the Arctic Flora and Fauna working group (CFFF) of the Arctic Council (Vongraven *et al.* 2009). The inconsistency of the schemes lies in the inclusion of the Faroe Archipelago area and the waters around the whole of Iceland, and the Norwegian shelf in the Arctic region, on one hand, and, from the other, in the establishment of a border in the Pacific sector corresponding to the Bering Strait, separating the Chukchi and Bering Seas, and excluding the latter basin from the Arctic.

The study area comprises shelf and upper slope of ten areas and seas of the Arctic Region (Table 1; Figure 1) with a depth range between intertidal and 800 m.

*Corresponding Author: ndenisenko@zin.ru

DOI: 10.1201/9781003454915-2

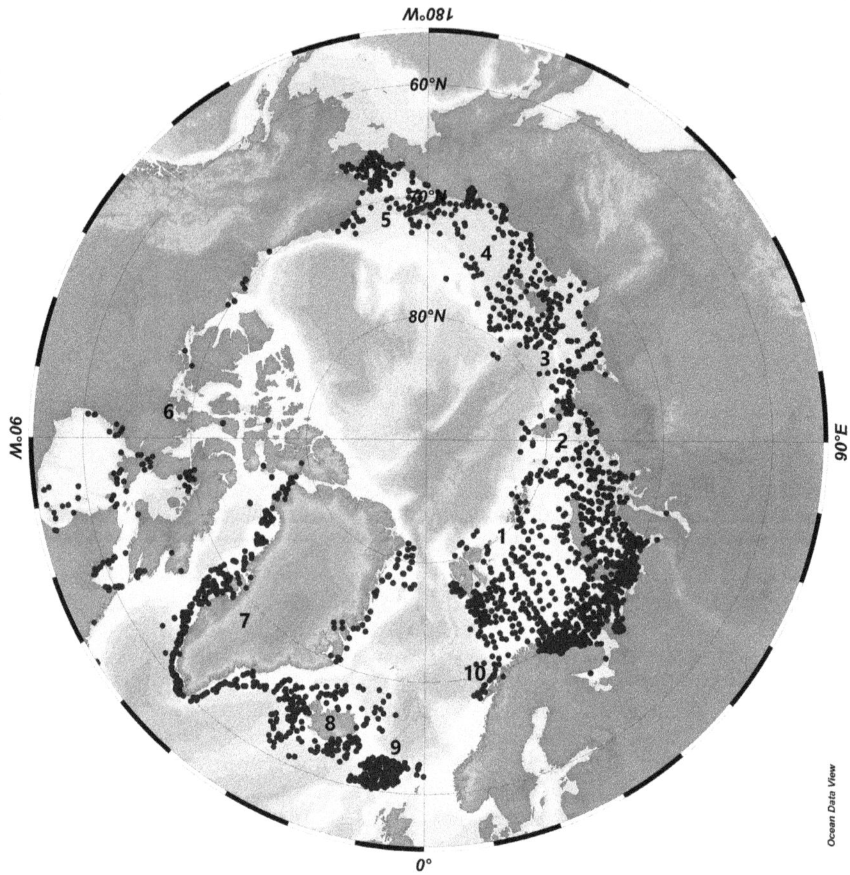

Figure 1. The Arctic region study areas and sampled stations within them. 1, Barents Sea; 2, Kara Sea; 3, Laptev Sea; 4, East Siberian Sea; 5, Chukchi Sea; 6, Canadian Arctic area; 7, Greenland area; 8, Iceland area; 9, Faroe Archipelago area; 10, Norwegian shelf.

3 MATERIAL AND METHODS

3.1 *Data sources*

The present study combined the results of analysis of the taxonomic catalogues of the Zoological Institute of the Russian Academy of Sciences, Polar branch of the Russian Federal Research Institute of Fisheries and Oceanography (PINRO) and Murmansk Marine Biological Institute of the Kola Branch of the RAS. The results of treatment of the recent bryozoan samples collected during the last three decades were also used. The sampling area encompassed the Eurasian Arctic seas (Norwegian shelf, Barents, Kara, Laptev, East-Siberian, and Chukchi Seas), Iceland, Greenland and the Faroe Islands areas. The newly obtained and archive data were added by the information from literature (Andersson 1901; Gontar & Denisenko 1989; Kluge 1962, 2009; Nordgaard 1916; Osburn 1919, 1936, 1955; Smitt 1868 ect.). In total, bryozoan samples were examined from more than 3100 stations (Table 1, Figure 1), and more than 20000 taxonomic identifications of bryozoan colonies were performed and subjected to the analysis.

3.2 *Data analysis*

The relationship between the diversity of bryozoan fauna and its change along longitude and latitude was assessed at the local (α-diversity) and sub-regional (γ-diversity) levels using the classification of diversity gradation suggested by Whittaker (1972). Iceland and Faroe Island faunas were excluded from the analysis of their changes along latitude-longitude range due to their narrow gradients within the areas. Analysis of relationships of the faunal richness and depth was carried out for all sub-regions involved in the study.

Pearson correlations (R) calculated by applying Software SyStat (2008; v. 8.0) were used to establish a level of the relationships between bryozoan species numbers and regarded parameters in all ten areas of the Arctic region. Significance of these

Table 1. Volume of analyzed material on the bryozoan fauna of the Arctic.

Sub-regions	Number of sampled stations
Barents Sea	1128
Kara Sea	401
Laptev Sea	177
East Siberian Sea	174
Chukchi Sea	151
Canadian Arctic	129
Greenland waters	314
Icelandic waters	295
Faroes area	364
Total	3133

Table 2. Statistic parameters of relationship of local ($\alpha-$) diversity of bryozoans along latitude gradient in different sub-regions of the Arctic.

Area	R	p-level
Canadian Arctic	−0.08	0.12
Greenland waters	0.026	0.09
Barents Sea	0.05	0.06
Kara Sea	0.013	0.14
Laptev Sea	0.05	0.15
East Siberian Sea	−0.36	0.13
Chikchi Sea	0.05	0.1

relationships was estimated by the null hypothesis values (p-level) which initially suggests the lack of relationship or correlation between studied variables.

4 RESULTS

The local or α-diversity of bryozoan species within the seas of the Arctic region, in our case the diversity of species in the sample, which reflects the diversity of habitats, varies considerably. The number of species in samples varied from 1 to 120, but at most collection points the number of species did not exceed five. The species representation in samples remained within specified limits regardless of the sampling gear. The Barents Sea and the Chukchi Sea showed the greatest range in the number of bryozoan species in the samples from 1 to 120 and from 1 to 89 respectively. Statistical analysis of variation in the number of species along the latitude gradient did not reveal any significant relationship between them (Table 2).

Analysis of species richness along the latitude gradient on sub-regional level was also carried out using the information including both previously published and newly collected data. In the Barents and in the Laptev Seas, the species richness of the bryozoans increased toward the north (Figure 2A, C). The correlations between changes between these characteristics at the sub-regional scale was moderate, and statistically insignificant (R = 0.45; $p = 0.12$ and R = 0.53; $p = 0.17$ respectively) due to high variance. In contrast to the listed Seas, the results of the analysis indicate a moderate, close to statistically significant, decrease of the bryozoan species richness toward north in the Kara Sea (R = −0.50; $p = 0.06$) (Figure 2B). In the East Siberian Sea, correlation between these parameters was very weak but this relationship was positive: species number increased with latitude (R = 0.34, $p = 0.31$) (Figure 2D). An insignificant increase in species richness within latitude was also observed in Greenland waters (R = 0.348, $p = 0.11$) (Figure 2E). No correlation between bryozoan species richness and latitude was observed in the Chukchi Sea (R = 0.06, $p = 0.16$) and in the Canadian Arctic (R = 0.05, $p = 0.83$).

No or very weak correlation was found between α-diversity and longitude (Table 3). But at sub-regional level, a relationship was statistically significant (Table 3).

Analysis of the variations in the species richness of the bryozoans with depth showed a statistically significant relationship between them in most of the defined sub-regions (Table 4). In the Barents Sea, this relationship was close to statistical significance, but in shallow shelf seas such as East-Siberian and Chukchi, the decrease of species number with depth can be considered as a trend only, because, despite high R, the correlation is not significant ($p > 0.05$) (Table 4).

5 DISCUSSION

The success of the analysis of variation in species richness along the longitudinal and latitudinal gradients is heavily reliant on the quality of the material under comparison, which, in turn, depends primarily on the collection tools and the evenness of distribution of sampling stations (Gray 1997, 2000). The data we have used for comparison had been collected during the expeditions over the continental shelf and the upper bathyal zone with relatively stable environmental conditions. Overall, compared data satisfied the requirements of Gray (2001, 2002) as sampling was made with similar gear and using regularly spaced station grids. The feasibility of studying bryozoan species richness along the longitudinal-latitudinal gradient is further strengthened by compliance of this study with the requirement proposed by Hillebrand (2004), according to which the taxon under study must belong to a

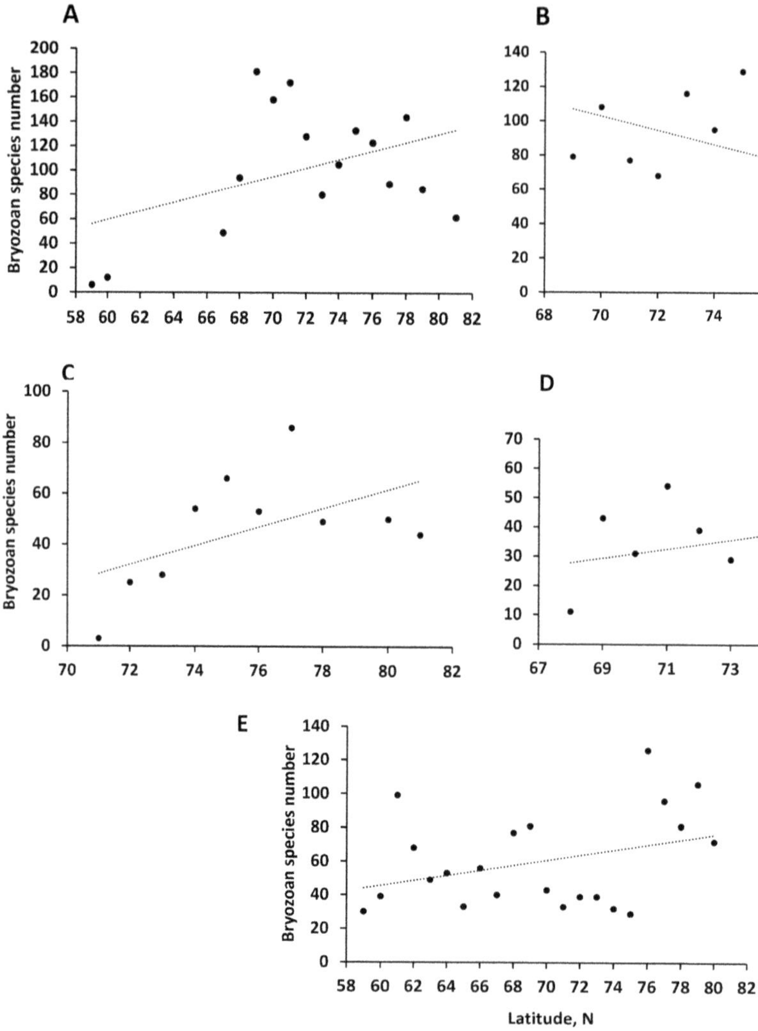

Figure 2. Variation of bryozoan species richness along latitude gradient in the Arctic subregions. A, Barents Sea; B, Kara Sea; C, Laptev Sea; D, East Siberian Sea; E, Greenland waters.

single ecological group. Since all these conditions have been met, the results of the study are likely to reflect the actual variation in species richness in the seas surveyed along the longitudinal and latitudinal gradients in different areas of the Arctic region.

The assessment of variation in bryozoan species richness along the longitudinal and latitudinal gradients performed at the local level (α-diversity) has shown that this variation is not related to either latitude or longitude. Bryozoan species richness has also been shown to exhibit no distinct patterns of variation within the local habitats, which indicates an aggregation of bryozoan settlements caused by the mosaic character of the environment, even within a single geographic degree. These results are in good agreement with the conclusions made earlier for the bryozoan fauna of the tropical and temperate zones of the Atlantic Ocean (Clarke & Lidgard 2001).

At the subregional level (γ-diversity), variation in bryozoan species richness with latitude showed more pronounced trends, but they had different directions. A decline in abundance of the bryozoan fauna with an increasing latitude has been found only in the Kara Sea, which indicates a decrease in species richness of this group in typically Arctic sub-region. In some other sub-regions, in contrast, the diversity increased with an increasing latitude. These regions included Barents, Laptev and East Siberian seas, as well as the waters around Greenland, where environmental heterogeneity was considerable. However, the analysis has

Table 3. Statistic parameters of relationship of local ($\alpha-$) and regional ($\gamma-$) diversity of bryozoan fauna along longitude in different sub-regions of the Arctic region. Significant values are in bold.

	$\alpha-$diversity		$\gamma-$diversity	
Area	R	$p-$level	R	$p-$level
Canadian Arctic	−0.1	0.25	**−0.62**	**0.049**
Barents Sea	−0.03	0.93	**−0.59**	**0.05**
Kara Sea	0.05	0.34	**−0.75**	**0.008**
Laptev Sea	−0.05	0.34	−0.09	0.79
East Siberian Sea	−0.21	**0.009**	0.042	0.92
Chukchi Sea	0.38	**<0.0001**	**0.52**	**0.01**

Table 4. Statistic parameters of the relationship between bryozoan species richness and depth on sub-regional level in the Arctic region. Significant values are in bold.

Area	Pearson correlation co-efficient (R)	p-level
Canadian Arctic	−0.81	**0.025**
Greenland waters	−0.96	**0.005**
Icelandic waters	−0.76	**0.006**
Faroese waters	−0.98	**0.001**
Barents Sea	−0.69	0.06
Kara Sea	−0.86	**0.001**
Laptev Sea	−0.58	**0.01**
East Siberian Sea	−0.84	0.16
Chukchi Sea	−0.51	0.16

revealed only trends toward an increase or decrease in faunal diversity, rather than statistically significant correlations. Moreover, these trends were only partially consistent with the trends of latitudinal variation in bryozoan species richness observed in the tropical and temperate waters of North Atlantic (Clarke & Lidgard 2001).

The results derived from data aggregation proved, in some cases, to be less reliable or even contradictory to the patterns previously observed for this taxonomic group. For instance, the pooled data for Greenland (combining that from East and West Greenland) shown that bryozoan species richness increases toward the North Pole, while the data analysis made for each of these two areas separately gave the opposite results (Denisenko & Blicher 2021). The observed discrepancy can be caused by differences in species composition between these two areas and by significant environmental differences between the western and eastern coasts of Greenland (Puerta *et al.* 2020).

A significant environmental heterogeneity also occurs in the Barents Sea, where depth differences of hundreds of meters occur within the same latitude in the open regions of the sea, and where shallow stretches alternate with deep depressions of the seafloor (Gorshkov 1980). This changes the velocity and warm capacity of Atlantic currents, which results in significant differences in temperature of near-bottom waters (Tantsyura 1973). The south-eastern part of the sea is heavily influenced by river run-off. The run-off causes poorer fauna in south compared to the nothern regions.

In Kara and Laptev seas, where the influence of the Atlantic Ocean is generally weak and the relief of the seafloor is fairly even, the bryozoan fauna declines in abundance as the depth increases and temperature decreases from south to north which restrict the distribution of stenobathic pacific species and shallow-water species with Atlantic origin (Denisenko 2021).

A northward increase in species richness in the East Siberian Sea is explained by the living conditions in the southern part of the sea, which are highly unfavorable for this taxonomic group because of the breakdown of coastal permafrost. This breakdown results in the formation of finely grained muddy sediments and increases turbidity of near-bottom waters in the coastal areas, which has an adverse effect on the growth of bryozoan colonies.

Despite its highly even seafloor relief, the Chukchi Sea (Dobrovolskiy & Zalogin 1982), like the Barents Sea, shows a significant variation in habitats within the same latitude, which prevents the emergence of any distinct patterns in the distribution of bryozoans.

All of the above suggests that environments are of major significance for the formation of bryozoan species richness and their changes in future can reflect on fauna composition. The role of the evolutionary factor that governs faunal variation with latitude in the regions of the world's oceans with a long geological history (Hillebrandt 2004) is fairly insignificant in the Arctic, because the shelves of the northern seas have been repopulated relatively recently, after they had been freed from glaciers during the Late Pleistocene to Early Holocene.

In contrast to the multidirectional trends observed in latitudinal variation, an increase in longitude is associated with a statistically significant decrease in bryozoan species richess in most subregions of the Eurasian Arctic, which has been observed not only in the present study, but also reported earlier (Denisenko 2021; Denisenko & Grebmeier 2015). Since the shelf zone and the existing system of currents in the Arctic basin have been shaped in a recent geological period after the quaternary glaciation, one might expect that this trend must be distinctly pronounced in this region, because even under favorable conditions the

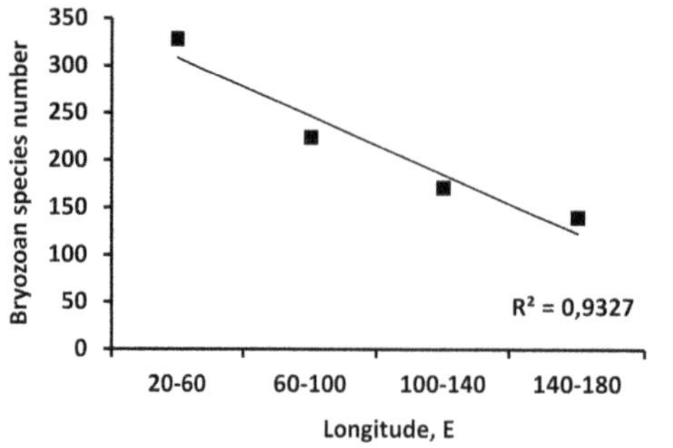

Figure 3. Variation of bryozoan species richness along longitude gradient in the Eurasian sector of the Arctic region.

expansion of the species requires a sufficiently long timespan. As the Atlantic waters undergo transformation, a process already presents in the Barents Sea (Tereschenko 1985), the bryozoans that are the inhabitants of the temperate latitudes of the Atlantic Ocean disappear already in the eastern parts of this sea (Denisenko 1990, 1996). The most substantial drop in bryozoan species richness occurs in the eastern direction in the Siberian seas, where, as the influence of the Atlantic water becomes significantly weakened, the predominance of the Arctic water mass becomes evident already in the Kara Sea (Prischepa 2018). The corresponding living conditions cause a decline in number of species of Atlantic origin with a wide geographic range and an increase in number of Arctic species, whose proportion in the overall bryozoan fauna is small (Denisenko 2022, in press). In general, despite distinctive subregional characteristics of the environment, warm Atlantic waters play major role in the formation of the fauna in most seas of the eastern part of the Arctic. In the Chukchi Sea, warm Bering-Sea waters also spread primarily in the eastern direction (Pisareva *et al.* 2015) and have a corresponding effect on species richness that decreases from west to east (Denisenko & Grebmeier 2015).

The results of the analysis of integral data on bryozoans for the whole Eurasian sector of the Arctic indicate the decline bryozoan fauna diversity with increasing distance from the Atlantic ocean (Figure 3). It does not contradict the existing opinion that species diversity decreases away from the speciation centers, which are located in North Atlantic and North Pacific. A high correlation between species richness and geographic longitude in the Arctic seas confirms the view that this trend is more pronounced in young, rather than old faunas, as was found when the Atlantic fauna was compared to that of the Pacific Ocean (Hillebrandt 2004). In modern studies, the assessment of longitudinal variation in diversity is usually performed to compare the faunas of different oceans in order to understand the history of their biota and the environmental changes in their geological past that could have influenced the transformation of the fauna (Poore & Wilson 1993).

Depth is the other crucial factor, beside geographic location, that determines species distribution within sea regions. As an ecological factor, depth, like latitude and longitude, reflects the combined effect of environmental parameters on biota. Depth determines the intensity of hydrodynamics, the associated structure of seafloor sediments and the number of substrates suitable for bryozoans as well as the temperature regime, the concentration of dissolved oxygen and the availability of food sources. This set of factors controls the dependence of bryozoan species richness on depth, although the extent of this dependence varies across different seas and regions. In the Barents Sea, this dependence approaches the statistical significance, which is explained by a complex relief of the seafloor with numerous depressions and seamounts, whose slopes form additional hard substrates as a result of an increased near-bottom hydrodynamic activity, and these substrates determine species richness of bryozoans at different depths. In the East Siberian and Chukchi seas, no significant variation in species richness with depth has been observed, because in most parts of these seas, depth differences do not exceed several dozen of meters.

The island regions of Greenland, Iceland and the Faroes have relatively narrow shelves with steep depth drops (Gougeon *et al.* 2017; ICES 2018; Nørrevang *et al.* 1994), which has an adverse impact on the number of substrates suitable for bryozoans, thereby causing their species richness.

It should also be noted that, despite the overall trend toward faunal impoverishment with depth, in all the regions studied the local maxima of species richness were observed at certain specific depths. The highest species richness in Greenland waters is currently found at depths of 150–200 m (Denisenko & Blicher 2021), while in the temperate regions of North Atlantic the maximum diversity of this group occurs at depths of 75–100 m (Clarke & Lidgard 2001). In the Barents Sea, the highest number of species has been reported at a depth range of 100–200 m (Denisenko 1990). In the Kara and East Siberian seas, the maximum diversity of bryozoan species has been observed at depths of 50–70 m (Denisenko 2010, 2021). The observed differences in bathymetric location of the maximum species richness are likely to be caused by interaction of Atlantic and Arctic waters, which occurs at different depths in different regions (McCartney *et al.* 1982), causing the overlap of the geographic ranges of species with different biogeographic origin (Denisenko & Blicher 2021; Denisenko *et al.* 2016; 2019). It is also possible that the occurrence of the local maxima of species richness at different depths in the Arctic seas and regions studied is caused by the geomorphological features of their seafloor (Gougeon *et al.* 2017).

6 CONCLUSION

From a geological point of view, the modern fauna of Bryozoa was formed relatively recently in the late Pleistocene – early Holocene, after the end of the Quaternary glaciation. Accordenly, the proportion of regional endemics in the fauna of this group is relatively small (Denisenko 2022b), and the wide connection of the Arctic Basin with the Atlantic Ocean contributed to the intensive distribution of the Atlantic allochthonous fauna. The prevalence of the latter species in the bryozoan fauna of the Arctic region suggests that the trends in the change in species richness along the gradients of latitude and longitude, as well as in depth, depend on the environmental conditions and their heterogeneity in the study area, whereas the evolutionary factor does not have a large importance.

ACKNOWLEDGEMENTS

The study was conducted as a part of the State Assignment to the Laboratory of Marine Research of Zoological Institute, Russian Academy of Sciences, no. 1021051402797-9 "Invertebrate Taxonomy, Diversity, and Ecology in the Russian and Neighboring Waters of the World Ocean, Inland Water Bodies, and Wet Territories"

REFERENCES

Andersson, K. 1902. Bryozoen, Während der Schwedischen Expeditionen 1898 und 1899 Unter Leitung von Prof. A.G. Nothorst und 1900 Unter Leitung von Conservator G. Kolthoff gesammelt. *Zoologische Jahrbücher. Abteilung für Systematik, Geographie und Biologie der Tiere* XV(5): 537–560.

Chaudhary, C., Saeedi, H. & Costello, M.J. 2017. Marine Species Richness: Is Bimodal with Latitude: A Reply to Fernandez and Marques. *Trends in ecology & evolution* 32(4): 234–237.

Clarke, A. & Lidgard, S. 2001. Spatial Patterns of Diversity in the Sea: Bryozoan Species Richness in the North Atlantic. *Journal of Animal Ecology* 69: 799–814.

Denisenko, N.V. & Blicher, M.E. 2021. Bryozoan Diversity, Biogeographic Patterns, and Distribution in Greenland Waters. *Marine Biodiversity* 51(4), article number 73 (2021).

Denisenko, N.V. & Grebmeier, J.M. 2015. Spatial Patterns of Bryozoan Fauna Biodiversity and Issues of Biogeographic Regionalization of the Chukchi Sea. *Oceanography (Washington D.C.)* 28(3): 134–145.

Denisenko, N.V. 1990. *Distribution and Ecology of Bryozoans of the Barents Sea.* Apatity, Isdatel'stvo KSC RAS. (in Russian)

Denisenko, N.V. 1996. Structure and Distribution of the Bryozoan Fauna of the South-western Part of the Barents Sea. In G.G. Matishov, G.A. Tarasov, S.G. Denisenko & K.V. Galaktionov (eds). *Biogeocenosy Glacial'nykh Shelfov Sapadnoy Arctiki*: 224–237. Apatity: Isdatel'stvo KSC RAS. (in Russian)

Denisenko, N.V. 2010. Bryozoans (Bryozoa) of the East Siberian Sea. In B.I. Sirenko & S.G. Denisenko (eds). *Fauna of the East Siberian Sea, Distribution Patterns and Structure of Bottom Communities.* Staint Petersburg: ZIN RAS. *Explorations of the Fauna of the Seas* 66 (74): 89–129. (in Russian)

Denisenko, N.V. 2021. Bryozoans of the Kara Sea: Species Diversity, Changes, and Peculiarities of Beogeographic Composition. *Proceedings of the Zoological Institute of the Russian Akademy of Sciences* 325(2): 217–234.

Denisenko, N.V. 2022a. Species Richness and Distribution Patterns of Bryozoans of the Arctic Region: List of the Bryozoan Fauna. In P.N. Wyse Jackson & M.E. Spencer Jones (eds.), *Annals of Bryozoology 7: Aspects of the History of Research on Bryozoans*: 121–143. Dublin: International Bryozoology Association.

Denisenko, N.V. 2022b. Taxonomic Structure and Endemism of the Bryozoan Fauna of the Arctic Region. *Paleontological Journal* 56: 774–792.

Denisenko, S.G., Denisenko, N.V., Chaban, E.M., Gagaev, S.Yu., Petryashov, V.V., Zhuravleva, N.E. & Sukhotin, A.A. 2019. The Current Status of the Macrozoobenthos Around the Atlantic Walrus Haul-outs in the Pechora Sea (SE Barents Sea). *Polar Biology* 42: 1703–1717.

Dobrovolsky, A.D. & Zalogin, B.S. 1982. *Seas of the USSR.* Moscow: Moscow State University. (in Russian)

Gontar, V.I. & Denisenko, N.V. 1989. Arctic Ocean Bryozoa. In Y. Herman (ed.) *The Arctic Seas*: 341–371. New York: Van Nestrand.

Gorshkov, G.P. 1980. *Atlas of Oceans. Arctic Ocean.* Moscow: Publishing house of the Main office of navigation and oceanography.

Gougeon, S., Kemp, K.M., Blicher, M.E. & Yesson, C. 2017. Mapping and Classifying the Seabed of the West Greenland Continental Shelf. *Estuarine, Coastal and Shelf Science* 187: 231–240.

Gray, J.S. 1997. Marine Biodiversity: Patterns, Threats and Conservation Needs. *Biodiversity and Conservation* 6: 153–175.

Gray, J.S. 2000. The Measurement of Marine Species Diversity, with an Application to the Benthic Fauna of the Norwegian Continental Shelf. *Journal of Experimental Biology and Ecology* 250: 23–49.

Gray, J.S. 2001. Antarctic Marine Benthic Biodiversity in a World-wide Latitudinal Context. *Polar Biology* 24: 633–641.

Gray, J.S. 2002. Species Richness of Marine Soft Sediments. *Marine Ecology Progress Series* 244: 285–297.

Hillebrand, H. 2004. Strength, Slope and Variability of Marine Latitudinal Gradients. *Marine Ecology Progress Series* 273: 251–267.

ICES. 2018 *ICES Ecosystem Overviews.* Report of ICES Advisory Committee, 2018. ICES Technical Guidelines, ICES Advice 2018, Section 16.2.

Kendall, M.A. & Aschan, M. 1993. Latitudinal Gradients in the Structure of Macrobenthic Communities: A Comparison of Arctic, Temperate

and Tropical Sites. *Journal of the experimental biology and ecology* 172: 157–169
Kluge, G.A. 1962. *Bryozoans of the Northern Seas of the USSR*. Moscow-Leningrad: Academy of Sciences of the USSR. (in Russian)
Kluge, G.A. 2009. *Ecology and Distribution of Bryozoans in the Barents Sea and in the Siberian seas*. Saint Petersburg: LULU Inc. (in Russian)
McArthur, R. & Wilson, E.O. 1967. *The Theory of Island Biogeography*. Princeton: Princeton University Press N.J.
McCartney, M.S. & Talley, L.D. 1982. The Subpolar Mode Water of the North Atlantic Ocean. *Journal of Physic Oceanography* 12(11): 1169–1188
Mokievskiy, V. & Azovsky, A. 2002. Re-evaluation of Species Diversity Patterns of Free-living Marine Nematodes. *Marine Ecology Progress Series* 238: 101–108.
Nordgaard, O. 1906. Bryozoa from 2nd "Fram" Expedition 1898–1902. *Report 2nd Norwegian Arctic Expedition of the Fram*: 1–44.
Nørrevang, A., Brattegard, T., Josefson, A.B., Sneli, J. & Tendal, O.S. 1994. List of BIOFAR Stations. *Sarsia* 79: 165–80.
Osburn, R.C. 1919. Bryozoa of the Crocker Land Expedition. *Bulletin of American Museum of Natural History* 41: 603–624.
Osburn, R.C. 1936. Bryozoa Collected in the American Arctic by Captain R. A. Bartlett. *Journal of Washington Academy of Sciences* 26 (12): 538–543.
Osburn, R.C. 1955. The Circumpolar Distribution of Arctic-Alaskan Bryozoa. In *Essays in the Natural Sciences in Honor of Captain Allan Hancock*. Los Angeles: The University of Southern California Press, 29–38.
Pianka, E.R. 1966. Latitudinal Gradients in Species Diversity: A Review of Concepts. *American Naturalist* 100(110): 3–46.
Poore, G.C.B. & Wilson, G.D.F. 1993. Marine Species Richness. *Nature* 361(6413): 597–598.
Prischepa, B.F. (Editor) 2008. *Ecosystem of the Kara Sea*. Murmansk: PINRO publisher.
Puerta, P., Johnson, C., Carreiro-Silva, M., Henry, L.-A., Kenchington, E., Morato, T., Kazanidis, G., Rueda, J.L., Urra, J., Ross, S., Wei, C.-L., González-Irusta, J.M., Arnaud-Haond, S. & Orejas, C. 2020. Influence of Water Masses on the Biodiversity and Biogeography of Deep-sea Benthic Ecosystems in the North Atlantic. *Frontiers in Marine Sciences* 00239.
Rex, M.A., Stuart, C.T., Hessler, R.R., Allen, J.A., Sander, H.L. & Wilson, G.D.F. 1993. Global Scale Latitudinal Pattern of Species Diversity in the Deep-sea Benthos. *Nature* 365: 636–639.
Roy, K., Jablonski, D. & Valentine, J.W. 2000. Dissecting Latitudinal Diversity Gradients: Functional Groups and Clades of Marine Bivalves. *Proceeding of Royal Society* B267: 293–299.
Sanders, H.L. 1968. Marine Benthic Diversity: A Comparative Study. *American Naturalist* 102: 243–282.
Smitt, F. 1868. Kritisk Förteckning öfver Skandinaviens Hafs-Bryozoer: Pt III. *Öfversigt af Kongliga Vetenskaps-Akademiens Förhandlingar* 24: 279–429.
Stehli, F.G. 1968. Taxonomic Diversity Gradients in Pole Location: The Recent Model. In E.T. Drake (ed.). *Evolution and Environment*: 163–227. New Haven: Yale University Press
Tantsyura, A.I. 1973. About Seasonal Changes of Currents of the Barents Sea. *PINRO Proceedings* XXXIV: 108–112. Murmansk: PINRO publisher (in Russian)
Tereschenko, V.V. 1985. *Seasonal and Inter-years Changes of Temperature and Salinity in the Barents Sea*. Murmansk: PINRO publisher. (In Russian)
Thorson, G. 1957. Bottom Communities (Sublittoral or Shallow Shelf). In J.W. Hedgepeth (ed). *Treatise on Marine Ecology and Paleoecology*: 461–534. New York: Geological Society of America.
Virnstein, R.W., Nelson, W.G., Graham Lewis III, F. & Howard, R.K. 1984. Latitudinal Patterns in Seagrass Epifauna: Do Patterns Exist, and Can They be Explained? *Estuaries* 7(4): 310–330.
Vongraven, D., Arneberg, P., Bysveen, I., Crane K., Denisenko, N.V. Gill M., Gofman, V., Grant-Freidman, A., Gudmundsson, G., Hindrum, R., Hopcroft, R., Iken, K., Labansen, A., Liubina, O.S., Moore, S.E., Melnikov, I.A., Reist, J.D., Stow, J., Tchernova, J., Ugarte, F. & Watkins, J. 2009. *Circumpolar Marine Biodiversity Monitoring Plan – Background Paper*. Akureyri: International secretariat, Iceland, CAFF CBMP. Report N 19.
Warwick, R.M. & Ruswahyuni. 1987. Comparative Study of the Structure of Some Tropical and Temperate Marine Soft-bottom Macrobenthic Communities. *Marine Biology* 95: 641–649.
Whittaker, R.H. 1972. Evolution and Measurement of Species Diversity. *Taxon* 21: 213–251.

Bryozoan Studies 2022 – Key, Porter & Wyse Jackson (eds)
© 2023 The Author(s), ISBN: 978-1-032-59487-3

Food for thought: investigating the impacts of feeding regime on the growth and survival of a locally invasive cheilostome bryozoan, *Watersipora subatra*

Tyler M. Feary*
Department of Marine Science, University of Otago, Dunedin, New Zealand

Abigail M. Smith*
Department of Marine Science, University of Otago, Dunedin, New Zealand
Orcid: 0000-0001-6468-9124

ABSTRACT: *Watersipora subatra* is an invasive cheilostome bryozoan found locally in shallow waters, especially on artificial substrates in Otago Harbour, Dunedin, southern Aotearoa New Zealand. The aim of this study is to assess the effect of six different feeding regimes on growth, morphology and survival of *W. subatra* in the laboratory, with the goal of finding a suitable regime for sustaining colonies in a laboratory setting. We tested feeding regimes of *Chaetoceros mulleri*, *Tetraselmis chuii*, *Pavlova lutea*, a mixture of the three, kelp flakes, and natural seawater. Growth was most rapid in the *Chaetoceros* and mixed treatments, whereas no growth was observed in the kelp flake or seawater treatments. Zooids and opercula were measured, but showed no discernible trends. This research contributes to our knowledge of *W. subatra* as a model species, allowing further laboratory-based studies into the effects of climate change on invasiveness.

1 INTRODUCTION

Bryozoans are a phylum of mostly colonial marine invertebrates, though there are some freshwater species (Ko 2019), and some non-sessile bryozoa (Hageman *et al.* 1997, 1998; Hermansen *et al.* 2001). Most bryozoans calcify, growing a biomineralized skeleton for protection and support (Hageman *et al.* 1998; Loxton *et al.* 2018; Steger & Smith 2005; Smith *et al.* 2006). Bryozoan colonies are made up of individual animals called zooids (Hageman *et al.* 1998; Lidgard & Jackson 1989). Autozooids are the most abundant of these, and are responsible for feeding (Okamura *et al.* 2014); in contrast heterozooids are non-feeding zooids that may play a role in reproduction, defense, or motility (Schack *et al.* 2019, 2020; Serova *et al.* 2017). Bryozoans are lophotrochozoans, meaning they collect suspended particles with tentacles on a muscular ring, called a lophophore (Okamura *et al.* 2014; Sun *et al.* 2009). Bryozoans usually grow on hard surfaces including rocks, macroalgae, hard-shelled animals, boats, wood, and man-made objects, preferring to grow on undersides and in darkness (Li *et al.* 2016; Mackie *et al.* 2014).

Bryozoans are essential components of biodiversity, particularly in the temperate Southern Hemisphere (Smith 2014). They provide both habitat stabilisation by encrusting over loose substrate (Hageman *et al.* 1998), and habitat space for other invertebrates and small fish in the form of bryozoan 'reefs' or thickets (Wood *et al.* 2013). In addition, bryozoans sequester carbon in their skeletons, which can contribute to the formation of sediment after death (Smith 2014). Our capacity to manage impacts and understand resiliency of these creatures is complicated by their colonial lifestyle. Even such basic things as "how old can they get" and "how fast do they grow" are both unknown and surprisingly difficult to ascertain.

Published bryozoan growth rates vary greatly (Amui-Vedel *et al.* 2007; Barnes *et al.* 2006). Coastal crustose Bryozoa can grow very quickly and live for only a few months. For example, *Membranipora membranacea* can grow as much as 12mm in one day (Smith 2014). Measuring bryozoan growth rate is difficult, not least because they are colonial, small, cryptic, and tend to be found at depth.

Different methods have been used for studying the growth rate of Bryozoa including calcein staining (Linard *et al.* 2011; Moran 2000; Smith 2014), but methods used are not comparable across the phylum, so that measured growth rates are also not compatible. Nevertheless, a basic understanding of growth is essential background to understand the effects of climate change and ocean acidification on the growth and survival of bryozoans (Lombardi *et al.* 2011; Seroy & Grünbaum 2018; Smith 2009).

Watersipora subatra, the red-ripple bryozoan, has the potential to be an effective "lab rat" model

*Corresponding authors: tyler.feary@gmail.com and abby.smith@otago.ac.nz

DOI: 10.1201/9781003454915-3

organism (Figure 1). It is invasive, fast-growing, and abundant (Culver *et al.* 2021; Reverter-Gil & Souto 2019; Viola *et al.* 2018). The area of origin of *W. subatra* is yet to be determined, although other members of the genus are thought to have originated in Asia, and have since invaded parts of Europe and the Pacific (Ferrario *et al.* 2015; Reverter-Gil & Souto 2019). The first record of *W. subatra* in Aotearoa New Zealand was in Dunedin in 1982, though it was initially misidentified as *Watersipora subtorquata* (Gordon & Mawatari 1992; Vieira *et al.* 2014). A population of *W. subatra* is conveniently located living on a set of tyres dangling off the pier at Portobello Marine Laboratory (Figure 2).

The opportunities offered by *Watersipora subatra*, and the genus *Watersipora* in general have been recognised, and it appears frequently in the literature (e.g., Ferrario *et al.* 2015; Kuhlenkamp & Kind 2013; Láruson *et al.* 2012; Mackie *et al.* 2006, 2012, 2014; Ng & Keough 2003; Sams *et al.* 2015; Sun *et al.* 2009). A thorough review, however, reveals gaps in laboratory culture protocols. In particular, what do we feed it, what is its wild growth rate, and how does laboratory feeding regime affect growth?

2 METHODS

2.1 *Feeding experiments*

Eight *Watersipora subatra* colonies were collected with a serrated knife from the tires at Portobello Marine Laboratory and cut into 48 fragments with diameters of approximately 0.5 – 0.9 cm. All fragments had live zooids and a defined growing edge. These fragments were glued to microscope slides in randomly assorted lines of three zooids with cyanoacrylate marine glue, yielding a total of sixteen microscope slides each with three colony fragments on them.

As a pilot study in winter 2021, two slides were placed into each of four feeding treatments: the diatom *Chaetoceros mulleri* (3–20 μm), *Tetraselmis chuii*, a green flagellate (12–14 μm), haptophyte *Pavlova lutea* (approximately 5 μm), and a mixture of roughly 1/3 of each, to investigate the effect of food type and size on survival and growth in *W. subatra*. For each treatment a 500ml beaker with two slides (i.e., six colonies) was kept in a flow-through dark-covered water bath. Input of 10ml of food occurred every three days for two weeks. Water was drained from the bath during feeding to prevent contamination among feeding regimes. Diameter of each of the colony fragments was measured at the beginning of the experiment and at the end of each week, using calipers.

Methodology was improved and refined for the summer feeding trial. Colonies of *W. subatra* were again collected from several tires at Portobello Marine Laboratory and divided into 72 fragments with a scalpel. Four fragments with diameters of approximately 0.5 – 0.9 cm were glued to each of 18 6-cm^2 settlement plates using cyanoacrylate glue (Figure 3A, B). They were then divided into each of six feeding treatments – *Chaetoceros mulleri*, *Tetraselmis chuii*, *Pavlova lutea*, a mixture of 1/3 each of the three, kelp flakes, and plain seawater. Seawater was collected locally at Portobello Marine Lab. Kelp flakes were cut from local *Macrocystis* sp. and minced small by knife. Colony plates were photographed weekly for three weeks. Fragments were photographed individually at the end of the experiment, and were subsequently bleached in a 10% solution and re-photographed to investigate zooid characteristics (Figure 3C). The characters measured included zooid length, zooid width, zooid morphology, operculum length, and operculum width. Fragment growth was defined as change in total area over time, measured with ImageJ.

2.2 *Wild growth*

To estimate the wild growth rate of *W. subatra,* 33 colonies on two tires at Portobello Marine Laboratory were labelled and digitally photographed monthly. Of these 33 colonies, seven were successfully rephotographed; the other colonies died or were overgrown. Photographs were analyzed using the imaging software ImageJ to find the change in colony diameter and area over time. The results were then compared with the results of the feeding experiments.

3 RESULTS

Laboratory-grown colonies in winter did not grow much, and in some cases they shrank; diameter change ranged from −0.190 to 0.095 mm/day (mean of 0.008 mm/day, SD = 0.06, n = 48). Wild growth on the tires at the same time was between 0 and 0.15 mm/day (mean of 0.08 mm/day, SD = 0.07, n = 7). While there was an overlap in the growth of wild and lab colonies, the wild colonies overall grew faster. Wild growth in winter is equivalent to a maximum annual growth rate of 5.5 cm/year.

The effect of feeding regime on *W. subatra* colonies varied between treatments during the winter trials (Figure 4). Change in colony diameter (mm/day) was observed to be greatest in colonies fed *Chaetoceros*, and the mixed treatment (equal parts *Chaetoceros*, *Pavlova*, and *Tetraselmis*). Change in colony diameter was observed to be either low or to show no change in colonies fed *Pavlova.* Colonies fed *Tetraselmis* showed substantial variation in response to the treatment with two showing an increase in colony

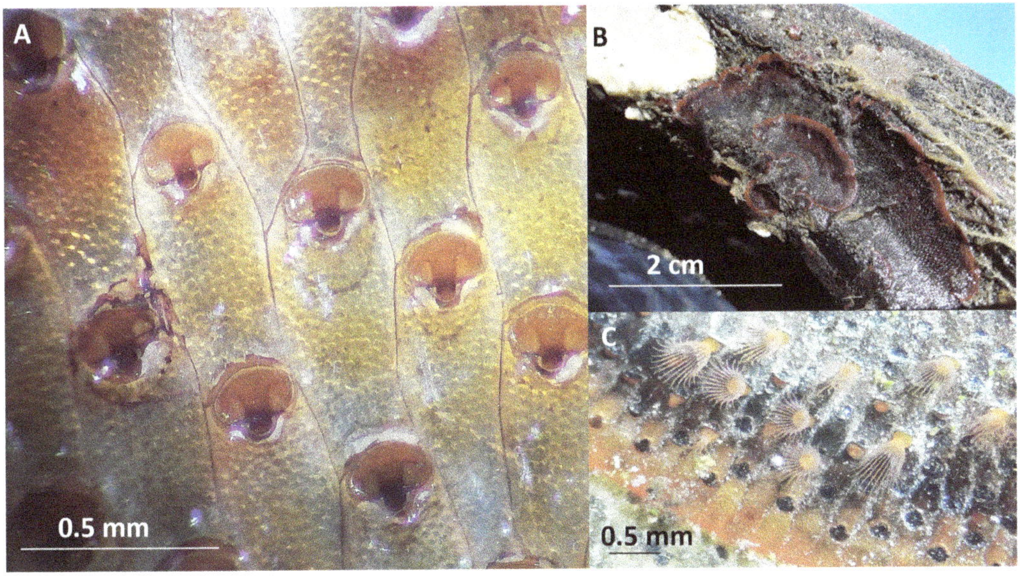

Figure 1. *Watersipora subatra* zooids under A. magnification, B. encrusting on a tire, and C. magnified with lophophores extended during feeding.

Figure 2. Location of the study population of *Watersipora subatra*. A. map highlighting Portobello Marine Lab in Otago Harbour, New Zealand; B. the pier at Portobello Marine Lab; and C. the colonized tire.

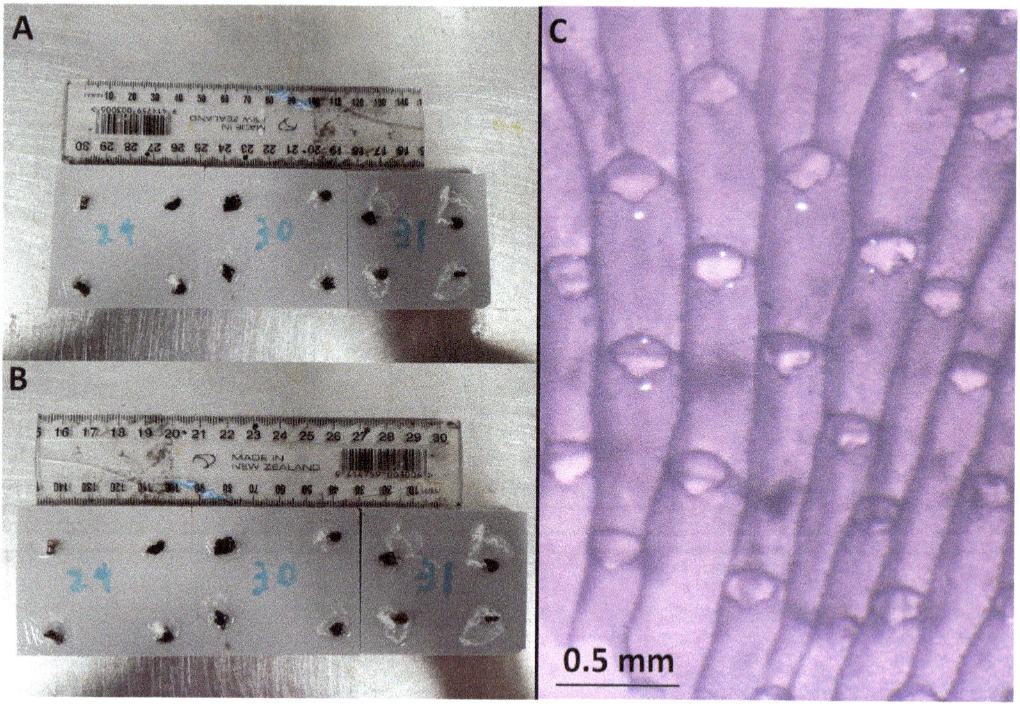

Figure 3. *Watersipora subatra* colony fragments adhered to A, B. settlement plates; and C. magnified after bleaching.

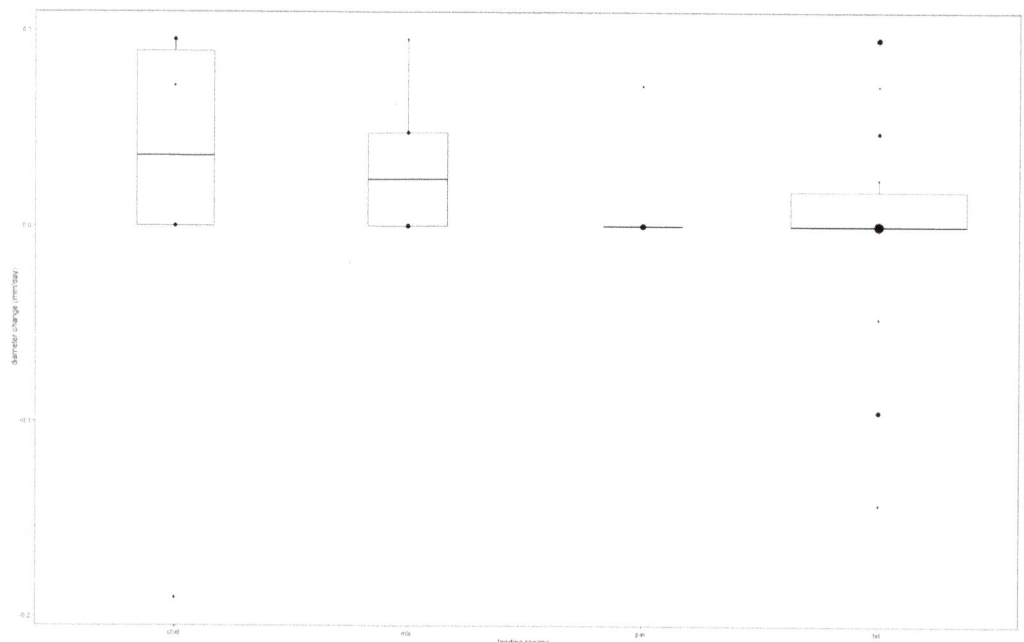

Figure 4. Graph comparing feeding regime with change in fragment diameter during winter 2021. Chet = *Chaetoceros*, mix = mixed treatment, pav = *Pavlova*, tet = *Tetraselmis*.

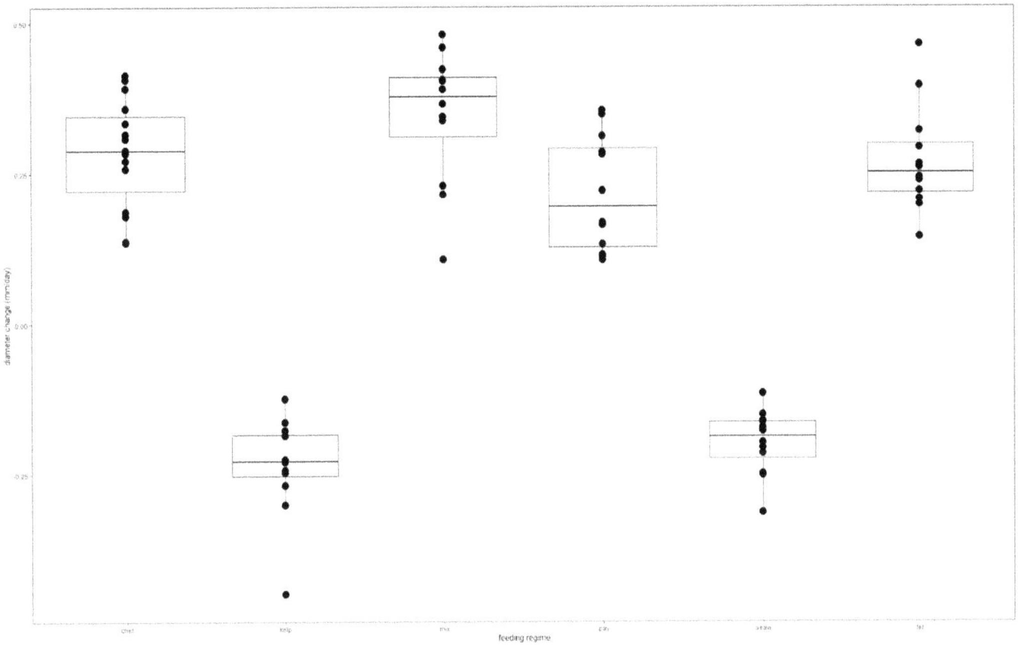

Figure 5. Graph comparing feeding regime with change in fragment diameter after three weeks during summer 2022. Chet = *Chaetoceros*, kelp = kelp flakes, mix = mixed treatment, pav = *Pavlova*, seaw = seawater treatment, tet = *Tetraselmis*.

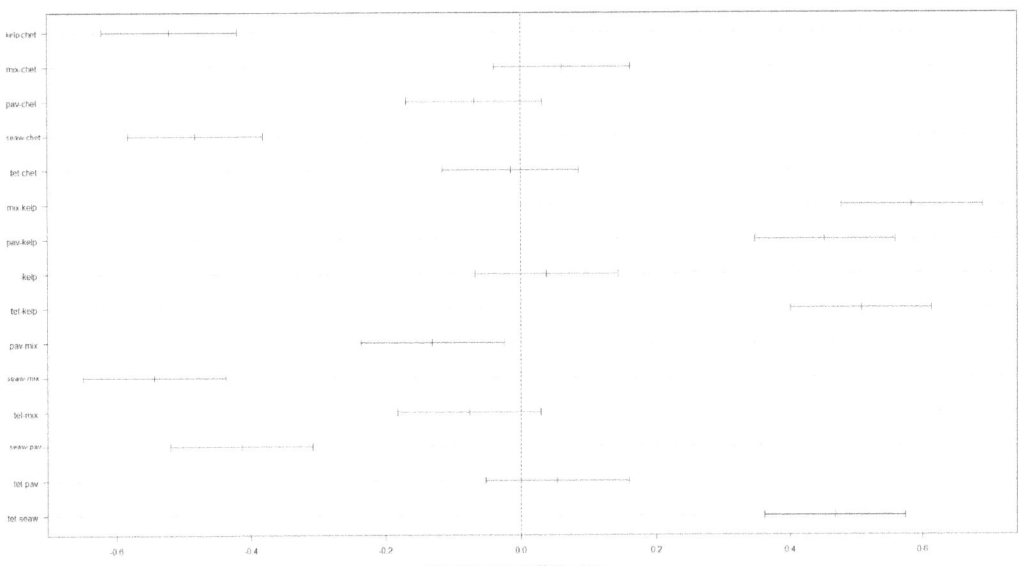

Figure 6. Graph showing the difference in means of feeding regime with 95% confidence intervals. Chet = *Chaetoceros*, kelp = kelp flakes, mix = mixed treatment, pav = *Pavlova*, seaw = seawater treatment, tet = *Tetraselmis*.

diameter, and the remainder showing either no change, or a negative change.

In the summer trials, colonies in the laboratory grew more. Total change in diameter ranged from −0.448 to 0.479 mm/day (mean of 0.12 mm/day, SD = 0.251, n = 72) whereas wild growth on the tires at the same time was between 3.03 and 13.48 mm/day (mean of 8.25 mm/day, SD = 7.38, n = 7). Although there was an overlap in the growth of wild and lab colonies, wild colonies in general grew more. Growth in summer is equivalent to a maximum annual growth rate of 30.4 cm/year.

The effect of feeding regime on *W. subatra* colonies varied between treatments during the summer trials (Figure 5). Change in colony diameter (mm/day) was again observed to be highest in colonies fed *Chaetoceros*, the mixed treatment, and *Pavlova*. Change in colony diameter was observed to be negative in the kelp flake and seawater treatments. ANOVA and post-hoc Tukey's tests were run to analyse the effect of feeding regime on change in colony diameter, yielding a significant difference (Figure 6).

4 DISCUSSION

4.1 *What does Watersipora like to eat?*

In winter, growth was observed in colonies fed *Chaetoceros, Pavlova, Tetraselmis,* and an even mixture of the three, suggesting that all these foods are acceptable for *W. subatra*. The highest growth was seen in the mixed treatment, suggesting that for *W. subatra* a mixed diet is optimal.

In summer, growth was observed in colonies fed *Chaetoceros, Tetraselmis,* and the mixed treatment. As with the winter results, this indicates *W. subatra* likes a mixed diet, but also supports the use of *Chaetoceros* and *Tetraselmis* as feeds. Unlike the summer trials, no growth was observed in the colonies fed *Pavlova*, possibly due to colony preference or a flaw in the earlier methodology as replication was lower. It may also be a case of seasonality, such that *W. subatra* exhibits greater diet selection in summer when growth is typically greatest. Presumably it is beneficial to eat everything offered in order to utilize optimal conditions for growth. It has been observed that *Fredericella sultana* is also a bryozoan with a generalist feeding strategy (Raddum & Johnsen 1983), suggesting that there is a benefit to feeding on a range of foods in interspecific competition among bryozoa (Allen *et al.* 2008; Best & Thorpe 1986; Comerford *et al.* 2020; Svensson & Marshall 2015). Furthermore, seasonal changes in feeding preferences have also been previously recorded in Bryozoa (Todini *et al.* 2018).

Chaetoceros mulleri is a diatom with an average length of 3–20 µm, *Tetraselmis chuii* is a green flagellate with an average length of 12–14 µm, and *Pavlova lutea* is a haptophyte with an average length of approximately 5 µm. The mixture of the three therefore contains a size range of 3–20 µm. The results of the feeding trials indicate that *W. subatra* can ingest particles of at least up to 20 µm, and shows little discrimination among phytoplankton groups.

Next steps worth trying include: both larger and smaller food, ciliates, and marine invertebrate food available in pet stores. Other foods successfully fed to Bryozoa include *Dunaliella, Cryptomonas,* and *Isochrysis* (Khale *et al.* 2003; McKenzie *et al.* 2012). *Thalassiosira, Nannochloropsis,* and *Brachionus* have also been fed to Bryozoa (including *Watersipora)* in the past (Svensson & Marshall 2015). *Synechoccus, Colpidium, Tetrahymena* have been fed to freshwater Bryozoa (Wood 2021), as has fish food and algal pellets, with varying results.

4.2 *How fast does Watersipora grow?*

Colonies in our laboratory trials grew at rates between −0.448 to 0.479 mm/day, with slower growth in winter (mean = 0.008 mm/day, SD = 0.060, N = 48) and faster in summer (mean = 0.120 mm/day, SD = 0.251, N = 72). Wild colonies grew at similar rates, on average 55 mm/y in winter (mean = 0.08 mm/day, SD = 0.07, N = 7) and 304 mm/y (mean = 8.25 mm/day, SD = 7.38, N = 7) in summer. These rates fall well within the normal range for encrusting bryozoans (Smith 2014).

When comparing between winter and summer trials, both wild and cultured growth was, as expected, higher during the summer. Similar trends in seasonality have been observed in *Watersipora* previously (Ng & Keough 2003; Sams *et al.* 2015).

4.3 *How does diet affect growth in Watersipora?*

Feeding regime was observed to have an effect on growth in *W. subatra* colonies. Change in colony diameter (mm/day) was highest in colonies fed *Chaetoceros*, one of the *Tetraselmis* treatments, and the mixed treatment (equal parts *Chaetoceros, Pavlova,* and *Tetraselmis*). Change in colony diameter was observed to be low or show no change in colonies fed *Pavlova* and *Tetraselmis*. Feeding *Chaetoceros, Pavlova, Tetraselmis,* and a mixture of the three promoted growth during summer, but during winter only *Chaetoceros, Tetraselmis,* and the mixture promoted growth. Feeding kelp flakes or seawater did not promote growth, and yielded a negative change in colony diameter. Other studies have noted that changes in diet may affect particle capture rates as well as growth (Okamura 1985; Riisgard & Manriquez 1997).

We show here that colonies of *Watersipora subatra* will grow in culture when fed *Chaetoceros* or a mixture of phytoplankton, suggesting that *W. subatra* is a generalist feeder rather than specializing in a single type of phytoplankton. It may also be that they grow better on a varied diet. In either case they appear to dislike or be unable to feed on *Pavlova*, at least in summer. The effects of feeding *Tetraselmis* varied among individuals. One difference between the two experiments is the inclusion of kelp flakes and filtered seawater as a control. *W. subatra* showed no growth when fed kelp flakes indicating they are unable to be sustained with this food source. The filtered seawater treatment acted as a control; as expected there was no growth in these treatments. Since growth in the wild is similar to maximum growth in the lab, our results support the feeding of *W. subatra* a varied mixed diet, but not including *Pavlova* or kelp flakes.

5 SUMMARY AND CONCLUSIONS

Feeding *W. subatra* a variety of cultured plankton enabled lab-based growth that was similar to wild growth nearby in several colonies. *W. subatra* thus demonstrates considerable potential as a "lab-rat" model species for experiments designed to explore environmental effects on growth.

ACKNOWLEDGEMENTS

This research was supported by the Department of Marine Science, University of Otago. The authors thank the International Bryozoology Association for a generous travel award to TMF.

REFERENCES

Allen, R.M., Buckley, Y.M. & Marshall, D.J. 2007. Offspring Size Plasticity in Response to Intraspecific Competition: An Adaptive Maternal Effect Across Life-history Stages. *The American Naturalist* 171(2): 225–237.

Amui-Vedel, A.-M., Hayward, P.J. & Porter, J.S. 2007. Zooid Size and Growth Rate of the Bryozoan *Cryptosula pallasiana* Moll in Relation to Temperature, in Culture and in its Natural Environment. *Journal of Experimental Marine Biology and Ecology* 353(1): 1–12.

Barnes, D., Webb, K. & Linse, K. 2006. Slow Growth of Antarctic Bryozoans Increases Over 20 years and is Anomalously High in 2003. *Marine Ecology Progress Series* 314: 187–195.

Best, M.A. & Thorpe, J.P. 1986. Effects of Food Particle Concentration on Feeding Current Velocity in Six Species of Marine Bryozoa. *Marine Biology* 9: 255–262

Comerford, B., Alvarez-Noreiga, M. & Marshall, D. 2020. Differential Resource Use in Filter-feeding Marine Invertebrates. *Oecologia* 194: 505–513.

Culver, C.S., Johnson, L.T., Page, H.M., Dugan, J.E. & Santschi, C.A.. 2021. Integrated Pest Management for Fouling Organisms on Boat Hulls. *North American Journal of Fisheries Management* 41(2): 301–328.

Ferrario, J., D'Hondt, J.-L., Marchini, A. & Occhipinti-Ambrogi, A. 2015. From the Pacific Ocean to the Mediterranean Sea: *Watersipora arcuata*, a New Non-indigenous Bryozoan in Europe. *Marine Biology Research* 11(9): 909–919.

Hageman, S.J., Bock, P.E., Bone, Y. & McGowran, B. 1998. Bryozoan Growth Habits: Classification and analysis. *Journal of Paleontology* 72(3): 418–436.

Hageman, S.J., Bone, Y., McGowran, B. & James, N.P. 1997. Bryozoan Colonial Growth-forms as Paleoenvironmental Indicators: Evaluation of Methodology. *PALAIOS* 12(5): 405–419.

Hermansen, P., Larsen, P.S. & Riisgård, H.U. 2001. Colony Growth Rate of Encrusting Marine Bryozoans (*Electra pilosa* and *Celleporella hyalina*). *Journal of Experimental Marine Biology and Ecology* 263(1): 1–23.

Kahle, J., Liebezeit, G. & Gerdes, G. 2003. Growth Aspects of *Flustra foliacea* (Bryozoa, Cheilostomata) in laboratory culture. *Hydrobiologia* 503: 237–244.

Ko, E.-J. 2019. Distribution and Growth of Non-native bryozoan *Pectinatella magnifica* (Leidy, 1851) in four large rivers in South Korea. *Aquatic Invasions* 14: 384–396.

Kuhlenkamp, R. & Kind, B. 2013. Arrival of the invasive *Watersipora subtorquata* (Bryozoa) at Helgoland (Germany, North Sea) on Floating Macroalgae (Himanthalia). *Marine Biodiversity Records* 6.

Láruson, Á.J., Craig, S.F., Messer, K.J. & Mackie, J.A. 2012. Rapid and Reliable Inference of Mitochondrial Phylogroups among *Watersipora* species, an Invasive Group of Ship-fouling Species (Bryozoa, Cheilostomata). *Conservation Genetics Resources* 4(3): 617–619.

Li, H.-X., Orihuela, B., Zhu, M. & Rittschof, D. 2016. Recyclable Plastics as Substrata for Settlement and Growth of Bryozoans *Bugula neritina* and Barnacles *Amphibalanus amphitrite*. *Environmental Pollution* 218: 973–980.

Lidgard, S. & Jackson, J.B.C. 1989. Growth in Encrusting cheilostome Bryozoans: I. Evolutionary trends. *Paleobiology* 15(3): 255–282.

Linard, C., Gueguen, Y., Moriceau, J., Soyez, C., Hui, B., Raoux, A., Cuif, J.P., Cochard, J.-C., Le Pennec, M. & Le Moullac, G. 2011. Calcein Staining of Calcified Structures in Pearl Oyster *Pinctada margaritifera* and the Effect of Food Resource Level on Shell Growth. *Aquaculture* 313(1–4): 149–155.

Lombardi, C., Rodolfo-Metalpa, R., Cocito, S., Gambi, M.C. & Taylor, P.D. 2011. Structural and Geochemical Alterations in the Mg Calcite Bryozoan *Myriapora truncata* under Elevated Seawater pCO_2 Simulating Ocean Acidification. *Marine Ecology* 32 (2): 211–221.

Loxton, J., Spencer Jones, M., Najorka, J., Smith, A.M. & Porter, J.S. 2018. Skeletal Carbonate Mineralogy of Scottish Bryozoans. *PLoS One* 13(6): e0197533.

Mackie, J.A., Darling, J.A. & Geller, J.B. 2012. Ecology of Cryptic Invasions: Latitudinal Segregation Among *Watersipora* (Bryozoa) Species. *Scientific Reports* 2 (1): 871.

Mackie, J.A., Keough, M.J. & Christidis, L. 2006. Invasion patterns Inferred from Cytochrome Oxidase I Sequences in Three Bryozoans, *Bugula neritina, Watersipora subtorquata*, and *Watersipora arcuata*. *Marine Biology* 149(2): 285–295.

Mackie, J.A., Wostenberg, D., Doan, M., Craig, S.F. & Darling, J.A. 2014. High-throughput Illumina

Sequencing and Microsatellite Design in *Watersipora* (Bryozoa), a Complex of Invasive Species. *Conservation Genetics Resources* 6(4): 1053–1055.
McKenzie, L.A., Johnston, E.L. & Brooks, R. 2012. Using Clones and Copper to Resolve the Genetic Architecture of Metal Tolerance in a Marine Invader. *Ecology and Evolution* 2(6): 1319–1329.
Moran, A.L. 2000. Calcein as a Marker in Experimental Studies Newly-hatched Gastropods. *Marine Biology* 137(5–6): 893–898.
Ng, T. & Keough, M. 2003. Delayed Effects of Larval Exposure to Cu in the Bryozoan *Watersipora subtorquata. Marine Ecology Progress Series* 257: 77–85.
Okamura, B. 1984. The Effects of Ambient Flow Velocity, Colony Size, and Upstream Colonies on the Feeding Success of Bryozoa I. *Bugula stolonifera* (Ryland), an arborescent species. *Journal of Experimental Marine Biology* 83: 179–193.
Okamura, B. 1985. The Effects of Ambient Flow Velocity, Colony Size, and Upstream Colonies on the Feeding Success of Bryozoa II. *Conopeum reticulum* (Linnaeus), an Encrusting Species. *Journal of Experimental Marine Biology* 89: 69–80.
Okamura, B., O'Dea, A. & Knowles, T. 2011. Bryozoan Growth and Environmental Reconstruction by Zooid Size Variation. *Marine Ecology Progress Series* 430: 133–146.
Pratt, M.C. 2008. Living Where the Flow is Right: How Flow Affects Feeding in Bryozoans. *Integrative and Comparative Biology* 48(6): 808–822.
Raddum, G.G. & Johnsen, T.M. 1983. Growth and feeding of *Fredericella sultana* (Bryozoa) in the Outlet of a Humic Acid Lake. *Hydrobiologia* 101: 115–120
Reverter-Gil, O. & Souto, J. 2019. Watersiporidae (Bryozoa) in Iberian waters: An Update on Alien and Native Species. *Marine Biodiversity* 49(6): 2735–2752.
Riisgård, H.U. & Manriquez, P. 1997. Filter-feeding in Fifteen Marine Ectoprocts (Bryozoa): Particle Capture and Water Pumping. *Marine Ecology Progress Series* 154: 223–239.
Riisgård, H.U., Okamura, B. & Funch, P. 2010. Particle Capture in Ciliary Filter-feeding Gymnolaemate and Phylactolaemate Bryozoans – A Comparative Study. *Acta Zoologica* 91(4): 416–425.
Sams, M., Warren-Myers, F. & Keough, M. 2015. Increased Larval Planktonic Duration and Post-recruitment Competition Influence Survival and Growth of the Bryozoan *Watersipora subtorquata. Marine Ecology Progress Series* 531: 179–191.
Schack, C.R., Gordon, D.P. & Ryan, K.G.. 2019. Modularity is the Mother of Invention: A Review of Polymorphism in Bryozoans. *Biological Reviews* 94 (3): 773–809.
Schack, C R., Gordon, D.P. & Ryan, K.G. 2020. Community Assembly in a Modular Organism: the Impact of Environmental Filtering on Bryozoan Colony Form and Polymorphism. *Ecology* 101(9): e03106.
Serova, K.M., Vishnyakov, A.E., Zaitseva, O.V., Kotenko, O.N. & Ostrovsky, A.N. 2017. Comparative Analysis of the Nervous System Structure of Polymorphic Zooids in Marine Bryozoans. *Doklady Biological Sciences* 475(1): 144–147.
Seroy, S. & Grünbaum, D. 2018. Individual and Population Level Effects of Ocean Acidification on a Predator-prey System With Inducible Defenses: Bryozoan-nudibranch Interactions in the Salish Sea. *Marine Ecology Progress Series* 607: 1–18.
Smith, A.M. 2009. Bryozoans as Southern Sentinels of Ocean Acidification: A Major Role for a Minor phylum. *Marine and Freshwater Research* 60(5): 475–482.
Smith, A.M. 2014. Growth and Calcification of Marine Bryozoans in a Changing Ocean. *The Biological Bulletin* 226: 203–210.
Smith, A.M. 2018. Addicted to Bryozoans. *Junctures: The Journal for Thematic Dialogue* 37–46.
Smith, A.M., Key, M.M., Jr. & Gordon, D.P. 2006. Skeletal Mineralogy of Bryozoans: Taxonomic and Temporal Patterns. *Earth-Science Reviews* 78(3–4): 287–306.
Steger, K.K. & Smith, A.M. 2005. Carbonate Mineralogy of Free-living Bryozoans (Bryozoa: Otionellidae), Otago Shelf, Southern New Zealand. *Palaeogeography, Palaeoclimatology, Palaeoecology* 218(3–4): 195–203.
Strathmann, R.R., Foley, G.P. & Hysert, A.N. 2008. Loss and Gain of the Juvenile Rudiment and Metamorphic Competence During Starvation and Feeding of Bryozoan Larvae. *Evolution & Development* 10(6): 731–736.
Sun, M.A., Wu, Z., Shen, X., Ren, J., Liu, X., Liu, H. & Liu, B. 2009. The Complete Mitochondrial Genome of *Watersipora subtorquata* (Bryozoa, Gymnolaemata, Ctenostomata) with Phylogenetic Consideration of Bryozoa. *Gene* 439(1–2): 17–24.
Svensson, J.R. & Marshall, D.J. 2015. Limiting Resources in Sessile Systems: Food Enchances Diversity and Growth of Suspension Feeders Despite Available Space. *Ecological Society of America.* 96(3): 819–827.
Todini, C., Elia, A.C., Selvaggi, R., Scoparo, M. & Taticchi, M.I. 2018. Food Selection by *Plumatella geimermassardi* Wood and Okamura, 2004 (Phylactolaemata, Bryozoa). *Knowledge and Management of Aquatic Ecosystems* 419(26).
Vieira, L.M., Spencer Jones, M. & Taylor, P.D. 2014. The Identity of the Invasive Fouling Bryozoan *Watersipora subtorquata* (d'Orbigny) and Some Other Congeneric Species. *Zootaxa* 3857(2): 151–182.
Viola, S.M., Page, H.M., Zaleski, S.F., Miller, R.J., Doheny, B., Dugan, J.E., Schroeder, D.M. & Schroeter, S.C. 2018. Anthropogenic Disturbance Facilitates a Non-native Species on Offshore Oil Platforms. *Journal of Applied Ecology* 55(4): 1583–1593.
Wood, A.C.L., Rowden, A.A., Compton, T.J., Gordon, D.P. & Probert, P.K. 2013. Habitat- Forming Bryozoans in New Zealand: their Known and Predicted Distribution in Relation to Broad-scale Environmental Variables and Fishing Effort. *PLoS One* 8(9): e75160.
Wood, T.S. 2020. Digestion and Nutrition in Freshwater Bryozoans (Phylactolaemata). *Invertebrate Biology* 140:e12314.

Bryozoan Studies 2022 – Key, Porter & Wyse Jackson (eds)
© 2023 The Author(s), ISBN: 978-1-032-59487-3

Pennsylvanian bryozoans from the Ixtaltepec Formation, Oaxaca, Mexico

Sergio González Mora*
Museo de Paleontología, Departamento de Biología Evolutiva, Facultad de Ciencias, Universidad Nacional Autónoma de México, Ciudad de México, México
ORCID ID: 0000-0001-9709-2033

Patrick N. Wyse Jackson
Department of Geology, Trinity College, Dublin, Ireland
ORCID ID: 0000-0001-5605-0670

Sara A. Quiroz Barroso & Francisco Sour Tovar
Museo de Paleontología, Departamento de Biología Evolutiva, Facultad de Ciencias, Universidad Nacional Autónoma de México, Ciudad de México, México
ORCID ID: 0000-0002-8864-6329

ABSTRACT: Eight species of bryozoans are described from Upper Carboniferous outcrops of the Ixtaltepec Formation, Nochixtlán Municipality, Oaxaca, Mexico: *Rhabdomeson* sp., *Rhombopora* sp. (Order Cryptostomata), *Fenestella s.l.* sp., *Spinofenestella* sp., *Ptylopora* sp. 1, *Ptylopora* sp. 2, *Acanthocladia* sp. and *Septopora* sp. (Order Fenestrata). The genera listed here are in addition to five more that have been previously described for this formation. All these specimens are preserved as molds in shale and were found in Pennsylvanian (Bashkirian-Moscovian) strata. The referred taxa support the similarity between the studied fauna and coeval ones from the Mid-Continent Province of the United States and support the hypothesis of the existence of an epicontinental sea that extended from the central-eastern of North America to southeastern Mexico. The occurrence of representatives of the genus *Ptylopora* in levels API-6 and API-7 of the Ixtaltepec Formation indicates a protected shallow reef environment with a muddy bottom.

1 INTRODUCTION

The region of Santiago Ixtaltepec is characterized by one of the most extensive sequences of Palaeozoic rocks in Mexico. Particularly, the Santiago and Ixtaltepec formations, previously assigned to the Mississippian (Tournaisian-Serpukhovian) and Pennsylvanian (Bashkirian-Moscovian), contain abundant fossil remains of sponges, conulariids, corals, trilobites, bivalves, gastropods, cephalopods, brachiopods, bryozoans, and crinoids. The fauna resembles those of the Mid-Continent Province (central-eastern United States) (Navarro-Santillán *et al.* 2002; Quiroz-Barroso & Perrilliat 1997, 1998; Quiroz-Barroso & Sour-Tovar 1995; Sour-Tovar 1994; Sour-Tovar & Martínez-Chacón 2004; Torres-Martínez 2014; Torres-Martínez & Sour-Tovar 2012; Torres-Martínez *et al.* 2008; Villanueva-Olea & Sour-Tovar 2014; Villanueva-Olea *et al.* 2011).

Although the fossil invertebrates of the Ixtaltepec Formation have been extensively studied, few studies have focused on bryozoans. Similarly, little research has been done on Upper Palaeozoic bryozoans in Mexico in general, despite the group being well represented (Quiroz-Barroso & Sour-Tovar 2006). One of these studies was carried out by Ernst & Vachard (2017), who described 10 bryozoan species from a rich bryozoan assemblage from the Middle Pennsylvanian of Cerros del Tule, Sonora, Mexico. On the other hand, for the Ixtaltepec Formation, González-Mora & Sour-Tovar (2014) described five taxa of fenestrate bryozoans assigning them to four genera. In addition, González-Mora *et al.* (2020) reported on rare palaeocorynid-type structures in fenestellid bryozoans from the same formation. Two palaeoecological and palaeoenvironmental studies of the Ixtaltepec Formation have utilized the morphology of the bryozoans to infer the conditions of the depositional environment (Morales-Soto 1984; Ortiz-Lozano 1996). As a result of the scarcity of studies on bryozoans, there is a need for research that corroborate the stratigraphic and palaeobiogeographic

*Corresponding Author: gioser@ciencias.unam.mx

DOI: 10.1201/9781003454915-4

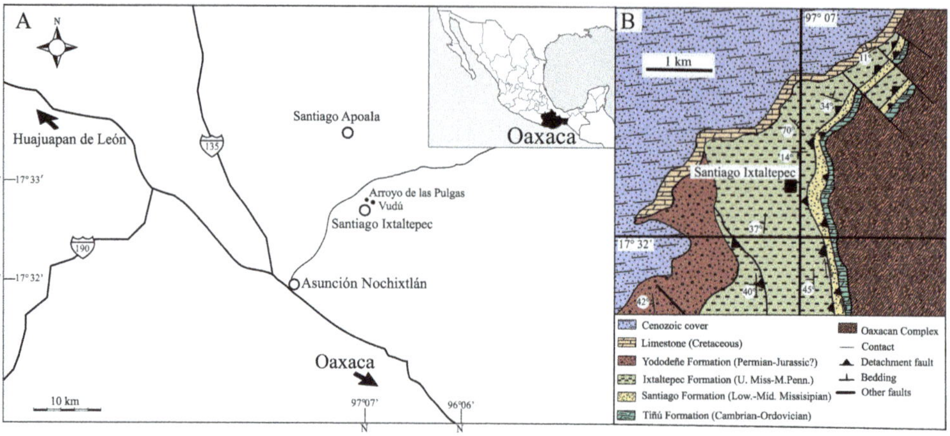

Figure 1. Location of the Ixtaltepec Formation outcrops examined in this study. A. Location of the Arroyo de las Pulgas and Vudú outcrops; B. Geologic map of the Santiago Ixtaltepec area (modified from Centeno-García & Keppie 1999; Navarro-Santillán *et al.* 2002; Pantoja-Alor 1970).

interpretations that have been made for the region. In the present work, we report the occurrence of eight species of bryozoans in outcrops of the Ixtaltepec Formation (Mississippian-Pennsylvanian), Nochixtlán Municipality, Oaxaca, Mexico.

2 GEOLOGICAL SETTING

The Santiago Ixtaltepec region is located 19 km northwest of the city of Asunción Nochixtlán, State of Oaxaca in southern Mexico. The specimens were collected in the Ixtaltepec Formation whose type section is located at Arroyo de las Pulgas, approximately 500 m north of the town of Santiago Ixtaltepec (17°32′–17°33′N; 97°06′–97°07′W), and Vudú outcrops, located approximately 800 m northeast of Santiago Ixtaltepec (17°33′15′′N, 97°06′ 36′′W) (Figure 1A).

The Carboniferous Santiago Ixtaltepec succession (Figure 1B) overlies the Tinú Formation (Cambrian-Ordovician), and it is divided into two formations (Pantoja-Alor & Robison 1967; Robison & Pantoja-Alor 1968). The 'Santiago Formation', an informally named unit that is Mississippian (Tournaisian–Serpukhovian) in age (Castillo-Espinoza 2013; Navarro-Santillán *et al.* 2002; Quiroz-Barroso *et al.* 2000). The Ixtaltepec Formation comprises 430 m of sedimentary rocks of Late Mississippian–Middle Pennsylvanian age. The base of this formation contains 90 m of siltstones, intercalations of shale, fine-grained calcareous sandstone, and thin layers of slightly clayish calcarenite, and it is overlain by units of slightly sandy shale with fine-grained sand beds and micaceous silts. Above these beds are thick layers of sandy shale interbedded with fine-grained shale. For palaeontological analysis, the Ixtaltepec Formation has been informally divided into eight levels (API-1 to API-8), each characterized by its fossil content and lithology (Quiroz-Barroso & Perrilliat 1997, 1998). Strata from levels API-1 to API-4 contain fossils of Mississippian age (Tournaisian–Serpukhovian) (Torres-Martínez & Sour-Tovar 2016a) and levels API-5 to API-8 yield a Pennsylvanian fauna (Bashkirian–Moscovian) (Torres-Martínez & Sour-Tovar 2016b, 2022) (Figure 2).

3 MATERIALS AND METHODS

The study of the bryozoans from the Pennsylvanian of Ixtaltepec Formation was performed from the observation of material using a binocular microscope. Due to the fact that these specimens are poorly preserved in shales and strongly weathered, use of thin sections was prohibited. However, the materials are preserved as molds and casts, in which are visible some taxonomically important external and internal features. Morphological character terminology and parameters measured are adopted from Hageman (1991) and Ernst & Minwegen (2006) for fenestrates and follow those utilized by Wyse Jackson & Bancroft (1995) for cryptostomes. All the specimens were measured from photographs using the Palaeontological Database System's PaleoTax's PaleoTax/Measure module (http://www.paleotax.de/measure). Measurements were summarised using the arithmetic mean, standard deviation, coefficient of variation, and minimum and maximum values. The material described herein is housed in the collection of the Museo de Paleontología, Facultad de Ciencias, Universidad Nacional Autónoma de México. Figured specimens are designated in the descriptions by the prefix FCMP.

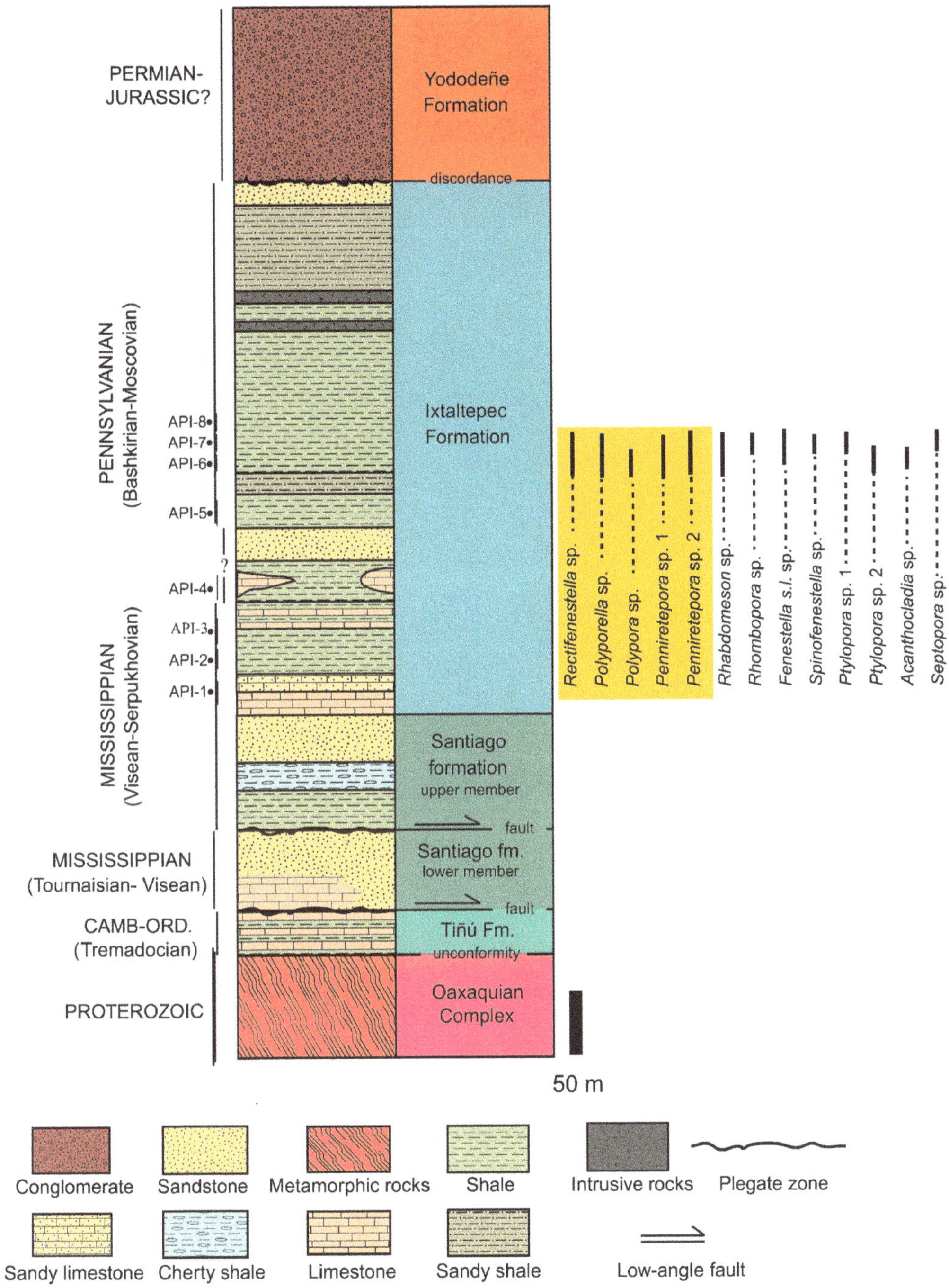

Figure 2. Stratigraphic column of the Palaeozoic region of Santiago Ixtaltepec, Mexico. The vertical bars indicate the levels where the bryozoan taxa in this study were found. In orange, the bryozoan taxa previously described by González-Mora & Sour-Tovar (2014) are shown.

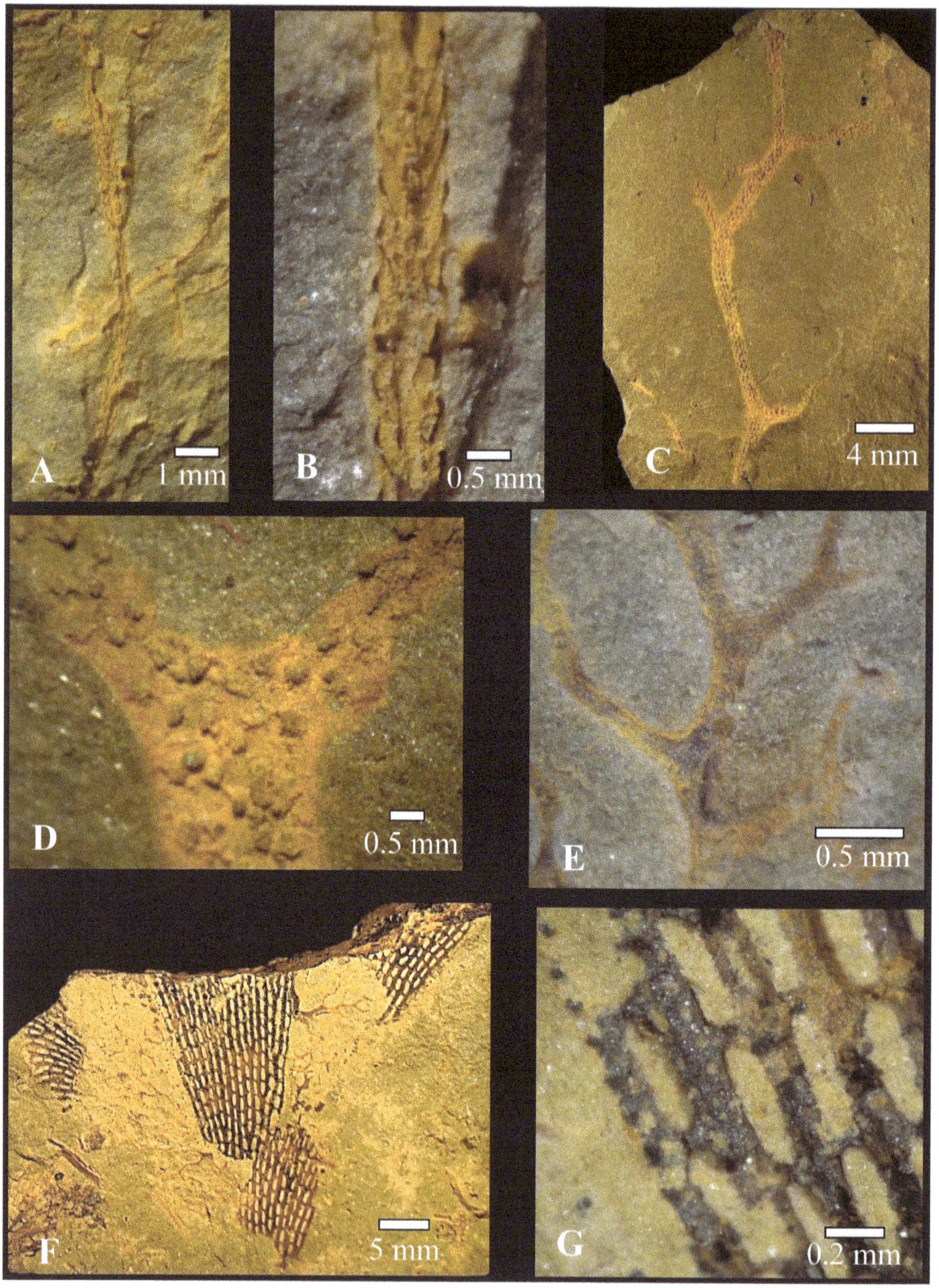

Figure 3. A-B. *Rhabdomeson* sp. (FCMP 1552): A. Zoarium; B. Autozooecial chambers and axial cylinder; C-D. *Rhombopora* sp. (FCMP 1553); C. Zoarium; D. Autozooecial apertures. *Rhabdomeson* sp. and *Rhombopora* sp. were collected in levels API-6 and API-7 of the Ixtaltepec Formation. E-G. *Fenestella s.l.* sp. (FCMP 1381) collected at the Vudú locality correlated to levels API-6 and API-7 of the Ixtaltepec Formation: E. Palaeocorynid-type structure associated; F. Zoarium; G. Apertures on the branches.

4 SYSTEMATIC PALAEONTOLOGY

Order Cryptostomata Vine, 1884
Suborder Rhabdomesina Astrova & Morozova 1956
Family Rhabdomesidae Vine, 1884
Genus *Rhabdomeson* Young & Young, 1874

Type species. Rhabdomeson progracile Wyse Jackson & Bancroft, 1995 from Hurst, North Yorkshire, England.

Diagnosis. Rhabdomesid with delicate dendroid colony with irregularly dichotomizing branches. Autozooecia regularly budded around axial cylinder in annual or spiral manner. Hemisepta common. Autozooecial apertures elliptical, pyriform or rhombic, closely spaced, arranged quincunxially on zoarial surface, of constant or variable dimensions around the branch. Stylets abundant and structurally diverse (Wyse Jackson & Bancroft 1995).

Geographic distribution. Cosmopolitan

Stratigraphic distribution. Middle Devonian – Upper Permian.

Rhabdomeson sp.
Figure 3A, B

Description. Erect, slightly curved colony with axial cylinder. The branch is 0.56–0.66 mm in diameter. The axial cylinder is circular and small, 0.11–0.19 mm in diameter. Autozooecia budded from its axis in spiral pattern. There are no bifurcations preserved. Measurements in Table 1.

Material. Specimen FCMP 1552. Colony preserved in shale as an internal and external mold.

Occurrence. Specimen collected at the Vudú locality; the bryozoan-bearing levels studied correlate to levels API-6 and API-7 (Pennsylvanian) of the type section of the Ixtaltepec Formation.

Discussion. Rhabdomeson progracile Wyse Jackson & Bancroft, 1995 from the Carboniferous of England and Ireland, has similar dimensions to the Oaxacan specimen of *Rhabdomeson* sp. as well as the same budding pattern of the autozooecia but differs in the distal part of the autozoocial chambers of *Rhabdomeson* sp. which is more elongate. *Rhabdomeson rhombiferum* (Phillips, 1836) from the Carboniferous of Lanarkshire, Scotland differs from the Oaxacan species in its smaller dimensions, branch diameter 0.41 – 0.86 mm, and in an annular budding pattern of the autozooecia. Moreover, the cylindrical axis in the Mexican material is straight whereas that in the Scottish material is irregularly curved. The morphology of *Rhabdomeson simulatum* Moore, 1929 from Pennsylvanian of Texas is similar to the material studied, but its diameter (1.00–1.15 mm) is approximately twice than in the Mexican material. The divergence angles of the zooids from the central axis are comparable, ranging from 60°–70°. The Ixtaltepec Formation material lacks features from its external morphology, so no further comparison is possible. Moore (1929) also described *Rhabdomeson tenerum* and *R. filum* for the Pennsylvanian of Texas, both with sizes comparable to *Rhabdomeson* sp. but he did not describe any of the internal features of the specimens.

Family Rhomboporidae Simpson, 1895
Genus *Rhombopora* Meek, 1872

Type species. Rhombopora lepidodendroides Meek, 1872, Upper Carboniferous of the United States.

Diagnosis. Colonies branched. Tube-shaped autozooecia intersect the surface of the colony at a low angle. Diaphragms may be present. Hemisepta absent. Autozooecial apertures are oval. Metazooecia are rare or absent. One or two acanthostyles at the distal end of each aperture. Exozonal walls with abundant paurostyles arranged in regular patterns around apertures (Gorjunova 1985).

Geographic distribution. Cosmopolitan.

Stratigraphic distribution. Devonian to Permian.

Rhombopora sp.
Figure 3C, D

Description. Erect, branched colony, 0.76 to 1.68 mm in diameter. It has five rows of autozooecial chambers growing around an axis, appearing in spiral arrangement on the surface. The chamber openings are oval and alternate along the branches. No details of the autozooecia are visible. Measurements in Table 2.

Material. Specimens FCMP 1553 (part and counterpart) and FCMP 1554. Colonies preserved in shale as an external mold.

Occurrence. Specimen FCMP 1553 was collected in unit API-6 of the type section of the Ixtaltepec Formation. FCMP 1554 occurred in the Vudú locality; the bryozoan-bearing levels correlate to the API-7 level.

Discussion. The genus *Rhombopora* is distinguished from *Rhabdomeson* by the absence of a central axial cavity in the former. The morphology of *Rhombopora* sp. from the Ixtaltepec Formation is similar to *Rhombopora corticata* Moore, 1929 from the Pennsylvanian of Texas as well as *Rhombopora lepidodendroides* Meek, 1872 from the Carboniferous of Alabama, but differs from the former because it is smaller in both branch and aperture diameter and from the latter because the thin exozone is not visible due to preservation. *Rhombopora blakei* Hageman *et al.*, 2011 from the Tournaisian of the Hook Head Formation in Ireland is similar to the Mexican specimen but is distinguished by a smaller and consistent, branch diameter, ranging 0.79–0.81 mm, throughout the zoarium, whereas the specimen studied shows abrupt changes in the width of its branches in the areas before the bifurcations (Figure 3C, D).

Order Fenestrata Elias & Condra, 1957
Family Fenestellidae King, 1849
Genus *Fenestella* Lonsdale in Murchison, 1839

Type species. Fenestella subantiqua d'Orbigny, 1849 from the Lower Silurian of Wenlock, England.

Diagnosis. The zoarium are fan-shaped or funnel-shaped; regular network with straight or slightly sinuous branches and with dissepiments without autozooecia. On the branches there are two rows of alternating thin autozooecia, which are tetragonal, usually parallelogram-shaped in tangential section. Hemisepta poorly developed or absent. The keel wall is straight and the nodes on the keel are held in a row (Morozova 2001).

Geographic distribution. Cosmopolitan.

Stratigraphic distribution. Upper Ordovician to Upper Permian.

Fenestella s.l. sp.
Figure 3E-G

Description. Reticulate colony with straight branches joined by thin dissepiments. On the branches are two slightly offset rows of autozooids. Circular to oval apertures occur on the obverse surface. Between the apertures is a row of nodes arranged in a regular pattern along the branches. The reverse surface is grooved and free of apertures. On the obverse of the zoarium is an associated structure consisting of a series of elongated spines that repeatedly bifurcate and merge or overlap to form a meshwork that lacks internal structures and is ornamented with a series of continuous striations that are identical over the entire surface. Measurements in Table 3.

Material. Specimen FCMP 1381. Colony associated with a reticulate structure preserved in shale as an external mold.

Occurrence. Specimen collected at the Vudú locality; the bryozoan-bearing levels studied correlate to levels API-6 and API-7.

Discussion. Fenestella is distinguished from the genus *Moorephylloporina* by the shape of its short autozooids and by the zoarial structure that forms a regular network of branches linked by dissepiments (Morozova 2001). More than 2000 of species *Fenestella* have been described (Hageman 1991; Morozova 1974), but Morozova (1974) revised the genus and established several genera based on internal morphology, however, there are still about 1000 unassigned species of *Fenestella sensu lato*. McKinney & Wyse Jackson in the forthcoming revision of the *Treatise on Invertebrate Paleontology* restrict the geological range of *Fenestella* Lonsdale in Murchison, 1839 from the Silurian to Devonian and so the taxon herein is probably not *Fenestella sensu stricto* Lonsdale in Murchison, 1839 and is left as *sensu lato* until unequivocal characters can be determined to allow a definite generic identification.

The specimen of *Fenestella s.l.* sp. has a skeletal reticulum free of autozooecia, which has been interpreted as a palaeocorynid-type appendage, possibly a defence structure (González Mora *et al.* 2020). Particularly, this kind of palaeocorynid structure has been reported for *Fenestella plebeia* M'Coy, 1844 from the UK (Bancroft 1988).

Genus *Spinofenestella* Termier & Termier, 1971

Type species. Fenestella spinulosa Condra, 1902 from the Upper Carboniferous of Roca, Nebraska, USA.

Diagnosis. Reticulate colonies with relatively broad, thick branches and relatively thin dissepiments. Autozooecia arranged in two rows on the branches. Autozooecia triangular in midtangential section, and their shape may vary from triangular to pentagonal when close to a bifurcation. They have a narrow keel with a single row of developed nodes (Morozova 2001).

Geographic distribution. Cosmopolitan.

Stratigraphic distribution. Mississippian to Upper Permian.

Spinofenestella sp.
Figure 4A, B

Description. Reticulate colony with thin, straight branches joined by thin dissepiments. Two rows of alternating autozooecia on the branches are maintained even before the bifurcations. Fenestrae are regular ovals close to the shape of a square with curved edges. Autozooecia are triangular in tangential section. Keel and nodes are not seen. Measurements in Table 4.

Material. Specimen FCMP 1555. Preserved colony as an internal mold of the branches, where the morphology of the autozooecial chambers is visible.

Occurrence. Spinofenestella sp. specimen is from unit API-7 of the type section of the Ixtaltepec Formation.

Discussion. Some consider *Alternifenestella* Termier & Termier, 1971 to be a synonym of *Spinofenestella* because of the similar morphological characteristics between the type species of both genera (Hageman & McKinney 2010; Suárez Andrés & Ernst 2015). The genus *Spinofenestella* differs from the species of the genus *Rectifenestella* Morozova, 1974 in the triangular shape of the autozooecia in tangential midsection (Ernst 2001). The Mexican specimen is similar to *Spinofenestella geinitzi* (d'Orbigny 1850) described by Ernst (2001) from the Upper Permian Zechstein Formation of Germany, as they share the same range of branch width dimensions, but in the Mexican material, the autozooecia are larger, and also differ in that their fenestrae have less elongation. *Spinofenestella*

Figure 4. A-B. *Spinofenestella* sp. were collected in the API-7 level of the Ixtaltepec Formation (FCMP 1555): A. Zoarium; B. Triangular autozooecia; C-E. *Ptylopora* sp. 1 from the API-7 level of the Ixtaltepec Formation (FCMP 1556): C. Zoarium; B. Main branch with lateral branches joined by dissepiments; E. Lateral branches.

nodosa Tolokonnikova *et al.*, 2015 described from the Lower Mississippian of the Namur-Dinant Basin of Belgium is comparable in size but differs in the presence of almost pentagonal autozooecia. *Alternifenestella vagrantia* Gilmour & Snyder, 2000 from the Permian of Washington displays a morphology characterized by its small oval fenestrae similar to *Spinofenestella* sp. from the Ixtaltepec Formation but is distinguished by the presence of triangular oval autozooeoecia in mid-tangential section.

Genus *Ptylopora* M'Coy, 1844

Type species. Ptylopora pluma M'Coy, 1844 from the Lower Carboniferous (Tournaisian) of Ireland.

Diagnosis. Fan-shaped, pinnate zoarium consisting of a centrally located broad branch, from which additional slender parallel lateral branches arise. Connected by dissepiments at regular intervals. Above the main and lateral branches, there are two rows of slightly alternating autozooecia, with a pentagonal outline in tangential section. Hemisepta absent. The rows of apertures are separated by a keel with a single row of nodes (Morozova 2001).

Geographic distribution. Cosmopolitan.

Stratigraphic distribution. Lower Devonian to Lower Permian.

Ptylopora sp. 1
Figure 4C-E

Description. Regularly pinnate zoarium, the lateral branches are connected by straight and slender dissepiments forming regularly oval and elongate fenestrae. The main branches are straight, and the secondary branches are slightly curved. The reverse side of the colony is striated on both the main branch and the pinnae. There are two rows of autozooecia per branch, the autozooecial apertures are rounded. Autozooecia are triangular to pentagonal in shape. No heterozooecia are present. Measurements in Table 5.

Material. Specimen FCMP 1556 (molds).

Occurrence. Specimen FCMP 1556 was collected from unit API-7 of the type locality of the Ixtaltepec Formation.

Discussion. The specimen differs from species of the genera *Rectifenestella* and *Ispanayella* by the presence of the pinnate zooecial structure with a wide central branch. From the genus *Polycladiopora*, it differs by the pentagonal outline of the autozooecia in tangential section and by the absence of autozooecia on the dissepiments (Morozova 2001). It resembles *Ptylopora pluma* M'Coy *parva* subsp. from the Visean of County Fermanagh, Ireland (Tavener-Smith 1973) in the morphology of the colony, including the alternate arrangement of autozooecia on the main branch and pinnae, the arrangement of ridges and grooves on the reverse surface are also similar. However, the autozooecia in the Mexican specimens develop a triangular to pentagonal outline whereas those of *P. pluma* M'Coy *parva* have an unusual shape (Figure 4D). Given the preservation, it is not possible to assess the nature of any the keel nodes and thus assign a species.

Ptylopora sp. 2
Figure 5A-C

Description. Zoarium regularly pinnate, lateral branches are connected by straight, slender dissepiments forming oval to semi-angular fenestrae. Some lateral branches develop a subordinate pinna; the main branch is slightly curved, and the secondary branches are straight. There are two rows of autozooecia per branch, the secondary branches have three before a bifurcation; the autozooecial apertures are rounded. Due to the preservation of the material, the shape of the autozooecia is not visible. Measurements in Table 6.

Material. Three colonies, specimens: FCMP 1557, FCMP 1558, and FCMP 1559. All preserved in shale as external molds without internal features.

Occurrence. The specimens were collected in API-6 level of the type section of the Ixtaltepec Formation.

Discussion. It differs from *Ptylopora* sp. 1 from the Ixtaltepec Formation in its larger dimensions, pinnae arise at a greater angle, 36°–60°, and they are straight rather than curved in shape. All specimens of *Ptylopora* sp. 2 lack internal features, which makes comparison with other species of the genus difficult.

Family Acanthocladiidae Zittel, 1880
Genus *Acanthocladia* King, 1849

Type species. Keratophytes anceps Schlotheim 1820 from the Upper Permian, Glücksbrunn, Thuringia, Germany.

Diagnosis. Pinnate zoarium, consisting of a thick central branch from which lateral branches emerge free and diverging in a single plane. Branches with three or more rows of autozooecia. Autozooecia with a well-developed vestibule and upper hemisepta. Circular apertures. Autozooecia elongate-hexagonal in mid tangential section, and rhomboidal in deep tangential section. Sometimes with nodes in the peristome. The longitudinal rows of apertures are separated by differentially projected rings on which nodes and capillaries develop. The reverse side of the branches has capillaries and nodes (Morozova 2001).

Geographic distribution. Cosmopolitan.

Stratigraphic distribution. Carboniferous to Permian.

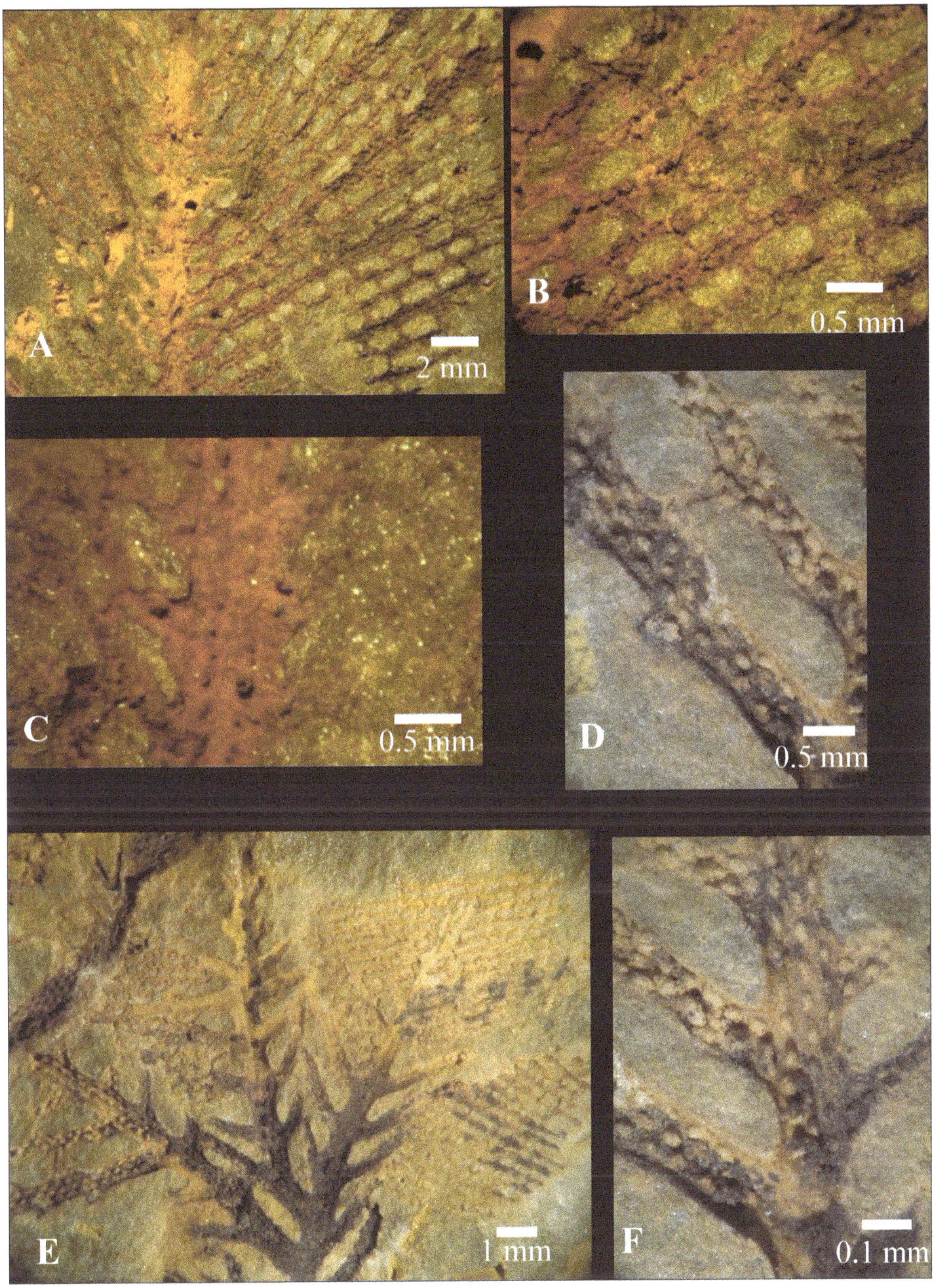

Figure 5. A-C. *Ptylopora* sp. 2 from the API-6 level of the Ixtaltepec Formation (FCMP 1557): A. Zoarium; B. Lateral branches joined by dissepiments; C. Autozooecia rows on the main branch; D-F. *Acanthocladia* sp. from the API-6 level of the Ixtaltepec Formation (FCMP 1560): D. Lateral branches; E. Zoarium; F. Autozooecial apertures on the main branch.

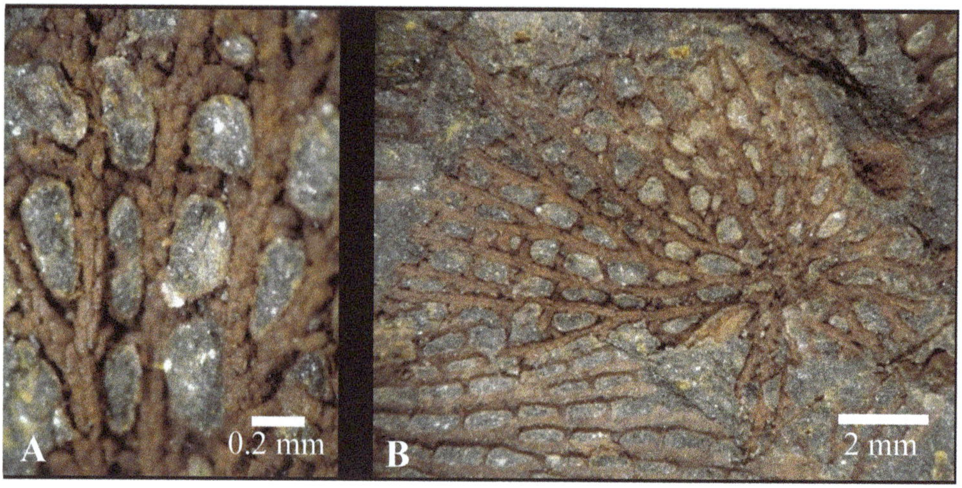

Figure 6. A-B. *Septopora* sp. from the API-7 level of the Ixtaltepec Formation (FCMP 1561): A. Autozooecia on the branches and dissepiments; B. Zoarium.

Acanthocladia sp.
Figure 5D-F

Description. Erect pinnate colony, with very frequent divergent branches, reaching fourth order branching. The number of autozooecia between branches varies according to the order of branching. The main branch has three rows of interspersed autozooecia, whereas the secondary branches have 2–3 autozooecia. Third-order branches are infrequently joined by dissepiments, whereas others coalesce. Most of the branches are distally curved, with a smaller proportion of straight branches. The autozooecial openings are round to oval. Hexagonal to triangular autozooecia can be inferred. Measurements in Table 7.

Material. Specimen FCMP 1560. Fragment of a colony preserved as an external mold.

Occurrence. Specimen collected from unit API-6 of the type section of the Ixtaltepec Formation.

Discussion. The genus *Kalvariella* has autozooecia similar to *Acanthocladia*, as well as their distribution on the branches and the shape of the zooids but differs in the number of rows of autozooecia on the main branch (Morozova, 2001). Likewise, *Acanthocladia* sp. from the Ixtaltepec Formation differs from *Penniretepora* sp. described from the same formation (González-Mora and Sour-Tovar, 2014), because the latter possesses two rows of autozooecia on all branches.

When comparing the Mexican material with the type species *Acanthocladia anceps* (Schlotheim, 1820) from the Permian of Western Europe, there is a great difference in size, as *Acanthocladia* sp. from the Ixtaltepec Formation is an order of magnitude smaller than the European material. The same situation exists for *Acanthocladia diffussa* (Korn, 1930) reported by Ernst (2001) from the Permian of Germany. Montgomery (2004) reported the presence of *Acanthocladia guadalupensis* from the Permian of Sierra Plomosa, Chihuahua, Mexico, but did not provide any description or illustration of this form that would allow for comparison of the currently studied material.

Family Septoporidae Morozova, 1962
Genus *Septopora* Prout, 1859

Type species. *Septopora cesteriensis* Prout, 1859 from the Lower Carboniferous of Illinois, USA.

Diagnosis. Colonies reticulate with moderately thick and broad branches. Frequent curved dissepiments. New branches originate by branch bifurcations or from dissepiments. Fenestrae irregular in shape. Autozooecia arranged in two rows on branches and dissepiments, rectangular to pentagonal in median tangential section. Upper hemisepta present; lower hemisepta absent. Keel broad, low, bearing a single row of nodes. Cyclozooecia irregularly spaced throughout the colony (modified from Morozova 2001).

Geographic distribution. North America, Europe, and Asia.

Stratigraphic distribution. Lower Carboniferous to Upper Permian.

Septopora sp.
Figure 6A, B

Description. Reticulate colony with undulating branches joined by dissepiments that are V-shaped and have autozooecia. The fenestrae are mostly oval. Autozooecia elongated pentagonal and alternating in two-row arrangement on branches and dissepiments, which display rows of two to four chambers. Given the preservation, no apertures are visible. Measurements in Table 8.

Table 1. Measurements of *Rhabdomeson* sp. 1. Abbreviations: N: Number of Measurements, Min: Minimum Value, Max: Maximum Value, X: Mean, SD: Standard Deviation, CV: Coefficient of Variation.

	N	Min	Max	X	SD	CV
Branch diameter, mm	10	0.56	0.66	0.60	0.04	6.30
Axial cylinder diameter, mm	10	0.11	0.19	0.16	0.03	15.95
Autozooecial chamber length, mm	10	0.32	0.49	0.39	0.06	14.76
Autozooecial aperture width, mm	10	0.19	0.41	0.31	0.06	18.85

Table 2. Measurements of *Rhombopora* sp. (abbreviations as for Table 1).

	N	Min	Max	X	SD	CV
Branch diameter, mm	10	0.76	1.68	1.15	0.29	25.00
Autozooecial aperture width, mm	10	0.09	0.14	0.11	0.02	15.45

Table 3. Measurements of *Fenestella s.l.* sp. (abbreviations as for Table 1).

	N	Min	Max	X	SD	CV
Branch width, mm	20	0.22	0.37	0.30	0.037	12.50
Fenestrule length, mm	20	0.75	0.92	0.84	0.052	6.19
Fenestrule width, mm	20	0.23	0.42	0.31	0.043	13.83
Dissepiment width, mm	20	0.13	0.21	0.16	0.023	14.20
Autozooecial aperture width, mm	15	0.08	0.13	0.11	0.015	13.64
Node width, mm	10	0.07	0.12	0.10	0.019	20.00
Node spacing, mm	10	0.27	0.34	0.32	0.024	7.52

Table 4. Measurements of *Spinofenestella* sp. (abbreviations as for Table 1).

	N	Min	Max	X	SD	CV
Branch width, mm	10	0.18	0.30	0.24	0.031	13.03
Fenestrule length, mm	10	0.44	0.57	0.53	0.040	7.58
Fenestrule width, mm	10	0.28	0.44	0.37	0.047	12.81
Dissepiment width, mm	10	0.10	0.19	0.14	0.026	18.44
Autozooecial chamber width, mm	10	0.10	0.16	0.13	0.018	13.43
Autozooecial chamber length, mm	10	0.17	0.23	0.20	0.018	9.00

Table 5. Measurements of *Ptylopora* sp. 1 (abbreviations as for Table 1).

	N	Min	Max	X	SD	CV
Branch width, mm	15	0.54	0.69	0.60	0.04	7.35
Fenestrule length, mm	20	0.66	0.91	0.77	0.07	8.43
Fenestrule width, mm	20	0.29	0.54	0.39	0.07	17.57
Dissepiment width, mm	10	0.09	0.15	0.11	0.02	18.02

(continued)

Table 5. Continued

	N	Min	Max	X	SD	CV
Aperture spacing across branch, mm	10	0.16	0.26	0.21	0.03	12.74
Aperture spacing along branch, mm	11	0.21	0.27	0.25	0.02	8.54
Autozooecial aperture width, mm	11	0.05	0.08	0.07	0.01	11.94
Lateral branch width, mm	20	0.21	0.36	0.29	0.04	15.07
Distance between lateral branch centres, mm	10	0.44	0.66	0.56	0.07	12.41
Aperture spacing across lateral branch, mm	10	0.11	0.21	0.16	0.04	22.64
Aperture spacing along lateral branch, mm	11	0.16	0.24	0.21	0.03	15.87
Autozooecial chamber width in lateral branch, mm	10	0.07	0.11	0.09	0.01	15.05
Autozooecial chamber length in lateral branch, mm	10	0.13	0.19	0.15	0.02	13.16
Lateral branch autozooecial chamber length, mm	10	0.16	0.24	0.19	0.03	13.47
Lateral branch diverging angle, °	10	24	35	31	3	11

Table 6. Measurements of *Ptylopora* sp. 2 (abbreviations as for Table 1).

	N	Min	Max	X	SD	CV
Branch width, mm	15	0.64	0.82	0.74	0.065	8.82
Fenestrule length, mm	30	0.37	0.89	0.66	0.127	19.13
Fenestrule width, mm	29	0.24	0.54	0.36	0.074	20.33
Dissepiment width, mm	15	0.13	0.40	0.20	0.064	32.00
Aperture spacing across branch, mm	13	0.17	0.46	0.33	0.081	24.85
Aperture spacing along branch, mm	10	0.22	0.32	0.27	0.039	14.34
Autozooecial aperture width, mm	13	0.05	0.14	0.09	0.027	28.72
Lateral branch width, mm	26	0.15	0.67	0.31	0.106	33.97
Distance between lateral branch centres, mm	20	0.42	0.83	0.59	0.098	16.61
Lateral branch diverging angle, °	16	36	60	45	7	16

Table 7. Measurements of *Acanthocladia* sp. (abbreviations as for Table 1).

	N	Min	Max	X	SD	CV
Branch width, mm	15	0.08	0.12	0.09	0.012	13.04
Autozooecial aperture width, mm	10	0.02	0.04	0.03	0.005	17.86
Lateral branch width, mm	10	0.05	0.09	0.06	0.013	20.31
Distance between lateral branch centres, mm	10	0.09	0.17	0.14	0.026	18.84
Lateral branch diverging angle, °	10	6	55	26	14	53

Table 8. Measurements of *Septopora* sp. (abbreviations as for Table 1).

	N	Min	Max	X	SD	CV
Branch width, mm	20	0.27	0.42	0.35	0.039	11.27
Fenestrule length, mm	20	0.56	1.18	0.81	0.199	24.69
Fenestrule width, mm	20	0.34	0.84	0.53	0.132	25.00
Dissepiment width, mm	20	0.21	0.51	0.30	0.067	22.11
Autozooecial chamber width, mm	10	0.14	0.20	0.17	0.018	10.91
Autozooecial chamber length, mm	10	0.20	0.35	0.28	0.042	15.22

Material. Specimen FCMP 1561. Colony preserved in shale as an internal mold.

Occurrence. The specimen was collected in unit API-7 of the type section of the Ixtaltepec Formation.

Discussion. The genus *Septopora* differs from *Synocladia* in that it possesses two rows of autozooecia on the branches and dissepiments. It also differs from *Ignotifenestella* and *Polyfenestella* in the presence of autozooecia and cyclozooecia on its dissepiments (Morozova, 2001). *Septopora blanda* Moore, 1929 described from the Pennsylvanian of Texas (Ernst *et al.* 2022; Moore, 1929) and Oklahoma (Ernst *et al.* 2016) has fenestrae with the same shape as the Mexican material. Branch width (0.55–0.60 mm) is in the range reported for *Septopora* sp. from the Ixtaltpec Formation. The Mexican material lacks external features, making comparison difficult. The dimensions of *Septopora alternata* Moore, 1929 and *Septopora robusta* Moore, 1929 are within a comparable range with the Mexican material, however, given the preservation, a deeper comparison is not possible. It should be noted that the material described by Moore (1929) displays detailed external features, but the specimens are highly fragmented. *Septopora* sp. is similar in dimensions as well as in its morphology to *Septopora carbonaria* (Etheridge, 1873) reported by Bancroft (1987) for the Carboniferous of Britain, but no external features such as cyclozooecia are observed to establish a specific assignment.

5 DISCUSSION

The present study of the bryozoans from the Ixtaltepec Formation led to the description of eight representatives of the orders Cryptostomata and Fenestrata with all of the specimens being determined to genus level: *Rhabdomeson* sp., *Rhombopora* sp., *Fenestella s.l.* sp., *Spinofenestella* sp., *Acanthocladia* sp., *Ptylopora* sp. 1, *Ptylopora* sp. 2, and *Septopora* sp. These taxa are in addition to the five previously described for the Pennsylvanian of the Ixtaltepec Formation: *Rectifenestella* sp., *Polyporella* sp., *Polypora* sp., *Penniretepora* sp. 1 and *P.* sp. 2 (González-Mora & Sour-Tovar 2014). All taxa described herein have been reported from other coeval localities in the Mid-Continent Province (East-Central region of United States), supporting previous arguments for the existence of an epicontinental sea that covered the Mid-Continent region during the Carboniferous, extending to what is currently the Oaxaca region (e.g., Navarro-Santillán *et al.* 2002; Quiroz-Barroso & Perrilliat 1998; Sour-Tovar & Martínez-Chacón 2004; Torres-Martínez & Sour-Tovar 2012).

Bryozoans have been considered very reliable proxies of palaeoenvironmental conditions because they are sessile organisms that feed by filtration and are very sensitive to the conditions of the environment, such as turbidity, illumination, depth, and currents (McKinney & Jackson 1991). In the case of Palaeozoic forms, McKinney & Gault (1980) studied bryozoan genera of the Order Fenestrata from different localities of the Chesterian of the North American Mid-Continent. They found environmental distributions of these genera through open-shelf, near shoal, and back-barrier environments. For example, the genus *Fenestella s.l.* was found in areas of moderately high to moderate energy, near shoals with protected muddy bottoms; colonies assigned to *Septopora* and *Polypora* in areas of moderate to low energy, near shoals; while erect and pinnate morphologies such as *Penniretepora* and *Ptylopora* were located exclusively in protected areas of low energy and muddy bottoms. Based on this, the occurrence of bryozoan genera of the Order Fenestrata: *Fenestella s.l.* sp., *Rectifenestella* sp., *Spinofenestella* sp., *Polypora* sp., *Polyporella* sp., *Ptylopora* sp. and *Penniretepora* sp. suggest a protected shallow reef environment with a muddy bottom, because of all these taxa have an environmental distribution close to the shoals. Particularly, both *Ptylopora* and *Penniretepora* have a very restricted distribution, especially the second genus, only reported from sheltered lagoonal environments located in the lee of the shoal. This is possibly due to their erect pinnate morphology is prone to fragmentation when faced with strong currents. The distribution of productid brachiopods in the Ixtaltepec Formation (Torres-Martínez & Sour-Tovar 2012) which are more abundant in calm and shallow waters (Fürsich & Hurst 1974), shows the same trend.

6 CONCLUSION

The presence of all these taxa is consistent with ages and palaeogeographical distributions that have been proposed in previous studies. The faunistic assemblages from API-6 and API-7 units and the bryozoan morphology are attributed to a protected shallow reef environment with a muddy bottom. Finally, the present work represents one of the first studies on Upper Palaeozoic bryozoans in Mexico.

ACKNOWLEDGEMENTS

We are grateful to C.M. Reid (University of Canterbury), A. Ernst (Universität Hamburg), and

M.M. Key, Jr. (Dickinson College) for their valuable comments and suggestions which improve our manuscript. We appreciate the facilities granted for field work by the authorities of Santiago Ixtaltepec town. The authors thank D. Navarro Santillán and M.I. Hernández Ocaña for collecting material and technical assistance. S. González Mora acknowledges Posgrado en Ciencias Biológicas, UNAM, and CONACYT for his graduate fellowship (No. 330554).

REFERENCES

Astrova, G. & Morozova, I. 1956. About Systematics of the Order Cryptostomata. *Doklady Akademii Nauk SSSR* 110(4): 661–664.

Bancroft, A.J. 1987. The Fenestrate Bryozoan *Septopora carbonaria* (Etheridge Jun. 1873a) From the Carboniferous of Great Britain. *Geological Journal* 22 (1): 1–10.

Bancroft, A.J. 1988. Palaeocorynid-type Appendages in Upper Palaeozoic Fenestellid Bryozoa. *Palaeontology* 31: 665–675.

Castillo-Espinoza, K.M. 2013. *Sistemática de Braquiópodos, Cefalópodos y Crinoideos del Misisípico Medio de la formación Santiago, Santiago Ixtaltepec, Oaxaca* [M.Sc. Thesis]. Facultad de Ciencias, Universidad Nacional Autónoma de México.

Centeno-García, E. & Keppie, J. D. 1999. Latest Paleozoic–early Mesozoic Structures in the Central Oaxaca Terrane of Southern Mexico: Deformation Near a Triple Junction. *Tectonophysics* 301(3–4): 231–242.

Condra, G.E. 1902. New Bryozoa from the Coal Measures of Nebraska. *The American Geologist* 30: 337–358.

d'Orbigny, A. 1849. Description de Quelques Genres Nouveaux de Mollusques bryozoaires. *Revue et Magasin de Zoologie* 2(1): 499–504.

d'Orbigny, A. 1850. *Prodrome de Paléontologie Stratigraphique Universelle des Animaux Mollusques & Rayonnés, Faisant Suite au Cours Élémentaire de Paléontologie et de Géologie Stratigraphiques (Vol. 1)*. Paris: Victor Masson.

Elias, M.K. & Condra, G.E. 1957. *Fenestella* from the Permian of West Texas. *Geological Society of America, Memoir* 70: 1–158.

Ernst, A. 2001. Bryozoa of the Upper Permian Zechstein Formation of Germany. *Senckenbergiana Lethaea* 81 (1): 135–181.

Ernst, A. & Minwegen, E. 2006. Late Carboniferous Bryozoans from La Hermida, Spain. *Acta Palaeontologica Polonica* 51(3): 569–588.

Ernst, A., Claussen, A. L., Seuss, B. & Wyse Jackson, P. N. 2022. Stenolaemate Bryozoans from the Graham Formation, Pennsylvanian (Virgilian) at Lost Creek Lake, Texas, USA. *Palaeontologia Electronica* 25(2): a15.

Ernst, A., Seuss, B., Taylor, P.D. & Nützel, A. 2016. Bryozoan fauna of the Boggy Formation (Deese Group, Pennsylvanian) of the Buckhorn Asphalt Quarry, Oklahoma, USA. *Palaeobiodiversity and Palaeoenvironments* 96(4): 517–540.

Ernst, A. & Vachard, D. 2017. Middle Pennsylvanian Bryozoans of Cerros de Tule, Sonora, Mexico. *Neues Jahrbuch für Geologie und Paläontologie-Abhandlungen* 285: 11–38.

Etheridge, R. 1873. I.—Description of *Carinella*, a New Genus of Carboniferous Polyzoa. *Geological Magazine* 10(112): 433–434.

Fürsich, F.T. & Hurst, J.M. 1974. Environmental Factors Determining the Distribution of Brachiopods. *Palaeontology* 17(4): 879–900.

Gilmour, E.H. & Snyder, E.M. 2000. Bryozoa of the Mission Argillite (Permian) Northeastern Washington. *Journal of Paleontology* 74(4): 545–570.

González-Mora, S. & Sour-Tovar, F. 2014. Briozoarios del Orden Fenestrida, Pensilvánico de la Formación Ixtaltepec, Municipio de Nochixtlán, Oaxaca; Consideraciones paleoambientales. *Boletín de La Sociedad Geológica Mexicana*, 66(3): 471–482.

González-Mora, S., Wyse Jackson, P.N., Bancroft, A.J., & Sour-Tovar, F. 2020. Palaeocorynid-type Structures in Fenestellid Bryozoa from the Carboniferous of Oaxaca, Mexico. In P.N. Wyse Jackson & K. Zágoršek (eds.), *Bryozoan Studies 2019*: 35–42. Prague: Czech Geological Survey.

Gorjunova, R. 1985. Morphology, System und Phylogeny of Bryozoa (Order Rhabdomesida). *Trudy Paleontologičeskogo Instituta Akademii Nauk SSSR* 208: 1–152.

Hageman, S.J. 1991. Approaches to Systematic and Evolutionary Studies of Perplexing Groups: an Example Using Fenestrate Bryozoa. *Journal of Paleontology*, 65(4): 630–647.

Hageman, S.J. & McKinney, F.K. 2010. Discrimination of Fenestrate Bryozoan Genera in morphospace. *Palaeontologia Electronica* 13: 1–41.

Hageman, S.J., Wyse Jackson, P.N., Abernethy, A.R. & Steinthorsdottir, M. 2011. Calendar Scale, Environmental Variation Preserved in the Skeletal Phenotype of a Fossil Bryozoan (*Rhombopora blakei* n. sp.), from the Mississippian of Ireland. *Journal of Paleontology* 85(5): 853–870.

King, W. 1849. XLIII.—On Some Families and Genera of Corals. *Annals and Magazine of Natural History* 3 (17): 388–390.

Korn, H. 1930. Die Cryptostomen Bryozoen des Deutschen Perms. *Leopoldina* 6: 341–375.

Lonsdale, W. 1839. Corals. In R. I. Murchison (ed.), *The Silurian System. Part 2. Organic Remains*: 675–694. London: John Murray.

McKinney, F.K. & Gault, H.W. 1980. Paleoenvironment of Late Mississippian Fenestrate Bryozoans, Eastern United States. *Lethaia* 13(2): 127–146.

McKinney, F.K. & Jackson, J.B.C. 1991. *Bryozoan Evolution*. Chicago: University of Chicago Press.

M'Coy, F. 1844. *A Synopsis of the Characters of the Carboniferous limestone Fossils of the Ireland*. Dublin: Dublin University Press.

Meek, F.B. 1872. *Report on the Paleontology of Eastern Nebraska. Final Report on the United States Geological Survey of Nebraska and Portions of Adjacent Territorie*. United States Government Printing Office, Washington: 81–239.

Montgomery, H. 2004. Deposition and Emplacement of Permian Reefs in Sierra Plomosa, Chihuahua, Mexico. *Revista Mexicana de Ciencias Geológicas* 21 (2): 236–246.

Moore, R.C. 1929. A Bryozoan Faunule from the Upper Graham Formation, Pennsylvanian, of North Central Texas. *Journal of Paleontology* 3(2): 1–27, 121–156.

Morales-Soto, S. 1984. *Estudio Paleoecológico del Paleozoico Superior (Pensilvánico) de Santiago Ixtaltepec, Oaxaca* [Undergraduate thesis]. Facultad de Ciencias, Universidad Nacional Autónoma de México.

Morozova, I.P. 1962. On the Systematics and Phylogeny of the Fenestelloidea. *Paleontological Journal* 4: 104–115.

Morozova, I.P. 1974. Reviziya Roda *Fenestella*. *Paleontologicheskij Zhurnal* 2: 54–67.

Morozova, I.P. 2001. Mshanki Otryada Fenestellida (Morfologiya, Sistema, Filogeniya, Istoricheskoe Razvitie). *Trudy Paleontologicheskogo Instituta Rossijskoi Akademiya Nauk* 277: 1–176.

Navarro-Santillán, D., Sour-Tovar, F. & Centeno-García, E. 2002. Lower Mississippian (Osagean) Brachiopods from the Santiago Formation, Oaxaca, Mexico: Stratigraphic and tectonic implications. *Journal of South American Earth Sciences* 15(3): 327–336.

Ortiz-Lozano, J.A. 1996. *Consideraciones Paleoambientales y Morfológicas de las Colonias de Briozoarios Pensilvánicos de la Formación Ixtaltepec en la Región de Nochixtlán, Oaxaca* [Undergraduate thesis]. Facultad de Ciencias, Universidad Nacional Autónoma de México.

Pantoja-Alor, J. 1970. Rocas Sedimentarias Paleozoicas de la Región Centro-septentrional de Oaxaca. In L.R. Segura & R. Rodríguez-Torres (eds), *Libro Gúia de la Excursión México-Oaxaca*: 67–84. Mexico: Sociedad Geológica Mexicana.

Pantoja-Alor, J. & Robison, R.A. 1967. Paleozoic Sedimentary Rocks in Oaxaca, Mexico. *Science* 157 (3792): 1033–1035.

Phillips, J. 1836. *Illustrations of the Geology of Yorkshire: The Mountain Limestone district (Vol. 2)*. London: J. Murray.

Prout, H.A. 1859. Third Series of Descriptions of Bryozoa from the Paleozoic Rocks of Western States and Territories. *Transactions of St. Louis Academy of Sciences* 1: 443–452.

Quiroz-Barroso, S.A. & Perrilliat, M. del C. 1997. Pennsylvanian Nuculoids (Bivalvia) from the Ixtaltepec Formation, Oaxaca, Mexico. *Journal of Paleontology* 71(3): 400–407.

Quiroz-Barroso, S.A. & Perrilliat, M. del C. 1998. Pennsylvanian Bivalves from the Ixtaltepec Formation, Mexico. *Journal of Paleontology* 72(6): 1011–1024.

Quiroz-Barroso, S.A., Pojeta, J., Sour-Tovar, F., & Morales-Soto, S. 2000. *Pseudomulceodens*: A Mississippian Rostroconch from Mexico. *Journal of Paleontology* 74(6): 1184–1186.

Quiroz-Barroso, S.A. & Sour-Tovar, F. 1995. *Nuevo Registro de Ofiuroideo (Ophiurinidae) Para el Pensilvánico de América del Norte, Proveniente de la Formación Ixtaltepec, Oaxaca. Memoria del V Congreso Nacional de Paleontología, Sociedad Mexicana de Paleontología*: 31.

Quiroz-Barroso, S.A. & Sour-Tovar, F. 2006. Fossil Record of Upper Paleozoic Marine Invertebrates from Mexico. In Vega, F.J., Nyborg, T.G., Perrilliat, M.del C., Cevallos-Ferriz, S. & S.A. Quiroz-Barroso (eds), *Studies on Mexican Paleontology*: 133–167. Heidelberg: Springer.

Robison, R.A. & Pantoja-Alor, J. 1968. Tremadocian Trilobites from the Nochixtlán Region, Oaxaca, Mexico. *Journal of Paleontology* 42(3): 767–800.

Schlotheim, E.F. 1820. *Die Petrefactenkunde auf Ihrem Jetzigen Standpunkte Durch die Beschreibung Seiner Sammlung Versteinerter und Fossiler Überreste des Thier und Pflanzenreichs der Vorwelt erläutert*: Mit 15 Kupfertafeln. Becker.

Simpson, G.B. 1895. A Handbook of the Genera of the North American Paleozoic Bryozoa. *14th Annual Report of the State Geologist (of New York) for the Year 1894*: 407–608.

Sour-Tovar, F. 1994. *Braquiópodos Pensilvánicos Del Area de Santiago Ixtaltepec, Municipio de Nochixtlán, Oaxaca* [M.Sc. thesis]. Facultad de Ciencias, Universidad Nacional Autónoma de México.

Sour-Tovar, F. & Martínez-Chacón, M.L. 2004. Braquiópodos Chonetoideos del Carbonífero de México. *Spanish Journal of Palaeontology* 19(2): 125–138.

Suárez Andrés, J.L. & Ernst, A. 2015. Lower–Middle Devonian Fenestellidae (Bryozoa) of NW Spain: Implications for Fenestrate Palaeobiogeography. *Facies* 61(1): 415.

Tavener-Smith, R. 1973. Fenestrate Bryozoa from the Visean of County Fermanagh, Ireland. *Bulletin of the British Museum (National History) Geology* 23(7): 1–493.

Termier, H. & Termier, G. 1971. Bryozoaires du Paléozoïque Supérieur de l'Afghanistan. *Travaux et Documents des Laboratoires de Géologie de Lyon* 47 (1): 1–52.

Tolokonnikova, Z., Ernst, A. & Poty, E. 2015. Tournaisian (Lower Mississippian) Bryozoans from Belgium. *Neues Jahrbuch für Geologie Und Paläontologie – Abhandlungen 278*: 23–45.

Torres-Martínez, M.A. 2014. *Braquiópodos Carboníferos del área de Santiago Ixtaltepec, Oaxaca. Implicaciones Paleoambientales, Estratigráficas y Paleobiogeográficas* [Ph.D. thesis]. Facultad de Ciencias, Universidad Nacional Autónoma de México.

Torres-Martínez, M.A. & Sour-Tovar, F. 2012. Nuevos Braquiópodos Prodúctidos (Rhynchonelliformea, Strophomenata) del Carbonífero de la región de Nochixtlán, Oaxaca. *Revista Mexicana de Ciencias Geológicas* 29(3): 696–712.

Torres-Martínez, M.A. & Sour-Tovar, F. 2016a. Braquiópodos discínidos (Lingulida, Discinoidea) de la Formación Ixtaltepec, Carbonífero del área de Santiago Ixtaltepec, Oaxaca. *Boletín de La Sociedad Geológica Mexicana* 68(2): 313–321.

Torres-Martínez, M.A. & Sour-Tovar, F. 2016b. New productide Brachiopods (Productoidea) from the Carboniferous of Ixtaltepec Formation, Oaxaca, Mexico. *Journal of Paleontology* 90(3): 418–432.

Torres-Martínez, M.A. & Sour-Tovar, F. 2022. New rhynchonellid and Spire-bearing Brachiopods from the Carboniferous of Mexico. Paleogeographical Significance of the Oaxacan Brachiopod Fauna Through the Serpukhovian–Moscovian. *Journal of Paleontology*: 1–22.

Torres-Martínez, M.A., Sour-Tovar, F. & Pérez-Huerta, A. 2008. Neospiriferinid Brachiopods (Spiriferida, Trigonotretidae) from Ixtaltepec Formation, Pennsylvanian of Oaxaca State, Southern Mexico: Fossils and Strata. *Lethaia Supplement* 54, 157–166.

Villanueva-Olea, R., Castillo-Espinoza, K.M., Sour-Tovar, F., Quiroz-Barroso, S.A. & Buitrón-Sánchez, B.E. 2011. Placas Columnares de Crinoides del Carbonífero de la Región de Santiago Ixtaltepec, Municipio de Nochixtlán, Oaxaca; Consideraciones Estratigráficas y Paleobiogeográficas. *Boletín de La Sociedad Geológica Mexicana* 63(3): 429–443.

Villanueva-Olea, R. & Sour-Tovar, F. 2014. A New Genus and Four New Species of Cladid Crinoids from the Carboniferous of Oaxaca State, Mexico. *Journal of Systematic Palaeontology* 13(7): 527–542.

Vine, G. 1884. Fourth Report of the Committee Consisting of Dr. HR Sorby and Mr. GR Vine, Appointed for the Purpose of Reporting on Fossil Polyzoa. *British Association Advancement of Science, Report 53d, Meeting (Southport, 1883)*: 161–209.

Wyse Jackson, P.N. & Bancroft, A.J. 1995. Generic Revision of the Cryptostome Bryozoan *Rhabdomeson* Young and Young, 1874, with Descriptions of Two Species from the Lower Carboniferous of the British Isles. *Journal of Paleontology* 69(1): 28–45.

Young, J. & Young, J. 1874. XLVI.—On a New Genus of Carboniferous Polyzoa. *Journal of Natural History*, 13(77): 335–339.

Zittel, K. A. von. 1880. *Handbuch der Palaeontologie. I. Band. Palaeozoologie. 1. Abtheilung, Protozoa, Coelenterata, Echinodermata und Molluscoidea*: München Oldenbourg.

Bryozoan Studies 2022 – Key, Porter & Wyse Jackson (eds)
 ISBN: 978-1-032-59487-3

Bryozoan dispersal using pumice rafting generated by the submarine eruption in the Ogasawara, Japan

Masato Hirose*
School of Marine Biosciences, Kitasato University, Sagamihara-Minami, Kanagawa, Japan
ORCID ID: 0000-0001-9991-7238

Sota Kaneko
Nakijin Elementary School, Nakijin, Kunigami, Okinawa, Japan

ABSTRACT: Fukutoku-Oka-no-ba, a submarine volcano located in the Ogasawara Islands, erupted in August 2021 and a large amount of pumice washed ashore on the Okinawa Islands in October of the same year. *Jellyella eburnea* was found on small pieces of pumice on the west coast of Okinawa Island in the October. Subsequently, some larger colonies were also found on more than half of the pumice of various sizes collected on the beach at the north shore of Nakijin Village. This is the first report of *J. eburnea* from Japanese waters, and it is considered also common around Japan. Although *J. eburnea* has been reported from various drift materials and several studies on pumice rafting reported bryozoans including *Jellyella* spp., this is the first report of *J. eburnea* from pumice. Pumice rafting produced by submarine volcanic eruptions is thought to provide a large volume of rafting substrates, and it may also play a significant role in the dispersal of this bryozoan species.

1 INTRODUCTION

Jellyella eburnea (Hincks, 1891) is often dispersed globally on planktonic mollusks, algae, and other debris (Gordon *et al.* 2007; Taylor & Monks 1997; Taylor & Tan 2015). Recently, *J. eburnea* has also been reported as an invasive species attached to Japanese tsunami marine debris (McCuller & Carlton 2018), but has never been reported formally from the Japanese waters.

Fukutoku-Oka-no-ba (24.285°N, 141.481°E), a submarine volcano located in the Ogasawara Islands, erupted on 13 August 2021 (Hydrographic and Oceanographic Department Japan Coast Guard, Maeno *et al.* 2022). Approximately two months later, a large amount of pumice was reported to have washed ashore in the Okinawa Island, which is situated in the Ryukyu Islands and approximately 1300 km distant from the submarine volcano (Figure 1). Cheilostome bryozoan colonies along with goose barnacles (*Lepas* sp.) and algae were found attached to those pumice pieces (Ohno *et al.* 2022). Bryozoans attached to these pumice pieces were identified as *J. eburnea*, and it is possible that pumice supplied in large quantities at irregular intervals also contributes to the dispersal of this species.

The Ryukyu Islands are strongly affected by the warm Kuroshio Current, which originates east of the Philippines and flows northeastward along the coast of southern Japan. A following current of the Kuroshio Current flows southward at southeast off Japan, and part of the following current connecting the Kuroshio Counter-Current flows southwestward in the middle of the northwest Pacific, passes the Ogasawara Islands, and flows to the Ryukyu Islands (Figure 1). The following current of the Kuroshio Current also connect the North Equatorial Current in the middle of the west Pacific near the equator, and the North Equatorial Current flows westward to the Philippines where is the origin of the Kuroshio Current.

In this study, we describe the morphology of *J. eburnea* attached to pumice that is thought to have been generated by the eruption in the Ogasawara Islands, collected from the western coast of Okinawa. We also discuss the timing and frequency of attachment based on the number and surface area of colonies and autozooid size variation within colonies for brief estimation of the number of days from the attachment. The importance of pumice in the dispersal of this species is also discussed based on the frequency of attachment of bryozoan colonies on pumice rafts.

*Corresponding Author: mhirose64@gmail.com

DOI: 10.1201/9781003454915-5

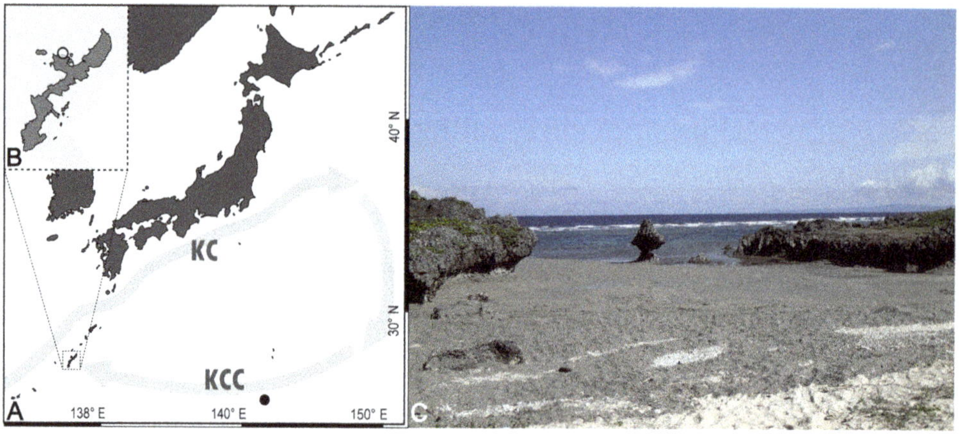

Figure 1. Map of the study area. A, map showing the locality of Fukutoku-Oka-no-ba submarine volcano located in the Ogasawara Islands (closed circle) and B, the locality where pumice were collected on Okinawa Island (open circle). Gray arrows indicate direction of the Kuroshio Current (KC) and the Kuroshio Counter-Current (KCC). C, the northern beach in Nakijin Village, Okinawa where the samples were collected. All of the grayish substrates shown in the photograph are drifted pumice.

2 MATERIALS AND METHODS

Drifted pumice samples were collected at the north shore of Nakijin Village in Okinawa, on 7 and 20 November 2021 and 10 August in 2022 (Figure 1). The maximum length of each pumice sample was measured using a ruler to the nearest 0.01 mm. Bryozoan colonies on pumice were observed under an Olympus SZX16 stereomicroscope and photographed with an Olympus DP73. Some colonies were also observed by using scanning electron microscope (SEM) (KEYENCE, VE-9800). The length and width of zooids in three randomly selected colonies collected in November 2021 was measured based on the SEM images. All measurements of bryozoans were taken to the nearest 0.01 mm from digital images using ImageJ 1.37v software (Image Processing and Analysis in Java, Wayne Rasband, National Institutes of Health, USA: http://rsb.info.nih.gov/ij/). Measurements are presented in millimeters as the range, followed by the average ± standard deviation (SD) with the sample size in parentheses using the following abbreviations: length of the zooid, ZL; width of the zooid, ZW; length of the opesia, OL; width of the opesia, OW; length of the operculum, OpL; width of the operculum, OpW.

3 RESULTS

3.1 *Description of the species*

Jellyella eburnea (Hincks, 1891)
(Figure 2)

Material examined. NSMT-Te1245, single small colony on drifted pumice, collected on 7 November 2021, shore in Nakijin Village, Okinawa (26°42′28.56″N, 127°56′59.52″E); NSMT-Te1246–1269, colonies on drifted pumice, collected on 20 November 2021 and 10 August 2022, shore in Nakijin Village, Okinawa.

Description. Colony encrusting, multiserial, unilamellar, white in color when dry. Autozooids irregularly rhomboidal, subhexagonal, or oval; ZL 0.36–0.74 (0.54 ± 0.10) mm, ZW 0.21–0.37 (0.29 ± 0.04) mm (n = 30), about 1.9 times longer than wide. Zooids arranged somewhat quincuncially, bifurcate preceded by a broader zooid followed by two or three narrower zooids. Zooidal boundaries in bleached specimens marked by a fine fissure. Gymnocyst variably developed; from just as a thin lateral and proximal rim (Figure 2C) to folded into frill-like projection or hollow spines slightly hanging over the proximal and lateral margins of opesia when it well-developed (Figure 2D), reached approximately 0.08 mm long with 0.03 mm diameter. In some zooids, gymnocyst well developed proximally and occupying about one-thrid of the proximal frontal area of the zooid. Cryptocyst forming a narrow crescent shelf at distal end of zooid.

Opesia longitudinally oval or subhexagonal in outline, occupying most of frontal surface; OL 0.26–0.54 (0.39 ± 0.08) mm, OW 0.19–0.30 (0.24 ± 0.03) mm (n = 30). Pore chambers absent. Vertical walls thin; lateral walls contaning approximately four uniporous or multiparous septulae, arranged almost equal distance at near the basal wall (Figure 2E); distal walls contaning more than 20 uniporous septulae, arranged densely (Figure 2F). Basal wall of zooids weakly calcified or uncalcified with large window. No ovicells, avicularia, or other heterozooids.

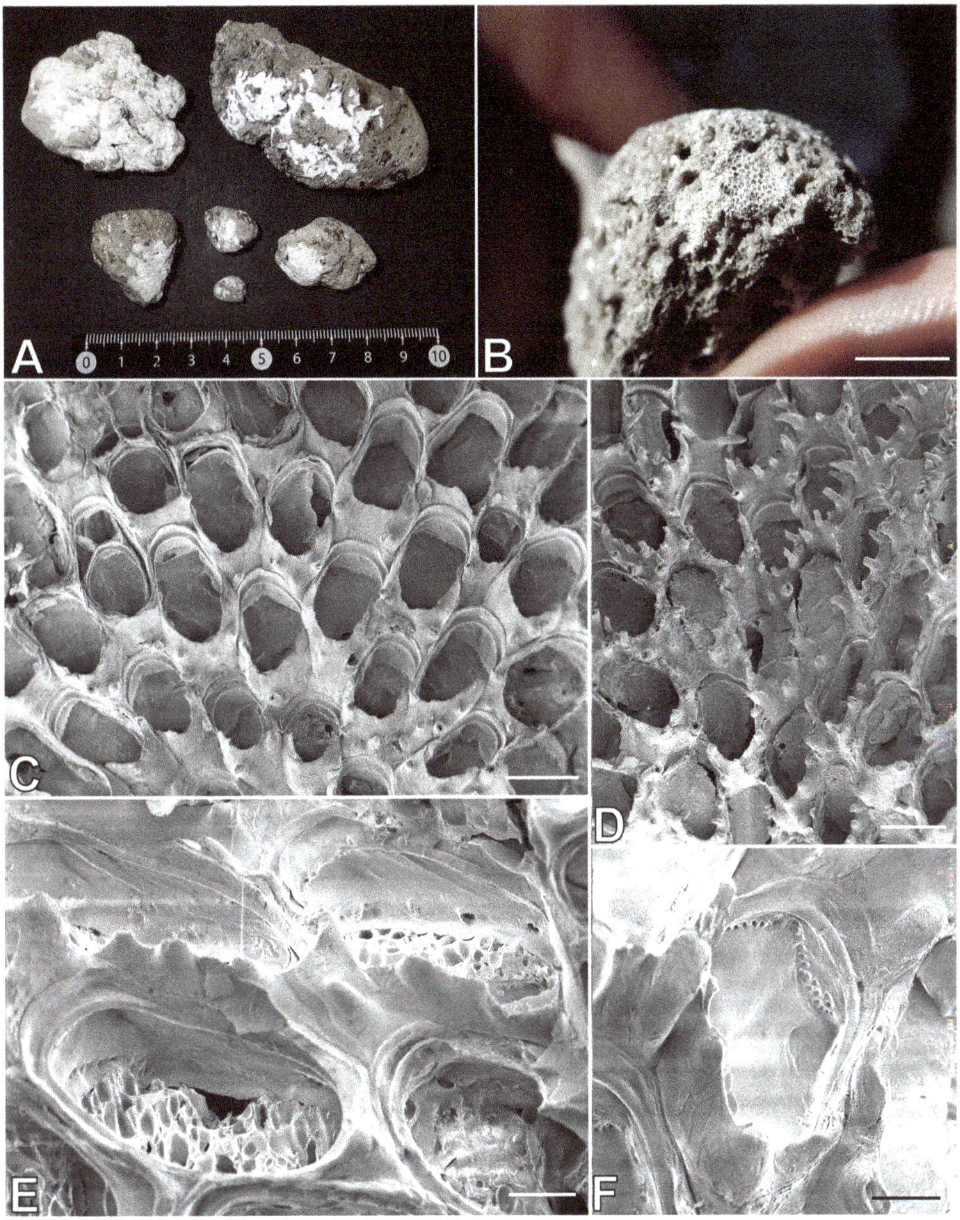

Figure 2. Photographs and scanning electron micrographs of *Jellyella eburnea* on pumice. A, B, colonies of *Jellyella eburnea* (white patches) on pumice (the numbers on ruler are shown in cm); C, autozooids in the younger part of colony; D, autozooids with a spiny rim of gymnocyst in the older part of colony; E, autozooids without a basal wall showing some septulae on the lateral wall; F, autozooids showing numerous uniporous septulae on the distal wall. Scale bars: B = 1 cm; C, D = 300 μm; E, F = 100 μm.

Frontal membrane translucent and shiny when in dry. Operculum semicircular, with brown marginal thickening at distal rim; OpL 0.07–0.10 (0.08 ± 0.01) mm, OpW 0.11–0.15 (0.13 ± 0.01) mm (n = 15), about 1.6 times broader than length. Polypide white to yellowish brown when in dry; tentacle number unknown.

Ancestrula twinned, small, irregularly oval, covered by the gymnocyst of the neighboring periancestrular zooids in the ancestrular complex and obscure; opesia occupying most of the surface.

Remarks. Wide morphological variations have been reported in *J. eburnea* (Taylor & Monks 1997; Taylor & Tan 2015). In some previous studies, it is thought several cryptic species were included in the previously reported nominal *J. eburnea* (McCuller & Carlton 2018). However, in this study, we treated all of the reports of the current species as a single species, on account of their rapid attachment and dispersal ability using various rafting objects such as floating shells of the squid *Spirula*, planktonic gastropod *Janthina*, plastic debris, and glass bottles (Goldstein *et al.* 2014; McCuller & Carlton 2018; Moyano 2005; Taylor & Monks 1997; Taylor & Tan 2015; Thiel & Gutow 2005). Further study based on a wider sample collection and study using both morphology and molecular data is needed to verify the presence or absence of cryptic species within this widely distributed species.

Distribution. Jellyella eburnea has a particularly wide distribution from Florida, West Atlantic, Straits of Magellan, South Africa, East Africa, Madagascar, Mauritius, Australia, Fiji, New Caledonia, New Zealand, Malaysia, and the Northwest Pacific (Dauphin 1979a, 1979b; Gordon *et al.* 2007; Stevens *et al.* 1996; Taylor & Monks 1997; Taylor & Tan 2015; Winston 1982).

3.2 *Number and size of colony*

3.2.1 *Number of colonies*

In total, at least 50 colonies were found on 22 of a total of 34 randomly collected pieces of pumice on 20 November 2021.The percentage of pumice encrusted by the bryozoan colonies was about 65%. On average, 2.3 colonies were found on each piece of pumice specimen (n = 22).

The size of the pumice pieces with bryozoan colonies was about 28.8 ± 12.6 mm (9.2–65.3 mm; n = 22) long in the maximum length. There was no significant correlation between the maximum length of the pumice pieces and the number of bryozoan colonies (Figure 3A).

In August 2022, 108 pumice specimens were encrusted by bryozoan colonies among the total 226 of randomly examined pumice pieces. The percentage of pumice piece encrusted by bryozoan colonies was about 48%. The maximum length of the pumice pieces encrustedby bryozoan colonies was about 24.8 ± 10.4 mm (9.0–67.0 mm; n = 108), whereas that of the pumice lacking bryozoan colonies were slightly smaller at 18.1 ± 8.4 mm (10.0–62.0 mm; n = 118) (Figure 4).

3.2.2 *Size of colonies and zooids*

The range of the sizes of observed colonies on the pumice collected in November 2021 is shown in Table 1. The surface area of colonies on 22 pumice pieces ranged from 5.5 to 1503.3 mm^2 (143.5 ± 254.8 mm^2; n = 50). There was no significant correlation between the surface area of colonies and the maximum length of the pumice pieces (Figure 3B).

The range of the sizes of observed zooids in each of the three randomly selected colonies are shown in Table 2. The length of zooids in a single colony ranged from 0.36 to 0.74 mm while the width ranged from 0.21 to 0.37 mm. There were no

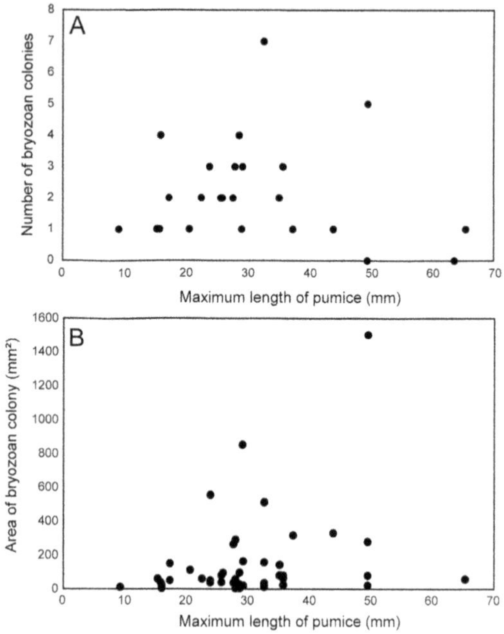

Figure 3. Maximum length of the examined pumice pieces and the number of bryozoan colonies (A) and the surface area of bryozoan colonies (B), collected in November 2021.

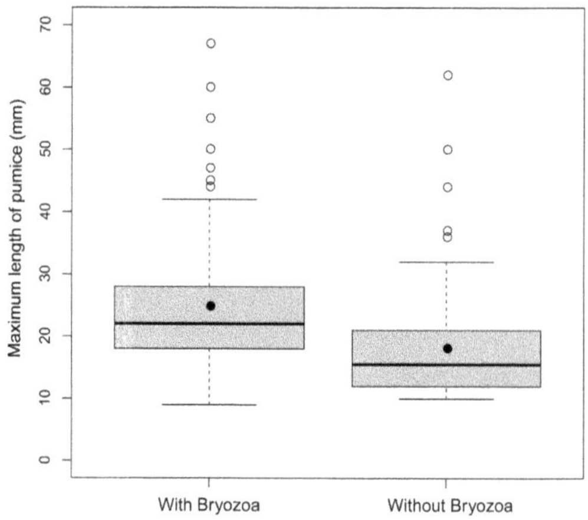

Figure 4. Maximum length of pumice pieces with bryozoans (n = 108) compared to those without bryozoan colonies (n = 118), collected in August 2022. Open circles indicate outlier, and closed circles indicate the mean value of each data.

Table 1. The sizes of observed colonies and pumice collected in November 2021.

	Min	Max	Mean	SD	N
Size of colony (mm^2)	5.52	1503.34	143.45	254.81	50
Maximum length of pumice (mm)	9.17	65.32	28.78	12.62	22

Table 2. The sizes of observed autozooids in each of three examined colonies collected in November 2021.

Colony #		1	2	3
Colony surface area (mm^2)		265.8	165.1	151.1
Zooid length	MIN (mm)	0.36	0.52	0.44
	MEAN (mm)	0.44	0.60	0.57
	MAX (mm)	0.51	0.74	0.69
	SD	0.05	0.07	0.08
	N	10	10	10
Zooid width	MIN (mm)	0.21	0.27	0.24
	MEAN (mm)	0.27	0.30	0.30
	MAX (mm)	0.37	0.34	0.37
	SD	0.05	0.02	0.04
	N	10	10	10

significant differences in zooid length and width between the colonies of different sizes (Kruskal-Wallis test, $p = 0.1468$).

4 DISCUSSION

In this study, more than half of the examined drifted pumice samples of various sizes were encrusted by colonies of *Jellyella eburnea* about three months after the eruption of Fukutoku-Oka-no-ba. The bryozoan colonies were not only attached on the larger pumice pieces but also on the smaller (< 1 cm in maximum length). If abundant larvae of this species are floating near the sea surface at a particular time, more and larger colonies should grow on larger pumice pieces. However, there were no significant correlations between the number or size of colonies and the size of the pumice. The results may indicate that a moderate number of larvae were available of multiple times, and the new colonies were always forming by nonselective attachment to pumice pieces by the larvae. The moderate to relatively high variability of autozooid size within a colony compared to that of the lectotype (0.51 ± 0.07 mm in length and 0.27 ± 0.02 mm in width) and one colony from Western Australia (0.61 ± 0.03 mm in length and 0.28 ± 0.03 mm in width) (from Taylor & Monks 1997) may also indicate the moderate duration of the growth during rafting. The percentage of bryozoan colony attachment to the drifted pumice was slightly decreased a year after the eruption in August 2022 (i.e. from 65% to 48%). Furthermore, pumice pieces that were not encrusted by bryozoan colonies tended to be slightly smaller overall. This is maybe due to the production of many smaller pumice fragments caused by the break up of larger pumice pieces on the beach.

Although the growth rate of the bryozoan colony varies depending on the food resources, substrate types, water temperature, flow velocity, and other environmental differences (e.g. Menon 1972; Okamura & Partridge 1999; Saunders & Metaxas 2009), some species such as the epifaunal encrusting colony of membraniporid bryozoans (e.g. *Membranipora membranacea*) generally grow rapidly (Eckman & Duggins 1993; Okamura & Partridge 1999; Seed 1976; Saunders & Metaxas 2009; Yoshioka 1973). Under the field experiments in a previous study of *M. membranacea*, growth rate of the younger colonies (5 to 20 mm^2/day) was smaller than that of the larger colonies (86 to 103 mm^2/day) (Eckman & Duggins 1993; Okamura & Partridge 1999). The observed mean colony size on pumice ranged from 5.5 to 1503.3 mm^2 in this study. The results suggest that the colonies on the examined pumice were quite young and probably attached at various different times during transport.

Pumice rafting is well known as an important process facilitating the distribution of marine organisms (Bryan *et al.* 2004, 2012; Jokiel 1990a, 1990b; Jokiel & Cox 2003; Ohno *et al.* 2022; Winston 2012). Submarine volcanic eruptions are low frequency events, and pumice is not always present on the sea surface since it disappears rapidly once beached. However, once pumice is supplied by a large submarine volcanic eruption, a huge volume of pumice can raft for more than a year and is transported over several thousand kilometers. In this study, most pumice was encrusted by various sizes of *J. eburnea* colonies. The results suggest that new larvae continuously attached and grew during the drifting period by larvae derived from a colony on another pieces of pumice or other floating objects. Therefore, if the area is regularly supplied with drift objects other than pumice (e.g. algae or artificial debris), pumice dispersal would be a factor in the rapid expansion of the bryozoan species distribution to other habitats, unlike the regular but small number of drift algae and drifting debris. The pumice from submarine volcanic eruptions may also be transported to more distant and different areas over time, either continuously or intermittently, by the complex ocean surface circulation patterns, as the marine debris originating from the Great East Japan Earthquake and Tsunami of 2011. Therefore, the supply of pumice by submarine volcanic eruptions may contribute significantly to the wide distribution of the bryozoan species.

ACKNOWLEDGEMENTS

We thank Atsushi Kaneko and Hiroko Takaoka for providing the opportunity to carry out this research and supporting it. We thank Koshiro

Hirose and Atsuki Hirose for cooperating the field survey. This study was partly supported by JSPS Grant-in-Aid for Scientific Research (C) Grant Number 22K05787.

REFERENCES

Bryan, S.E., Cook, A., Evens, J.P., Colls, P.W., Wells, M.G., Lawrence, M.G., Jell, J.S., Greig, A. & Leslie, R. 2004. Pumice Rafting and Faunal Dispersion During 2001–2002 in the Southwest Pacific: Record of a Dacitic Submarine Explosive Eruption from Tonga. *Earth and Planetary Science Letters* 227: 135–54.

Bryan, S.E., Cook, A.G., Evans, J.P., Hebden, K., Hurrey, L., Colls, P., Jell, J.S., Weatherley, D. & Firn, J. 2012. Rapid, Long-Distance Dispersal by Pumice Rafting. *PLoS One* 7 (7): e40583.

Dauphin, Y. 1979a. Contribution à l'étude de la Formation des Gisements de Céphalopodes. I. Les coquilles de Spirules (Dibranchiata) de Nouvelle Calédonie. *Cahiers Indo-Pacifique* 1 (2): 165–194.

Dauphin, Y. 1979b. Contribution à L'étude de la Formation des Gisements de Céphalopodes. II. Les Coquilles de Spirules (Dibranchiata, Cephalopoda) de l'ile Maurice (Océan Indien). *Cahiers Indo-Pacifique* 1 (3): 335–351.

Eckman, J.E. & Duggins, D.O. 1993. Effects of Flow Speed on Growth of Benthic Suspension Feeders. *The Biological Bulletin* 185: 28–41.

Goldstein, M.C., Carson, H.S. & Eriksen, M. 2014. Relationship of Diversity and Habitat Area in North Pacific Plastic-Associated Rafting Communities. *Marine Biology* 161: 1441–1453.

Gordon, D.P., Hossain, M.M. & Wood, T. 2007. The Known and Anticipated Bryozoan Diversity of Bangladesh. *Journal of Taxonomy and Biodiversity Research* 1: 45–58.

Hincks, T. 1891. Contributions Towards a General History of the Marine Polyzoa. XV. South-African and other Polyzoa. *Annals and Magazine of Natural History, Series 6*, 7: 285–298.

Hydrographic and Oceanographic Department Japan Coast Guard: Fukutoku-Oka-no-Ba submarine volcano information. https://www1.kaiho.mlit.go.jp/GIJUTSUKOKUSAI/kaiikiDB/kaiyo24-2.htm

Jokiel, P.L. & Cox, E.F. 2003. Drift Pumice at Christmas Island and Hawaii: Evidence of Oceanic Dispersal Patterns. *Marine Giology* 202: 121–133.

Jokiel, P.L. 1990a. Transport of Reef Corals onto the Great Barrier Reef. *Nature* 347: 665–667.

Jokiel, P.L. 1990b. Long-distance Dispersal by Rafting: Reemergence of an Old Hypothesis. *Endeavour* 14: 66–73.

Maeno, F., Kaneko, T., Ichihara, M., Suzuki, Y., Yasuda, A., Nishida, K. & Ohminato, T. 2022. First timeseries Record of a Large-scale Silicic Shallow-sea Phreatomagmatic Eruption. *Research Square*.

McCuller, M.I. & Carlton, J.T. 2018. Transoceanic Rafting of Bryozoa (Cyclostomata, Cheilostomata, and Ctenostomata) Across the North Pacific Ocean on Japanese Tsunami Marine Debris. *Aquatic Invasions* 13: 137–162.

Menon, N.R. 1972. Heat Tolerance, Growth and Regeneration in Three North Sea Bryozoans Exposed to Different Constant Temperatures. *Marine Biology* 15: 1–11.

Ohno, Y., Iguchi, A., Ijima, M., Yasumoto, K. & Suzuki, A. 2022. Coastal Ecological Impacts from Pumice Rafts. *Scientific Reports* 12 (1): 11187.

Okamura, B. & Partridge, J.C. 1999. Suspension Feeding Adaptations to Extreme Flow Environments in a Marine Bryozoan. *The Biological Bulletin* 196: 205–215.

Saunders, M.I. & Metaxas, A. 2009. Effects of Temperature, Size, and Food on the Growth of *Membranipora membranacea* in Laboratory and Field Studies. *Marine Biology* 156: 2267–2276.

Seed, R. 1976. Observations on the Ecology of *Membranipora* (Bryozoa) and a Major Predator *Doridella steinbergae* (Nudibranchiata) Along the Fronds of *Laminaria saccharina* at Friday Harbor, Washington. *Journal of Experimental Marine Biology and Ecology* 24: 1–17.

Stevens, L.M., Gregory, M.R. & Foster, B.A. 1996. Fouling Bryozoa on Pelagic and Moored Plastics from Northern New Zealand. In: D.P. Gordon, A.M. Smith & J.A. Grant-Mackie (eds), *Bryozoans in Space and Time*: 321–340. Wellington: NIWA.

Taylor, P.D. & Monks, N. 1997. A New Cheilostome Bryozoan Genus Pseudoplanktonic on Molluscs and Algae. *Invertebrate Biology* 116 (1): 39–51.

Taylor, P.D. & Tan, S.H.A. 2015. Cheilostome Bryozoa from Penang and Langkawi, Malaysia. *European Journal of Taxonomy* 149: 1–34. d

Thiel, M. & Gutow, L. 2005. The Ecology of Rafting in the Marine Environment. II. The Rafting Organisms and Community. *Oceanography and Marine Biology: An Annual Review* 43: 279–418.

Winston, J.E. 2012. Dispersal in Marine Organisms without a Pelagic Larval Phase. *Integrative and Comparative Biology* 52: 447–457.

Yoshioka, P.M. 1973. The Population Dynamics and Ecology of the Encrusting Ectoproct *Membranipora serrilamella*. PhD Thesis, University of California, San Diego, p. 143 + xii.

Bryozoan Studies 2022 – Key, Porter & Wyse Jackson (eds)
© 2023 The Author(s), ISBN: 978-1-032-59487-3

Fouling of the slipper lobster, *Scyllarides latus*, by cyclostome and ctenostome bryozoans in the Mediterranean Sea off Malta

Marcus M. Key, Jr.*
Department of Geosciences, Dickinson College, Carlisle, PA USA
ORCID ID: 0000-0003-4097-0143

Sebastian H. Decker
Department of Evolutionary Biology, University of Vienna, Vienna, Austria
ORCID ID: 0000-0001-9029-8717

ABSTRACT: A Mediterranean slipper lobster, *Scyllarides latus* (Latreille, 1802), was found near Malta in the Mediterranean Sea. It was conspicuously fouled by bryozoans which is a rare occurrence for lobsters in general, including this species. Based on its size, it was estimated to be an adult with an age of eight years and is eight months post-molt. Five erect branching *Crisia* sp. cyclostome bryozoan colonies, three encrusting lichenoporid cyclostome bryozoan colonies, and a few small ctenostome colonies were found on the dorsal surface of the lobster. The potential costs and benefits of this rare epibiotic relationship for the lobster and the bryozoans are reviewed. The goal of this paper is to describe this rare relationship between epizoic bryozoans and a slipper lobster.

1 INTRODUCTION

Epibiosis is the ecological relationship between organisms growing attached to a living surface such as the bryozoans on the lobster in this study. We follow the terminology of Wahl (1989) and refer to the slipper lobster as the basibiont (i.e. the motile host arthropod substrate) and the bryozoans as the epibiont (i.e. the sessile organism attached to the basibiont's exoskeleton). Fouling refers to the more general colonization process of a solid surface, living or dead, by epibionts (Wahl 1989). Here we use epibiosis and fouling interchangeably. Some epibionts are opportunists that simply require a hard substrate, others are obligate symbionts. Epibiosis is important because fouling epibionts degrade the functionality of ship hulls, heat exchangers, and water intake pipes (Hellio & Yebra 2009). The presence of epibionts like bryozoans can cover hosts' eyes, inhibit hosts' wound healing, decrease hydrodynamic efficiency of hosts, and result in an unappealing appearance of the affected lobster tails (the main consumable part of all lobsters), which lowers their commercial value (Zha *et al.* 2017).

A variety of organisms can foul the exoskeleton of lobsters, including bryozoans, hydroids, barnacles, serpulid worms, molluscs and many microorganisms (Key *et al.* 2021). Epibionts on lobsters are not as common as on other basibionts. This has been attributed to grooming by the host to remove settling larvae (Phillips *et al.* 1980; Quinn *et al.* 2009) and molting. Because lobsters typically do not have a terminal molt, they continue to molt throughout life, casting off any accumulated epibionts (Shields 2011). Thus, fouling of lobsters, such as reported here, is unusual.

The objective of this study is to describe a rare occurrence of bryozoans fouling the Mediterranean slipper lobster *Scyllarides latus* (Latreille, 1802).

2 MATERIALS AND METHODS

A Mediterranean slipper lobster, *Scyllarides latus* (Latreille, 1802) was noticed during a SCUBA dive on 27 July 2021 as it was conspicuously fouled by bryozoans (Figure 1A). It was found 40–50 m off the north coast of the island of Gozo, Malta in the Mediterranean Sea (36.080874°N, 14.241772°E) at a depth of 21.3 m (Figure 2). The water temperature was 23°C. For photographic documentation, a Canon PowerShot G7 X Mark II (Canon, Tokyo, Japan) in a Fantasea (Fantasea Photo and Marketing Ltd., Blaine, Washington, United States) underwater housing was used with an additional +7 diopter for higher magnification.

*Corresponding Author: key@dickinson.edu

DOI: 10.1201/9781003454915-6

Figure 1. The Mediterranean slipper lobster, *Scyllarides latus*, examined in this study. A, bryozoan colonies on host's carapace circled. B, closeup of rectangle in A showing the diversity of epibionts. C, close up of erect *Crisia* sp. colony in B. D, close up of encrusting lichenoporid colony in B.

This lobster's main defense mechanism is their fortress strategy (Lavalli *et al.* 2007). This involves clinging tenaciously with their sharp pointed stout legs to a rough substrate protected by their broad flattened heavily armored carapace, similar to a limpet (Barshaw & Spanier 1994a, 1994b; Spanier & Lavalli 1998). Their clinging force is up to 150 N, equivalent to 29 times their body weight (Spanier & Lavalli 1998). Therefore, to prevent damage to the delicate erect bryozoan colonies, the lobster was not pried off its rocky substrate. Thus, we were unable to photograph the ventral surface

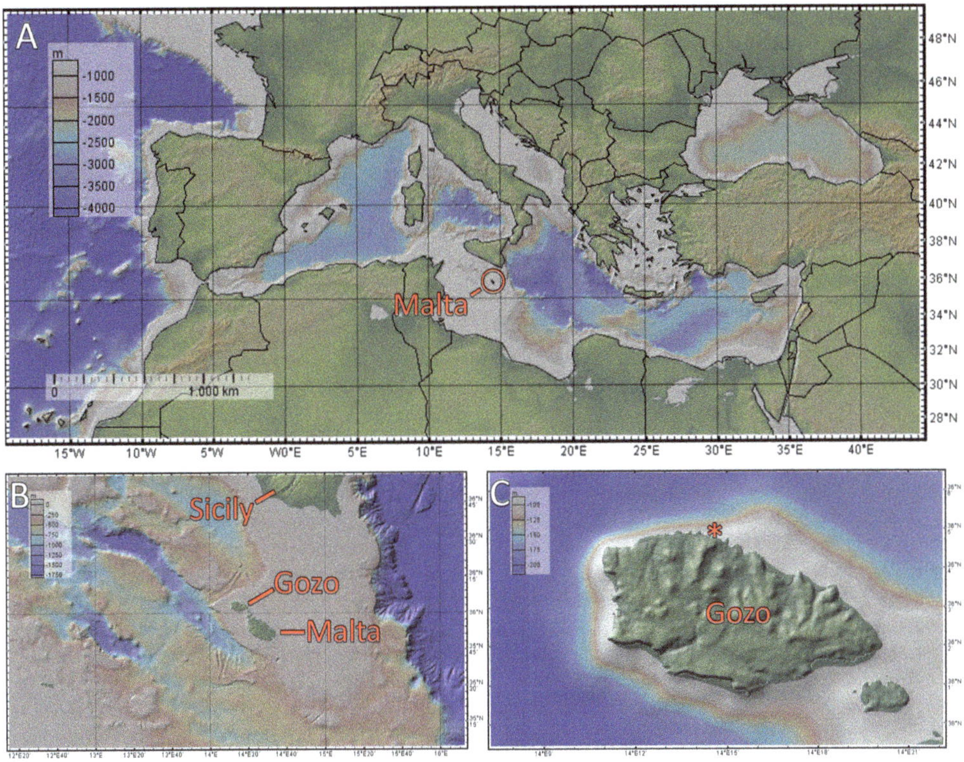

Figure 2. Bathymetric maps of the Mediterranean Sea. A, general location of Malta marked by a red circle. B, close-up of the Strait of Sicily showing islands of Malta and Gozo relative to Sicily. C, close-up of the Island of Gozo with the exact locality of the investigated slipper lobster (*Scyllarides latus*) marked by a red asterisk (36.080874°N, 14.241772°E). Maps were generated with GeoMapApp 3.6.14 using the Global Multi-Resolution Topography synthesis by Ryan *et al.* (2009).

and to observe the sexual orifices and the sexually dimorphic shapes of the pereiopods on the ventral surface (Romeo *et al.* 2004; Spanier & Lavalli 1998). Therefore, we did not determine the sex of the lobster.

3 RESULTS

Compared to other lobsters in the vicinity, the host lobster was fully mature and 25–30 cm in total length. Total length is not the usual size metric in lobsters; carapace length is the standard metric (Phillips *et al.* 1980). Using Martins (1985, fig. 3) to predict carapace length from total length in *S. latus*, our lobster would have had a carapace length of ~10 cm.

We counted five erect branching *Crisia* sp. cyclostome bryozoan colonies, three encrusting lichenoporid cyclostome bryozoan colonies, some small unidentified ctenostome bryozoan colonies, three encrusting serpulids, and some small hydrozoans (Figure 1B). All were growing on the dorsal surface of the host lobster, specifically the carapace and the enlarged flattened left second antennae. As stated above, we did not have access to the ventral surface. With only field photographs and no prepared SEM images, we could not determine the exact species of the bryozoans. The erect *Crisia* sp. colonies were up to 1 cm tall (Figure 1C). The *Crisia* colonies were either *C. sigmoidea* Waters 1916 or *C. oranensis* Waters 1916. The disc shaped lichenoporids were either *Disporella* sp. or *Patinella radiata* (Audouin 1826) and were up to 3 mm by 4 mm (Figure 1D).

4 DISCUSSION

A carapace length of 10 cm indicates an individual of 8 years of age (Bianchini & Ragonese 2007, fig. 9.1). An 8 year old lobster is an adult, and adult individuals of *S. latus* typically molt annually over the winter months, November to February (Bianchini *et al.* 2001; Martins 1985; Spanier *et al.* 1988; Spanier & Lavalli 1998). Therefore, based on the July date of our lobster, it was probably seven months, but up to eight months, post molt. The more recently *S. latus* has molted, the "cleaner" the carapace is (i.e. free from epibionts) (Bianchini *et al.* 2001; Spanier & Lavalli 1998). The lobster in this study was likely 58–67% of its way through its molt cycle, and thus an older, more fouled host.

We used Smith's (1992) experimental data on bryozoan colony growth rates to estimate the age of the largest (i.e. 1 cm tall) *Crisia* colony in this study. Smith (1992) did not include this weakly calcified genus in her study but did include another weakly calcified erect cheilostome species, *Caberea zelandica*. A 1 cm tall colony, weighing 0.020 g (Smith 1992, tbl. A1.3), and growing at 0.030 g/year (Smith 1992, tbl. 1D.2) would have been 8 months old. Even though these species are from different clades and environments, it adds credence to the estimate of the host lobster being up to 8 months post molt, assuming the host lobster was fouled by the bryozoan larvae soon after molting. This is possible as the carapace of *S. latus*, like other slipper lobsters without defensive claws and spines, hardens within 3–24 hours as compared to the prolonged post-molt soft condition of clawed lobsters (Barshaw *et al.* 2003; Horne & Tarsitano 2007).

The key for a successful epibiotic relationship, from the bryozoan perspective, is that the epibiont must reach reproductive maturity in less time than it takes for the basibiont to complete its molt cycle. Were the bryozoan colonies big enough to have sexually reproduced to make this host lobster basibiont a viable substrate? There is no data relating erect colony size to timing of sexual maturity, and no ovicells or embryos were visible on the colonies. But there is such data for encrusting cheilostome species, and it ranges widely. For example, *Rallocytus ridiculs* reaches sexual maturity at only four zooid size, many interstitial species reach sexual maturity by <10 zooids, *Drepanophora* sp. by 30 zooids, *Parasmittina* sp. and *Stylopoma spongites* by 150 zooids, but *Stylopoma* sp. not until it has 4600 zooids (Grischenko *et al.* 2018; Herrera *et al.* 1996; Jackson & Wertheimer 1985; Winston & Hakansson 1986). The encrusting cyclostome colonies were ~100 zooids in size. Thus, it is possible that these bryozoan colonies may not have reached sexual maturity before their host molted.

Most of the epizoic bryozoans in this study were cyclostomes. Cyclostomes are not as common as epizoans compared to cheilostomes and ctenostomes. For example, in a study of bryozoan fouling of crabs, cyclostomes were reported in only 10% of the cases of syn-vivo bryozoan fouling of extant crabs (Key *et al.* 2017, tbl. 3). Dick *et al.* (1998) reported an unidentified lichenoporid cyclostome and *Crisia* sp. fouling the snow crab *Chionoecetes bairdi*. The lichenoporids were found encrusting the dorsal and ventral surfaces, whereas the *Crisia* colonies were only on the ventral legs (Dick *et al.* 1998). In the crab fossil record, cyclostomes made up 16% of the cases (Key *et al.* 2017, tbl. 4). Cyclostomes have also been reported fouling fossil nautiloids (Turek 1987; Wyse Jackson & Key 2014) and trilobites (Tetreault 1992, 1997).

It is not surprising that one of the cyclostomes fouling the lobster was *Crisia* sp. as *Crisia* is one of the most species rich bryozoan genera in the Mediterranean (Harmelin 1968; Rosso 2003; Rosso & Di Martino 2016). In the Mediterranean, *Crisia sigmoidea* and *C. oranensis* are known to encrust rocks, sea grasses, sponges, and corals (Harmelin 1968) but not motile animals such as lobsters. In the south Pacific, erect crisiid colonies were reported on the red rock lobster *Jasus edwardsii* (Key *et al.* 2021).

Disporella and *Patinella* are just two of the many encrusting bryozoan genera in the Mediterranean (Rosso 2003; Rosso & Di Martino 2016). Encrusting bryozoans are much more common epibionts than erect bryozoans (Key *et al.* 2021). So, why were there more erect colonies (i.e. five) than encrusting colonies (i.e. three) on the host lobster in this study? It may be a function of the closely spaced setal tufts of the conspicuous and high tubercles that cover the surface of this lobster's carapace (Figure 1C) (Spanier & Lavalli 2013, fig. 13.18). These would have made an erect growth habit easier with its smaller surface area of attachment on the host lobster's carapace.

The incidence of epibionts on crustaceans is also a function of host grooming (Fernandez-Leborans 2010). *Scyllarides latus* does groom itself using mouthparts and legs to clean different parts of its body (Barshaw & Spanier 1994b; Lavalli *et al.* 2007, 2018; Spanier & Lavalli 1998). But those parts are generally ventral appendages, not the dorsal carapace where the bryozoans in this study were found.

What are the costs of this epibiotic relationship to the lobster and the bryozoans? Possible detrimental effects to the host include: (1) increased weight; probably not a major problem due to the almost neutral buoyancy of the bryozoans, (2) increased drag; probably not a major problem due to the encrusting nature of some of the colonies and the slow walking movement of the host, (3) shared doom (i.e. when a predator of either the host or the epibiont kills both (Wahl & Hay 1995)) (e.g. predation of bryozoans probably not a major problem as the epizoic bryozoan lifestyle may reduce susceptibility to predation via camouflage), and (4) increased susceptibility to predation (e.g. if the bryozoan colonies increase the host's weight and drag, decrease hosts functionality by overgrowing eyes or decreasing its flexibility, and attract predators of the bryozoans which may also eat the host) (Key & Barnes 1999; Key *et al.* 1995, 1996a, 1996b, 1999, 2000, 2010).

Possible detrimental effects to the bryozoans include: (1) being cast of by molting of the host, (2) abrasion of colonies by host crawling into crevices, (3) abrasion of colonies during mating, and (4) shared doom (e.g. predation of host) (Key & Barnes 1999; Key *et al.* 1995, 1996a, 1996b, 1999, 2000, 2010).

What are the benefits of this epibiotic relationship to the lobster and the bryozoans? The bryozoans may have provided camouflage to the host lobster as suggested by Martins (1985). For example, Vance (1978) documented a decrease in predation by starfish on clams, which were covered by bryozoans. Some crustaceans such as the decorator crabs purposefully camouflage themselves with bryozoans (Parapar *et al.* 1997; Stachowicz & Hay 2000). Camouflage is important for this species of lobster as it has no defensive claws or spines like other lobsters (Barshaw *et al.* 2003). The main predators of *S. latus* are humans through the over exploitative commercial fishery (Fisher 1973; Fisher *et al.* 1981; Martins 1985; Spanier 1991) and a variety of bony fishes, especially triggerfishes, *Balistes* spp. (Barshaw & Spanier 1994a, fig. 2; Barshaw *et al.* 2003; Lavalli & Spanier 2001; Martins 1985, Spanier & Lavalli 1998). Slipper lobsters such as *S. latus* are already camouflaged due to their flattened morphology and coloration (Barshaw & Spanier 1994a; Ogren 1977). This camouflage is enhanced in *S. latus* by the presence of its numerous attached epizoans including serpulids, hydroids, barnacles, bivalves, bryozoans, and foraminiferans (Martins 1985; Spanier & Lavalli 1998). Due to the minimal surface area coverage of the host by the bryozoans, the benefit of camouflage to the host is minimal.

Are the bryozoans receiving any benefit? There are a variety of benefits for a sessile encrusting animal living on a motile host. They include increased geographic range and substrate space. Living on a motile host provides the sessile bryozoans with free transport for avoiding predators, improve gamete dispersal, and increase geographic range. Hard substrate space is a limiting factor for bryozoans, especially encrusting bryozoans (Jackson 1977; Lidgard & Jackson 1989; McKinney 1995; Taylor 2016). Therefore, any increase in hard substrate space (e.g. motile host basibionts) should reduce competition for substrate space and result in an increase in bryozoan diversity as documented by Balazy & Kuklinski (2013). Finally, the host's carapace setae may provide camouflage for the small encrusting lichenoporids.

An additional benefit accrues to the bryozoans if the host is the encrusting animal's predator such as the bryozoans that foul predatory pycnogonids (Key *et al.* 2013). Other lobsters eat bryozoans, including the clawed lobster *Homarus* (Lawton & Lavalli 1995) and the spiny lobsters *Jasus* and *Panulirus* (Castaneda-Fernandez-de-Lara *et al.* 2005; Edmunds 1995; Fielder 1965; Lindberg 1955). The diet of *S. latus* consists of molluscs, mainly bivalves, but also including gastropods, but not bryozoans (Barshaw & Spanier 1994b; Martins 1985; Spanier 1987, 1991; Spanier *et al.* 1988), so it is doubtful this benefit accrues to the bryozoans here.

Weighing all these potential costs and benefits to the host lobster and the fouling bryozoans, the relationship is best described as commensalism. Overall, this is a one-sided commensal symbiotic relationship, best described as phoretic (i.e. hitchhiking relationship) for the bryozoans. This is the most commonly documented type of symbiosis between lobsters and bryozoans (Key & Hendrickx 2022; Key *et al.* 2021).

ACKNOWLEDGEMENTS

Simone Raus (University of Vienna, Austria) found the lobster during the dive. Abby Smith (University of Otago, New Zealand) brought Sebastian Decker's Twitter posting about this discovery to my attention. The following bryozoologists helped identify the bryozoans: Thomas Schwaha (University of Vienna, Austria), Dennis Gordon (National Institute of Water and Atmospheric Research, New Zealand), and Jean-Georges Harmelin (Centre d'Océanologie de Marseille, France). We thank Antonietta Rosso (University of Catania, Italy) and Hirose Masato (Kitasato University, Japan) for constructive feedback on the original manuscript.

REFERENCES

Audouin, J.V. 1826. Explication Sommaire des Planches de Polypes de l'Egypte et de la Syrie, Publiees par Jules-Cesar Savigny. In J.V. Audouin (ed.), *Description de l'Egypte, ou Recueil des Observations et des Recherches Qui ont été Faites en Egypte Pendant L'expédition de l'armée Française ... Histoire naturelle*, 1 (4): 225–244. Paris: Imprimerie Impériale.

Balazy, P. & Kuklinski, P. 2013. Mobile Hard Substrata – an Additional Biodiversity Source in a High Latitude Shallow Subtidal System. *Estuarine, Coastal and Shelf Science* 119: 153–161.

Barshaw, D.E. & Spanier, E. 1994a. Anti-predator Behaviors or the Mediterranean Slipper Lobster *Scyllarides latus*. *Bulletin of Marine Science* 55: 375–382.

Barshaw, D.E. & Spanier, E. 1994b. The Undiscovered Lobster: A First Look at the Social Behaviour of the Mediterranean Slipper Lobster, *Scyllarides latus* (Decapoda, Scyllaridae). *Crustaceana* 67: 187–197

Barshaw, D.E., Lavalli, K.L. & Spanier, E. 2003. Offense Versus Defense: Responses of Three Morphological Types of Lobsters to Predation. *Marine Ecology Progress Series* 256: 171–182.

Bianchini, M.L., Bono, G. & Ragonese, S. 2001. Long-term Recaptures and Growth of Slipper Lobsters, *Scyllarides latus*, in the Strait of Sicily (Mediterranean Sea). *Crustaceana* 74: 673–680.

Bianchini, M.L. & Ragonese, S. 2007. Growth of Slipper Lobsters of the Genus *Scyllarides*. In K.L. Lavalli & E. Spanier (eds), *The Biology and Fisheries of the Slipper Lobster*: 199–220. Boca Raton: CRC Press.

Castañeda-Fernández-de-Lara, V., Serviere-Zaragoza, E., Hernández-Vázquez, S. & Butler IV M.J. 2005. Feeding Ecology of Juvenile Spiny Lobster, *Panulirus interruptus*, on the Pacific coast of Baja California Sur, Mexico. *New Zealand Journal of Marine and Freshwater Research* 39: 425–435.

Dick, M.H., Donaldson, W.E. & Vining, I.W. 1998. Epibionts of the tanner crab *Chionoecetes bairdi* in the Region of Kodiak Island, Alaska. *Journal of Crustacean Biology* 18: 519–528.

Edmunds, M. 1995. The Ecology of the Juvenile Southern Rock Lobster, *Jasus edwardsii* (Hutton, 1875) (Palinuridae): 1–164. Tasmania: University of Tasmania, PhD dissertation.

Fernandez-Leborans, G. 2010. Epibiosis in Crustacea: an Overview. *Crustaceana* 83: 549–640.

Fielder, D.R. 1965. The Spiny lobster, *Jasus lalandei* (H. Milne-Edwards), in South Australia. III. Food, Feeding, and Locomotor Activity. *Australian Journal of Marine and Freshwater Research* 16: 351–367.

Fischer, W. (ed.) 1973. FAO Species Identification Sheets for Fishery Purposes. *Mediterranean and Blacks Sea (fishing area 37)*. Volume 2. Rome: Food and Agricultural Organization.

Fischer, W., Bianchi, G. & Scott, W.B. (eds) 1981. FAO Species Identification Sheets for Fishery Purposes. *Eastern Central Atlantic (fishing areas 34, 47 in part)*. Volume 5. Ottowa: Department of Fisheries and Oceans.

Grischenko, A.V., Gordon, D.P. & Melnik, V.P. 2018. Bryozoa (Cyclostomata and Ctenostomata) from Polymetallic Nodules in the Russian Exploration Area, Clarion-Clipperton Fracture Zone, Eastern Pacific Ocean – Taxon Novelty and Implications of Mining. *Zootaxa* 4484: 1–91.

Harmelin, J.G. 1968. Contribution à l'étude des Bryozoaires Cyclostomes de Méditerranée: Les *Crisia* des côtes de Provence. *Bulletin du Muséum National d'Histoire Naturelle* 40: 413–437.

Hellio, C. & Yebra, D. 2009. *Advances in Marine Antifouling Coatings and Technologies*. Oxford (UK): Woodhead Publishing.

Herrera, A., Jackson, J.B.C., Hughes, D.J., Jara, J. & Ramos, H. 1996. Life-history Variation in Three Coexisting Cheilostome Bryozoan Species of the Genus *Stylopoma* in Panamá. *Marine Biology* 126: 461–469.

Horne, F.R. & Tarsitano, S.F. 2007. The Mineralization and Biomechanics of the Exoskeleton. In K.L. Lavalli & E. Spanier (eds), *The Biology and Fisheries of the Slipper Lobster*: 183–198. Boca Raton: CRC Press.

Jackson, J.B.C. 1977. Competition on Marine Hard Substrata: The Adaptive Significance of Solitary and Colonial Strategies. *The American Naturalist* 111: 743–767.

Jackson, J.B.C. & Wertheimer, S.P. 1985. Patterns of Reproduction in Five Common Species of Jamaican Reef-associated Bryozoans. In C. Nielsen & G.P. Larwood (eds), *Bryozoa: Ordovician to Recent*. 161–168. Fredensborg: Olsen & Olsen.

Key, M.M., Jr. & Barnes, D.K.A. 1999. Bryozoan Colonization of the Marine Isopod *Glyptonotus antarcticus* at Signy Island, Antarctica. *Polar Biology* 21: 48–55.

Key, M.M., Jr. & Hendrickx, M.E. 2022. *Biflustra irregulata*: A Tsunami Debris Rafted Indo-Pacific Bryozoan Found in Eastern Pacific. *Zootaxa* 5128: 340–354.

Key, M.M., Jr., Hyžný, M., Khosravi, E., Hudáčková, N., Robin, N. & Mirzaie Ataabadi, M. 2017. Bryozoan Epibiosis on Fossil Crabs: A Rare Occurrence From the Miocene of Iran. *Palaios* 32: 491–505.

Key, M.M., Jr., Jeffries, W.B. & Voris, H.K. 1995. Epizoic Bryozoans, Sea Snakes, and Other Nektonic Substrates. *Bulletin of Marine Science* 56: 462–474.

Key, M.M., Jr., Jeffries, W.B., Voris, H.K. & Yang, C. M. 1996a. Epizoic Bryozoans and Mobile Ephemeral Host Substrata. In D.P. Gordon, A.M. Smith & J.A. Grant-Mackie (eds), *Bryozoans in Space and Time*: 157–165. Wellington: National Institute of Water and Atmospheric Research.

Key, M.M., Jr., Jeffries, W.B., Voris, H.K. & Yang, C. M. 1996b. Epizoic Bryozoans, Horseshoe Crabs, and Other Mobile Benthic Substrates. *Bulletin of Marine Science* 58: 368–384.

Key, M.M., Jr., Jeffries, W.B., Voris, H.K. & Yang, C. M. 2000. Bryozoan Fouling Pattern on the Horseshoe Crab *Tachypleus gigas* (Müller) from Singapore. In A. Herrera Cubilla & J.B.C. Jackson (eds), *Proceedings of the 11th International Bryozoology Association Conference*: 265–271. Balboa: Smithsonian Tropical Research Institute.

Key, M.M., Jr., Knauff, J.B. & Barnes, D.K.A. 2013. Epizoic Bryozoans on Predatory Pycnogonids from the South Orkney Islands, Antarctica: "If You Can't Beat Them, Join Them". In A. Ernst, P. Schäfer & J. Scholz (eds), *Bryozoan Studies 2010*: 137–153. Berlin: Springer-Verlag.

Key, M.M., Jr., Schumacher, G.A., Babcock, L.E., Frey, R.C., Heimbrock, W.P., Felton, S.H., Cooper, D.L., Gibson, W.B., Scheid, D.G. & Schumacher, S.A. 2010. Paleoecology of Commensal Epizoans Fouling *Flexicalymene* (Trilobita) from the Upper Ordovician, Cincinnati Arch Region, USA. *Journal of Paleontology* 84: 1121–1134.

Key, M.M., Jr., Smith, A.M., Hanns, B. & Kane-Sanderson, P. 2021. Rare Report of Bryozoan Fouling of Rock Lobsters (*Jasus edwardsii*: Decapoda: Palinuridae) from the North Island of New Zealand. *New Zealand Journal of Marine and Freshwater Research* 1–13 (online). https://doi.org/10.1080/00288330.2021.1977344

Key, M.M., Jr., Winston, J.E., Volpe, J.W., Jeffries, W. B. & Voris, H.K. 1999. Bryozoan Fouling of the Blue Crab, *Callinectes sapidus*, at Beaufort, North Carolina. *Bulletin of Marine Science* 64: 513–533.

Latreille, P.A. 1802. Histoire Naturelle, Générale et Particulière des Crustacés et des Insectes: Ouvrage Faisant Suite Aux Oeuvres de Leclerc de Buffon, et Partie du Cours Complet D'histoire Naturelle Rédigé par C.S. *Sonnini* 3: 1–468.

Lavalli, K.L., Malcom, C.N. & Goldstein, J.S. 2018. Description of Pereiopod Setae of Scyllarid Lobsters, *Scyllarides aequinoctialis*, *Scyllarides latus*, and *Scyllarides nodifer*, with observations on the Feeding During Consumption of Bivalves and Gastropods. *Bulletin of Marine Science* 94: 571–601.

Lavalli, K.L. & Spanier, E. 2001. Does Gregariousness Function as an Antipredator Mechanism in the Mediterranean Slipper Lobster, *Scyllarides latus*? *Marine and Freshwater Research* 52: 1133–1143.

Lavalli, K.L., Spanier, E. & Grasso, F. 2007. Behavior and Sensory Biology of Slipper Lobsters. In K.L. Lavalli & E. Spanier (eds), *The Biology and Fisheries of the Slipper Lobster*: 133–181. Boca Raton: CRC Press.
Lawton, P. & Lavalli, K.L. 1995. Postlarval, Juvenile, Adolescent, and Adult Ecology. In J.R. Factor (ed.), *Biology of the Lobster: Homarus americanus*: 47–88. San Diego: Academic Press.
Lidgard, S. & Jackson, J.B.C. 1989. Growth in Encrusting Cheilostome Bryozoans: I. Evolutionary Trends. *Paleobiology* 15: 255–282.
Lindberg, R.G. 1955. Growth, Population Dynamics and Field Behaviour in the Spiny Lobster, *Panulirus interruptus* (Randall). *University of California Publications in Zoology* 59: 157–248.
Martins, H.R. 1985. Biological Studies of the Exploited Stock of the Mediterranean Locust Lobster *Scyllarides latus* (Latreille, 1803) (Decapoda: Scyllaridae) in the Azores. *Journal of Crustacean Biology* 5: 294–305.
McKinney, F.K. 1995. One Hundred Million Years of Competitive Interactions Between Bryozoan Clades: Asymmetrical But Not Escalating. *Biological Journal of the Linnean Society* 56: 465–481.
Ogren, L.H. 1977. Concealment Behavior of the Spanish Lobster, *Scyllarides nodifer* (Stimpson), with Observations on its Diel Activity. *Northeast Gulf Science* 1: 115–116.
Parapar, J., Fernandez, L. Gonzalez-Gurriaran, E. & Muino R. 1997. Epibiosis and Masking Material in the Spider Crab *Maja squinado* (Decapoda: Majidae) in the Ria de Arousa (Galicia, NW Spain). *Cahiers de Biologie Marine* 38: 221–234.
Phillips, B.F., Cobb, J.S. & George, R.W. 1980. General Biology. In J.S. Cobb & B.F. Phillips (eds), *The Biology and Management of Lobsters. vol. I, Physiology and Behavior*: 1–82. New York: Academic Press.
Quinn, R.A., Smolowitz, R. & Chistoserdov, A. 2009. Eukaryotic Communities in Epizootic Shell Disease Lesions of the American Lobster (*Homarus americanus* H. Milne Edwards). *Journal of Shellfish Research* 28: 913–922.
Romeo, T., Florio, G., Lentini, F., Castriota, L., Falautano, M., Consoli, P., Pelusi, P. & Greco, S. 2004. Morphometric Aspects of *Scyllarides latus*. *Mediterranean Marine Science* 5: 65–71.
Rosso, A. 2003. Bryozoan Diversity in the Mediterranean Sea. *Biogeographia* 24: 227–250.
Rosso, A. & Di Martino, E. 2016. Bryozoan Diversity in the Mediterranean Sea: an Update. *Mediterranean Marine Science* 17: 567–607.
Ryan, W.B.F., Carbotte, S.M., Coplan, J.O, O'Hara, S., Melkonian, A., Arko, R., Weissel, R.A., Ferrini, V., Goodwillie, A., Nitsche, F., Bonczkowski, J. & Zemsky, R. 2009. Global Multi-resolution Topography (GMRT) Synthesis Data Set. *Geochemistry, Geophysics, Geosystems* 10(3): Q03014, https://doi.org/10.1029/2008GC002332
Shields, J.D. 2011. Diseases of Spiny Lobsters: A Review. *Journal of Invertebrate Pathology* 106: 79–91.
Smith, A.M. 1992. *Aspects of the Sedimentology of New Zealand Bryozoans and Mixed Carbonate-clastic Deposits: A Contribution to the Temperate Shelf Carbonate Model [dissertation]*. Waikato: University of Waikato.
Spanier, E. 1987. Mollusca as Food for the Slipper Lobster *Scyllarides latus* in the Coastal Waters of Israel. *Levantina a Journal of Malacology* 68: 713–716.
Spanier, E. 1991. Artificial Reefs to Insure Protection of the Adult Mediterranean Slipper Lobster, *Scyllarides latus* (Latreille, 1803). In C.F. Boudouresque, M. Avon & V. Gravez (eds), *Les Espèces Marines Protégées en Méditerranée*: 179–185. France: GIS Posidonia.
Spanier, E. & Lavalli, K.L. 1998. Natural History of *Scyllarides latus* (Crustacea: Decapoda): A Review of the Contemporary Biological Knowledge of The Mediterranean Slipper Lobster. *Journal of Natural History* 32: 1769–1786.
Spanier, E. & Lavalli, K.L. 2013. Commercial Scyllarids. In B.F. Phillips (ed.), *Lobsters: Biology, Management, Aquaculture and Fisheries*: 414–466. Oxford: Blackwell.
Spanier, E., Tom, M., Pisanty, S. & Almog, G. 1988. Seasonality and Shelter Selection by the Slipper Lobster *Scyllarides latus* in the Southeastern Mediterranean. *Marine Ecology Progress Series* 42: 247–255.
Stachowicz, J.J. & Hay, M.E. 2000. Geographic Variation in Camouflage Specialization by a Decorator Crab. *American Naturalist* 156: 59–71.
Taylor, P.D. 2016. Competition Between Encrusters on Marine Hard Substrates and its Fossil Record. *Palaeontology* 59: 481–497.
Tetreault, D.K. 1992. Paleoecologic Implications of Epibionts on the Silurian Lichid Trilobite *Arctinurus*. In S. Lidgard & P.R. Crane (eds), *North American Paleontological Convention Abstracts and Program*: 289. Knoxville: Paleontological Society and University of Tennessee.
Tetreault, D.K. 1997. Trilobite Behavior in the Silurian. *Second International Trilobite Conference, Abstracts with Programs*: 48–49. St. Catharines: Brock University.
Turek, V. 1987. On Some Epizoans of Mollusk Shells from the Upper Silurian (Pridoli) of the Barrandian area. *Vestnik Ustredniho Ustavu Geologickeho* 62: 105–111.
Vance, R.R. 1978. A Mutualistic Interaction Between a Sessile Marine Clam and its Epibionts. *Ecology* 59: 679–685.
Wahl, M. 1989. Marine Epibiosis. 1. Fouling and Antifouling – Some Basic Aspects. *Marine Ecology Progress Series* 58: 175–189.
Wahl, M. & Hay, M.E. 1995. Associational Resistance and Shared Doom: Effects of Epibiosis on Herbivory. *Oecologia* 102: 329–340.
Waters, A.W. 1916. Some Species of *Crisia*. *Annals and Magazine of Natural History*, 8th series 18: 469–477
Winston, J.E. & Håkansson, E. 1986. The Interstitial Fauna of the Capron Shoals, Florida. *American Museum Novitates* 2865: 1–98.
Wyse Jackson, P.N. & Key, M.M. Jr. 2014. Epizoic Bryozoans on Cephalopods Through the Phanerozoic: a Review. *Studi Trentini di Scienze Naturali* 94: 283–291.
Zha, H., Lewis, G., Alfaro, A., Wang, S., Dong, Y., Grandiosa, R. & Jeffs, A. 2017. Immune Status of the Spiny Lobster *Jasus edwardsii* with Tail Fan Necrosis. *Diseases of Aquatic Organisms* 126: 229–238.

Bryozoan Studies 2022 – Key, Porter & Wyse Jackson (eds)
 ISBN: 978-1-032-59487-3

Three-dimensional imaging of fossil cheilostome bryozoans in Eocene chert by Synchrotron Radiation Micro-Computed Tomography

Marcus M. Key, Jr.*
Department of Geosciences, Dickinson College, Carlisle, PA, USA
ORCID ID: 0000-0003-4097-0143

Jeremy Shaw
Centre for Microscopy Characterisation and Analysis (M010), University of Western Australia, Perth, WA, Australia
ORCID ID: 0000-0002-5140-457X

Ingrid Ward
School of Social Sciences, University of Western Australia, Perth, WA, Australia
ORCID ID: 0000-0002-0566-1479

ABSTRACT: The goal of this project is to describe the application of Synchrotron Radiation Micro-Computed Tomography (SRμCT) for three-dimensional imaging of fossil cheilostome bryozoan colonies. The technology is applied to the challenging problem of 3D visualization of silicified bryozoan fossils in Eocene cherts from southern Western Australia. The reconstructed colonies from Aboriginal chert artifacts and chert nodules from well cuttings allowed assignment to the following genera: *Quadricellaria*, *Siphonicytara*, *Trigonopora*, and *Reteporella*. The benefits of SRμCT include effective imaging of bryozoans in dense cherts non-destructively and the integration of 2D and 3D internal zooecium- and colony-level morphologic characters. The disadvantages of SRμCT include large data handling and analysis, relatively high cost, and low resolution in chert.

1 INTRODUCTION

It is a challenge to image species-diagnostic internal morphology of fossil bryozoans when they are embedded in rock, and even more so when they are in valuable cultural artifacts. An analysis of the three-dimensional (3D) morphology is a necessary part of species identification and typically requires the destructive process of thin sectioning which involves cutting, grinding, and polishing the rock to expose a bryozoan colony (Nye *et al.* 1972). The final product is a two-dimensional thin slice through the rock (i.e. a thin section) mounted on a glass slide that can be imaged with a microscope. Thin sections can yield high resolution images. Acetate peels are made similarly. After polishing, the colony is etched with acid, so any internal structures stand out in relief. Then a sheet of acetate is partially chemically dissolved onto that surface with acetone (Boardman & Utgaard 1964; Wilson & Palmer 1989). The acetate is then pulled off the colony and can be imaged with a microscope. Acetate peels allow image resolutions almost as good as thin sections, but they fall short in two areas. 1) They do not have the longevity that thin sections have. 2) Skeletal wall structure is not as well resolved as in thin sections.

Both techniques have always been destructive and labor-intensive (Boardman 2008). Intersecting a bryozoan colony inside the rock is not guaranteed. With both of these methods, accurately oriented sections (i.e. mutually perpendicular longitudinal, transverse, and tangential) are hard to achieve (Wyse Jackson & Buttler 2015). Fortunately, both methods allow for serial sectioning which can then be used to make three-dimensional reconstructions (Key *et al.* 2011). The grinding needed for serial sectioning is destructive and results in the complete loss of the specimen. The spacing is coarser with thin sections than acetate peels, and computer stitching of digital 2D slices is labor-intensive (Sutton *et al.* 2001).

The latest technique for three-dimensional imaging is laboratory-based X-ray Micro-Computed Tomography (μCT). By laboratory-based, we mean it is an accessory for an SEM or a small stand-alone cabinet-based system. X-ray

*Corresponding Author: key@dickinson.edu

DOI: 10.1201/9781003454915-7

absorption is the main source of contrast in μCT imaging. The ability to differentiate between morphological features depends on variations in mineral composition and/or void space/porosity, which have a bearing on X-ray attenuation through a colony. The first use of μCT was by Taylor *et al.* (2008) who imaged the internal morphology of Recent and fossil cyclostomes. Since then there have been dozens of publications using μCT 3D imaging technology in bryozoology (Key & Wyse Jackson 2022) . For example, the μCT lab at the Paleontological Institute of the Russian Academy of Sciences in Moscow has been very productive (Koromyslova & Pakhnevich 2016). Roughly twice as many of these studies are on fossil bryozoans, probably in response to the labor involved in making serial thin sections and acetate peels to create 3D reconstructions (e.g. Snyder 1991).

μCT 3D imaging works best when the mineralized skeleton and zooidal cavities have very different X-ray absorption properties. For example, Matsuyama *et al.* (2015) produced beautiful images of the autozooid chambers, suboral avicularian chambers, orifices, and frontal pores in Recent cheilostomes from marine colonies off Mauritania. Schwaha *et al.* (2018) used μCT to create 3D renderings of the various parts of cyclostome polypides, sampled from marine colonies off New Zealand. It is more of a challenge to obtain contrast for fossils in limestone when both the specimen, surrounding sediment matrix, and infilling cements are all composed of the same carbonate mineral and therefore share more similar X-ray attenuation (Keklikoglou *et al.* 2019; Sutton 2008). This homogenous carbonate mineralogy is often the case with a calcite bryozoan colony in a limestone (Key *et al.* 2005). If the densities of the bryozoan skeleton and infilling cements are too similar, imaging fails (Buttler *et al.* 2012).

Fortunately, in some carbonate sedimentary rocks, there is often enough variation between the fossil skeletons, the surrounding sediment matrix, and the secondary cements that sufficient contrast is achieved with μCT. For example, Viskova & Pakhnevich (2010) used μCT to image soft bodied Jurassic ctenostome colonies boring in mollusc shells. Wyse Jackson & McKinney (2013) applied this technology to a study of type material of a Mississippian fenestrate bryozoan in limestone to resolve the nature of its polymorphs along with the help of thin sections and SEM images. The application of μCT technology has not been limited to visualizing internal taxonomic characters. David *et al.* (2009) used this technique to document marine bryozoans encrusting Recent sea urchin spines from Antarctica. Heřmanová *et al.* (2020) used μCT to reveal commensal ecological relationships in Ordovician bryozoans encrusting conulariids and hemispherical colonies bored by sponges.

This problem of 3D imaging of fossilized bryozoans using μCT is accentuated when the fossilized remains have been replaced by new minerals. For example, Zhang *et al.* (2021) used μCT to show the first Cambrian bryozoans were soft-bodied. They were preserved through phosphatization which revealed bilaminate autozooids with possible interzooidal connections through the mesotheca. The challenge of 3D imaging of bryozoans is most difficult when there is not enough X-ray contrast to distinguish skeletal features from cement infilling. This is especially true in two situations. First, in Paleozoic palaeostomes, there is normally not enough contrast to identify features required for taxonomic identification where there are calcified walls and calcite infilling of zooecial chambers. Second, imaging is also a challenge when the original calcite or aragonite bryozoan skeleton has been replaced by silica during the diagenetic process creating chert (Maliva & Siever 1988; Murray *et al.* 1992). If the contrast between the bryozoan skeleton and its infilling cements and surrounding sediment matrix is too low, synchrotron phase-contrast imaging may be a better solution for 3D visualisation than μCT (Sutton *et al.* 2014).

Following the suggestion by Schmidt (2013), Ward *et al.* (2019a) were the first to try to image silicified fossil bryozoans in chert using Synchrotron Radiation Micro-Computed Tomography (SRμCT). SRμCT possesses some advantages over standard μCT. A synchrotron source provides a higher-flux, higher-intensity X-ray beam, allowing acquisition of potentially higher-resolution 3D images with a higher signal-to-noise ratio (Salomé *et al.* 1999). A synchrotron is ~100 m diameter dedicated facility that produces many times more flux than a laboratory-based μCT instrument, so the exposure times are smaller, and therefore data collection is faster (Betz *et al.* 2007). Because of the higher flux/brightness, the technicians can tune the energy (i.e. discard certain energy ranges leaving potentially complimentary energies to the sample chemistry) to try to obtain better contrast in the samples. Because of the faster scan times, multiple fields of view of larger/thicker samples can be acquired in a realistic timeframe. This is a key advantage over laboratory-based systems. Without the high-flux we could never have imaged the bryozoans in these cherts due to the size of the samples.

The goal of this project is to document the application of SRμCT technology for 3D imaging of bryozoans. In particular, we focus on silicified fossil bryozoans in Eocene cherts. The cherts came from well cuttings and from irreplaceable Australian Aboriginal prehistoric artifacts that prohibit destructive serial sectioning. One of the challenges of using fossil bryozoans to source artifacts is the need to image the internal morphology of the colonies to identify the exact species. This is typically done through destructive analysis in two dimensions from thin section and SEM images of the rock containing the fossil bryozoan (Wyse Jackson & Buttler 2015). Here we demonstrate that SRµCT technology offers a viable alternative.

2 MATERIALS

The samples in this study come from an ongoing geoarcheology project using fossil bryozoans to determine the source of Aboriginal prehistoric chert artifacts from Western Australia (Key *et al.* 2019; O'Leary *et al.* 2017; Ward *et al.* 2019a, 2019b, 2021). Most of the artifacts were chert cutting tools. The SRµCT technology was tested on fossil cheilostome bryozoans in Eocene cherts. Colonies of *Quadricellaria* were identified in Eocene chert Aboriginal artifacts (UWA-74610 Pinnacles and UWA-74781 Mandurah) collected on the Swan Coastal Plain around Perth and archived in the John Glover collection at the University of Western Australia museum. Colonies of *Siphonicytara*, *Trigonopora*, and *Reteporella* were identified from a 60.10 g chert nodule from well cuttings from 180 m deep in well FOR004 (Key *et al.* 2019). Well FOR004 (latitude 31.28008°S, longitude 128.55396°E) was drilled as part of the Eucla basement stratigraphic drilling program (Spaggiari & Smithies 2015). Stratigraphically, the well cuttings came from the Eocene Wilson Bluff Limestone (Key *et al.* 2019) which is a likely source of the chert for the artifacts (O'Leary *et al.* 2017; Ward *et al.* 2019a, 2019b). The size range of the chert samples ranged from 10–40 mm in maximum dimension. In all cases, samples that had bryozoans visible at or near the surface were deliberately targeted to increase the chances of finding embedded fossils. The specimen numbers and repositories are listed in Ward *et al.* 2019a).

3 METHODS

SRµCT requires a synchrotron that produces intense light. It does this by accelerating high-energy electrons in a circular orbit inside the synchrotron's tunnels by 'synchronized' application of strong magnetic fields (Betz *et al.* 2007). The light is channeled down beamlines to experimental workstations where it is used for research. We used the Australian Synchrotron which is located outside of Melbourne. It has nine beamlines. We used enclosure 3B of the Imaging and Medical BeamLine (IMBL) which provides a sample view area of up to 30 x 40 mm. In contrast, a typical laboratory-based µCT system can handle specimens up to 2 mm. Newer medical µCT systems can handle human skull-sized specimens. The IMBL delivers the world's widest synchrotron X-ray beam which allows high contrast 3D X-ray imaging at high resolution so as to hopefully reveal minute differences at the interfaces of the fossil bryozoan skeleton, sediment matrix, and secondary cements.

The resolution of each voxel (3D pixel) was 9.7 µm x 9.7 µm x 9.7 µm. A typical bryozoan zooecium is on the scale of 1 mm^3 (Ryland 2005), so a 9.7 µm resolution will allow up to 100 slices through a typical-sized zooecium. Each scan consisted of 1800 projections acquired as the sample is rotated at 0.1° steps through 180°. A standard filtered back-projection algorithm coupled with ring artifact removal was used to reconstruct the X-ray projection images into a stack of 2D slices, creating an isotropic voxel. Each voxel has a specific 16 bit grey value representing the linear X-ray attenuation coefficient of the composite material, which is dependent on density and chemical composition.

The first steps of image processing, such as flat-field correction, noise suppression, stitching and CT reconstruction were performed using the ASCI high performance cluster at the Australian Synchrotron. The subsequent 3D volumes could then be rendered for visualization and undergo subsequent post-processing and analysis. For further details on µCT scanning see Ngan-Tillard & Huisman (2017).

The synchrotron images were numerous and data rich. The number of synchrotron image files for each sample ranged from 340 to 6050. The size of each image file ranged from 0.64 to 20.3 MB. For the samples imaged, the total data set was > 800 GB. Hence, the first step involved reducing the data set to a more manageable size by cropping to specific regions of interest (ROIs) (i.e. bryozoan colonies) using Fiji (ImageJ) software (Schindelin *et al.* 2012). The ROIs were then analyzed using Avizo 9.2.0 (Avizo 2016) and Drishti 2.6.2 (Limaye 2012) software to assess presence or absence of individual bryozoan colonies. The xyz-position of individual colonies were recorded and examples of different bryozoans were selected to render in 3D in the Avizo software. Other 3D modelling software (e.g., Dragonfly) and freeware (e.g. Seg3D, 3D Slicer) are also available. Each voxel's X-ray attenuation coefficient (i.e. contrast value) was then color-coded to enhance visualization. This color coding (a.k.a. segmentation/labelling) is the most complex and labor-intensive aspect of the process. Owing to the sheer volume of data, only a small number of key fossil bryozoans from each region of interest were chosen for full 3D visualization. Normally, the more you look the more you find. Searching 3D data using slices can mean that, because of the oblique angles of the cuts through randomly oriented samples and the shape of the colony, much can be missed.

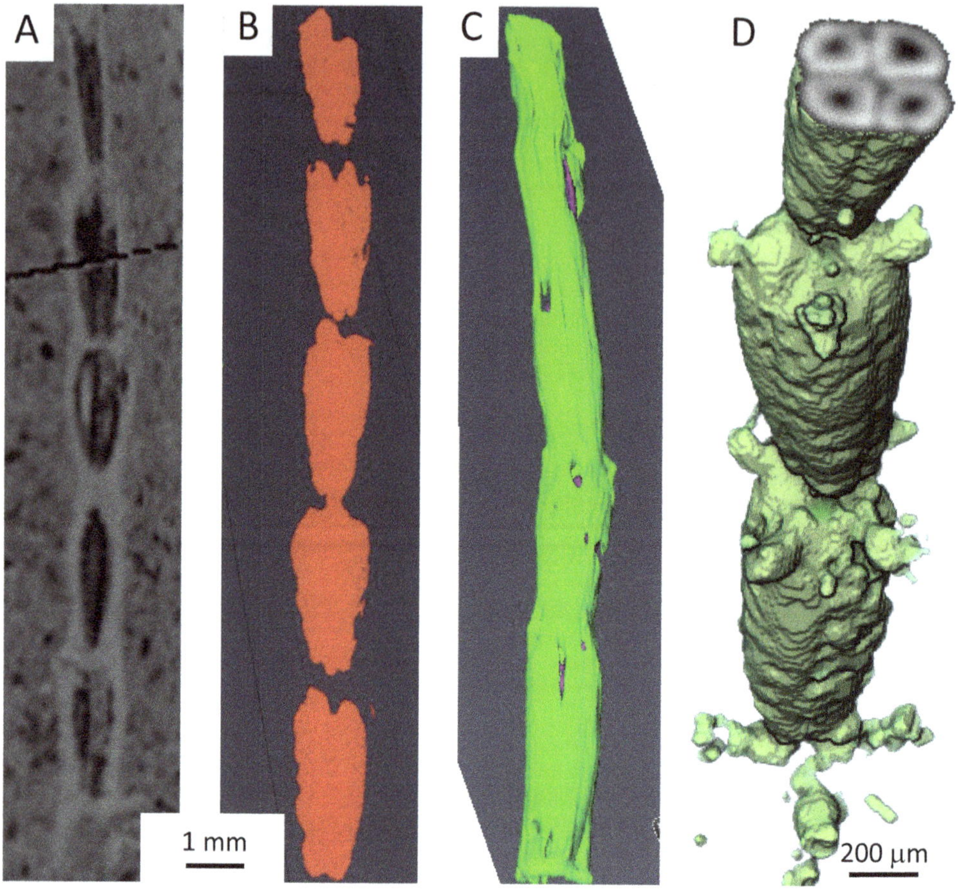

Figure 1. SRµCT images of the cheilostome bryozoan *Quadricellaria*. A, 2D grayscale image showing brighter skeletal walls surrounded by darker sediment matrix and even darker cement infilling zooecial cavities. B, 3D reconstruction of zooecial cavities in red. C, 3D reconstruction of exterior frontal walls of zooecia in green with opesia in purple. D, close up of B with exterior surface of zooecial cavities in green and a transverse 2D section at the top through the colony in grayscale.

4 RESULTS

What follows are the more successful 3D imaging attempts. Figure 1A shows a *Quadricellaria* colony with the skeletal walls in the grayscale 2D image clearly brighter than the surrounding sediment matrix and the cement infilling the zooecial cavities. In the resulting 3D reconstruction, the zooecial cavities are colored red (Figure 1B). Most of the exterior frontal walls of the zooecia (except for what remains of the opesia (colored purple in Figure 1C) have been covered by secondary calcification (colored green in Figure 1C). This allows us to infer that the the imaged fragment is from the proximal part of a colony because in some anascan cheilostomes, secondary calcification occurs in the older proximal parts of colonies (Sandberg 1983). In a close up of Figure 1B, the exterior surface of the zooecial cavities is colored green, a transverse 2D section through the colony is in grayscale (Figure 1D). The protuberances on the zooecial cavities (Figure 1D) are interpreted as opesia.

Figure 2 shows a 3D rendering of a branching *Siphonicytara* colony. The colony surface is colored in grayscale with orifices in green (Figure 2A) and the whorled zooecial cavities in green revealing opesia (Figure 2B). One of the zooids in the right-hand branch (just above the bifurcation) appears to show an orifice and latero-oral avicularium (Figure 2B arrow).

Figure 3 demonstrates the usefulness of the SRµCT technology by integrating traditionally oriented 2D sections with a 3D reconstruction of a *Trigonopora* colony. The colony surface is colored yellowish-brown with darker orifices (Figure 3A) with standard longitudinal (Figure 3B) and transverse (Figure 3C) sections in grayscale with skeletal walls brighter than surrounding sediment matrix and infilling cements.

The most challenging 3D rendering was a colony of *Reteporella*. The reticulate zoarial habit made it hard to isolate the target colony from other colony fragments in the sediment matrix

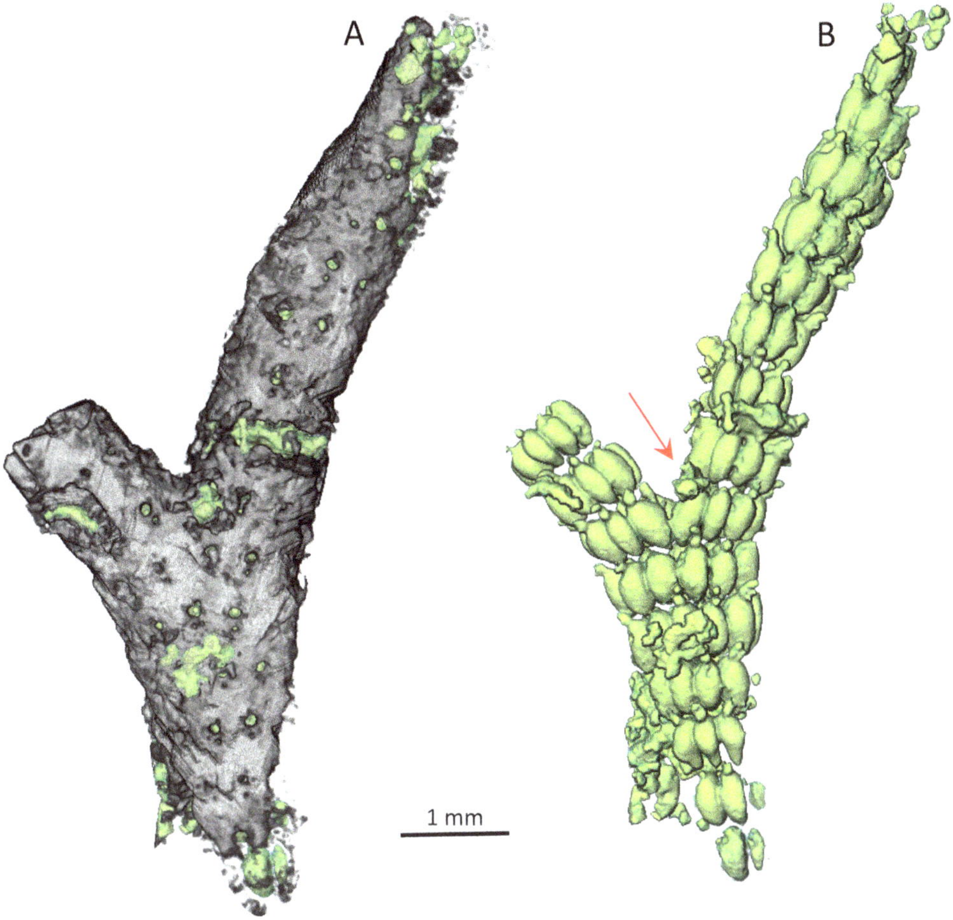

Figure 2. SRμCT 3D renderings of a branching colony of the cheilostome bryozoan *Siphonicytara*. A, colony surface is colored in grayscale with orifices in green. B, whorled zooecial cavities in green revealing opesia. One of the zooids in the right-hand branch (just above the bifurcation) appears to show an orifice and latero-oral avicularium (arrow).

(Figure 4A). The target colony surface is colored yellow with green orifices (Figure 4B) connected to the systematically arranged zooecial cavities in green (Figure 4C).

5 DISCUSSION

There are several benefits to SRμCT imaging of bryozoan fossils embedded in chert compared to more traditional thin sectioning and acetate peel technologies:

- It is non-destructive. Archaeologists, Traditional Owners, and museum curators of type specimens are reluctant to allow any destructive analysis of irreplaceable chert artifacts and type material such as by thin sectioning or acetate peels.
- The data processing makes creating 3D reconstructions of bryozoan colonies equally labor-intensive as serial sectioning, but the advantage is that you have a fully digital serial sectioned volume to explore in one dataset.
- A digital volumetric dataset can be sectioned and explored in any orientation non-destructively, something that is virtually impossible with thin sections. Therefore, it is easier to rotate the bryozoan colony into exact orthogonal section orientation (i.e. longitudinal vs. tangential vs. transverse) than with thin sections or acetate peels.
- It is easier to relate 2D images to 3D colony renderings as the technology provides both stand-alone 2D images and integrated 3D rendered colonies.
- Can make useful fly-through videos which facilitate relating 2D thin section views to 3D morphology and makes the results more useful for teaching and more accessible for other interested parties.
- Can easily distinguish skeletal walls and zooecial cavities.

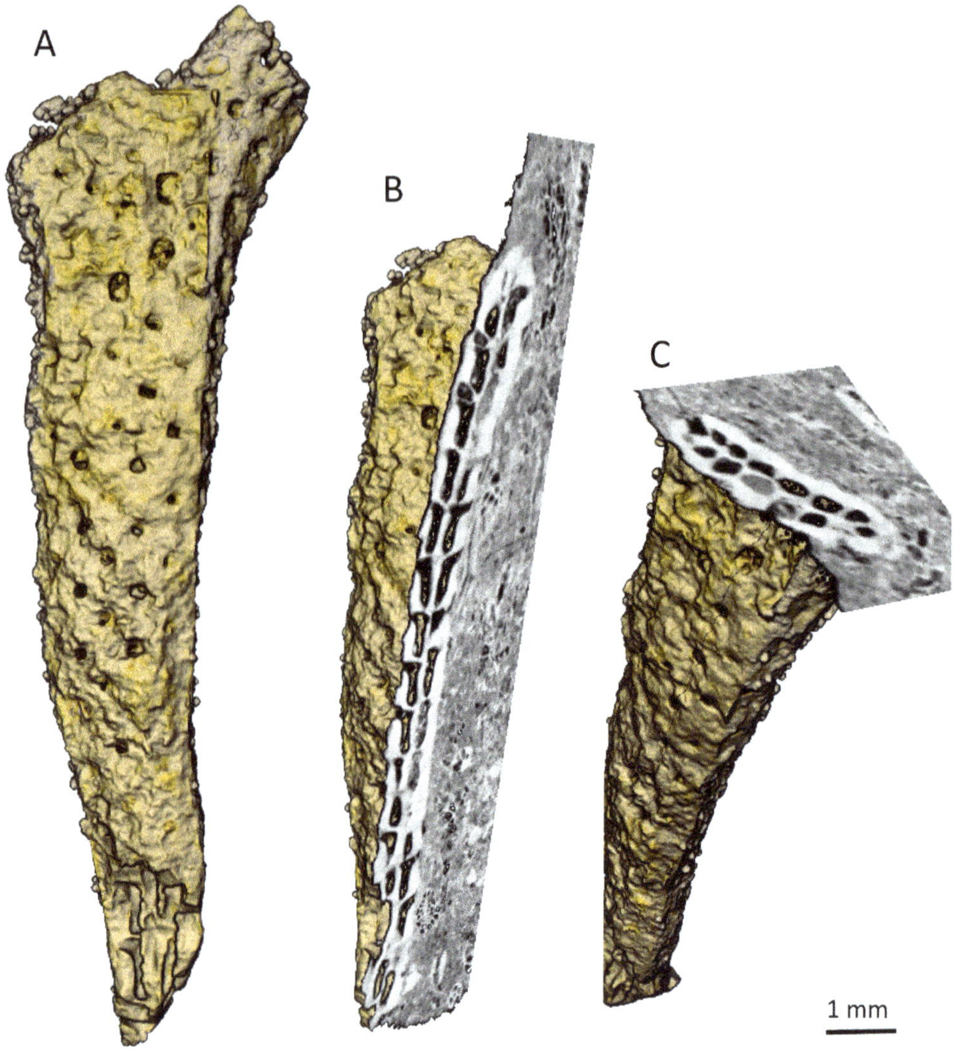

Figure 3. SRμCT 3D renderings of a blade-like colony of the cheilostome bryozoan *Trigonopora*. A, colony surface is colored yellowish-brown with darker orifices. Longitudinal (B) and transverse (C) 2D sections in grayscale with skeletal walls brighter than surrounding sediment matrix and infilling cements.

- 3D imaging allows easier inclusion of species-diagnostic internal zooecial morphology into taxonomic descriptions. Such characters are generally absent in bryozoology studies due to the lack of access to the zooecial interiors.
- 3D imaging allows easier inclusion of colony morphology characters into taxonomic descriptions. In fossils, the 3D colony morphology of a colony is rarely preserved due to burial, compaction, and fragmentation (Key *et al.* 2016). For example, the anastomosing branches of a net-like erect rigid fenestrate (reteporiform) colony of *Reteporella* would not be preserved in a fossil unless it was preserved on a bedding plane after burial (e.g. Suárez Andrés & Wyse Jackson 2015, fig. 3B).
- Compared to laboratory-based μCT systems, SRμCT 1) can typically image specimens an order of magnitude larger, and 2) image acquisition is much faster due to the greater X-ray flux.

Compared to more traditional thin sectioning and acetate peel technologies, there are several disadvantages to using SRμCT technology to image bryozoan fossils embedded in chert:

- Access to SRμCT technology is more geographically restricted (i.e. only ~90 synchrotrons globally). For example, not all artifacts can be

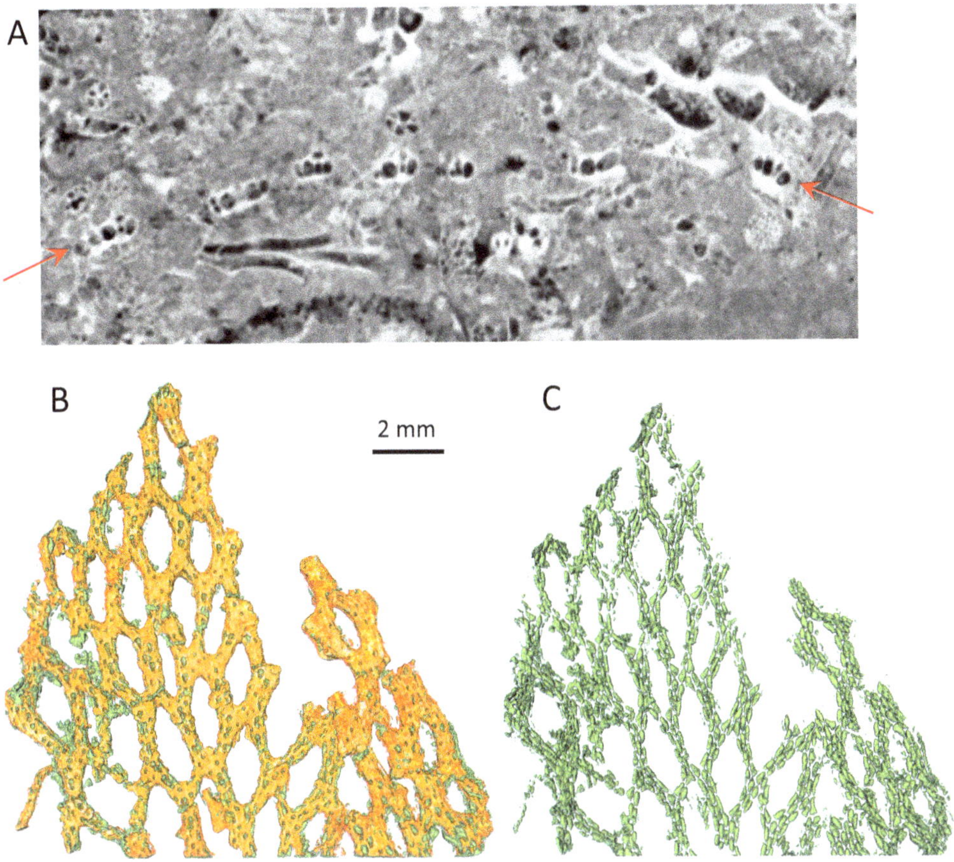

Figure 4. SRμCT images of the cheilostome bryozoan *Reteporella*. A, reticulate zoarial habit in cross section indicated by red arrows at each end of the colony. Brighter skeletal walls are surrounded by darker sediment matrix and even darker cement infilling zooecial cavities. B, 3D rendering with colony surface colored orange. C, systematically arranged zooecial cavities in green.

transported interstate or even overseas, without traditional owner permissions.

- Not all synchrotrons have beamlines suitable for imaging relatively large and dense objects such as chert artifacts with embedded bryozoans. The Australian synchrotron IMBL can handle specimens up to 30 x 40 mm.
- It is expensive, although access grants are often available. For example, the Australian synchrotron costs $800 AU ($575 US) per hour with a four-hour minimum usage.
- The post-processing of 3D tomographic data is time consuming due to the large amount of data. It may be as time consuming as thin sectioning and is best done by experts. Fortunately, this processing is getting faster due to machine learning and semi-automatic segmentation methods.
- Poor preservation of the fossils themselves is still a potential problem.
- The resolution is still not as good as thin section photographs (Ward *et al.* 2019a, fig. 10) or SEM images (Martha *et al.* 2019), which is required for species level identification. Higher resolution sub-micron X-ray CT scanning facilities exist as both synchrotron and laboratory-based X-ray sources and could provide greater spatial resolution, thereby providing more detailed taxonomic identification of embedded bryozoan fossils (Huisman *et al.* 2014). However, most laboratory-based μCT sources have limited flux, making the imaging and analysis of large/dense samples impractical. Sub-micron scale X-ray CT beamlines are now available. The one at the Australian Synchrotron is under construction and is planned to be available from 2022. Another solution is to scan samples before thin sectioning and combine the data from both into a single 3D data set.

6 CONCLUSIONS

This is a useful technique for 3D imaging of bryozoans in artifacts when destructive methods are not an option due to their cultural value. This study provides only the second known non-destructive synchrotron imaging of bryozoan fossils preserved in chert. SRµCT 3D imaging is sufficient for some genus level identification, but not for the species level. The resolution should be better in non-silicified fossil bryozoans where the colony skeleton has not been replaced by secondary minerals as in this study. There are pros and cons to using SRµCT technology to image 3D fossil bryozoans. In total, the benefits outweigh the disadvantages.

ACKNOWLEDGEMENTS

Dennis Gordon (National Institute of Water & Atmospheric Research, New Zealand) and Rolf Schmidt (Museum Victoria, Australia) helped with bryozoan identifications. This project was made possible by an Australian Synchrotron grant (No. AS171/IM/11589). The authors acknowledge the facilities and the scientific and technical assistance of Microscopy Australia at the Centre for Microscopy, Characterisation & Analysis, The University of Western Australia, a facility funded by the University, State and Commonwealth governments. We thank Caroline Buttler (National Museum of Wales, UK) and Thomas Schwaha (University of Vienna, Austria) for constructive feedback on the original manuscript.

REFERENCES

Avizo, 2016. *3D Analysis Software for Scientific and Industrial Data, version 9.2.0*, https://assets.thermofisher.com/TFS-Assets/MSD/Product-Updates/release-notes-avizo-920.pdf.

Betz, O., Wegst, U., Weide, D., Heethoff, M., Helfen, L., Lee, W.-K. & Cloetens, P. 2007. Imaging Applications of Synchrotron X-ray Phase-contrast Microtomography in Biological Morphology and Biomaterials Science. I. General aspects of the Technique and its Advantages in the Analysis of Millimetre-sized Arthropod Structure. *Journal of Microscopy* 22: 51–71

Boardman, R.S. & Utgaard, J. 1964. Modifications of Study Methods for Paleozoic Bryozoa. *Journal of Paleontology* 38: 768–770.

Boardman, R.S. 2008. Some early American History of the sectioning of Paleozoic Bryozoa: A Personal View. In P.N. Wyse Jackson & M.E. Spencer Jones (eds), *Annals of Bryozoology 2: Aspects of the History of Research on Bryozoans*: 1–8. Dublin: International Bryozoology Association.

Buttler, C.J., Rahman, I.A. & Slater, B. 2012. Three-dimensional Computer Reconstruction of a Lower Palaeozoic Trepostome bryozoan. *Palaeontological Association Newsletter, Abstracts Annual Meeting* 81: 62.

David, B., Stock, S.R., De Carlo, F., Hétérier, V. & De Ridder, C. 2009. Microstructures of Antarctic Cidaroid Spines: Diversity of Shapes and Ectosymbiont Attachments. *Marine Biology* 156: 1559–1572.

Heřmanová, Z., Bruthansová, J., Holcová, K., Mikuláš, R., Kočová Veselská, Tomáš, K., Dudák, J. & Vohník, M. 2020. Benefits and Limits of x-ray Micro-computed Tomography for Visualization of Colonization and Bioerosion of Shelled Organism. *Palaeontologia Electronica* 23: a23.

Huisman, D.J., Ngan-Tillard, D., Tensen, M.A., Laarman, F.J. & Raemaekers, D.C.M. 2014. A question of scales: Studying Neolithic Subsistence Using Micro CT Scanning of Midden Deposits. *Journal of Archaeological Science* 49: 585–594.

Keklikoglou, K., Faulwetter, S., Chatzinikolaou, E., Wils, P., Brecko, J. Kvaček, J., Metscher, B. & Arvanitidis, C. 2019. Micro-computed Tomography for Natural History Specimens: A Handbook of Best Practice Protocols. *European Journal of Taxonomy* 522: 1–55.

Key, M.M., Jr. & Wyse Jackson, P.N. 2022. History of Micro-Computed Tomographic Three-dimensional Imaging in Bryozoology. In P.N. Wyse Jackson & M.E. Spencer Jones (eds), *Annals of Bryozoology 7: Aspects of the History of Research on Bryozoans*: 57–72. Dublin: International Bryozoology Association.

Key, M.M., Jr., Burkhart, M.S. & O'Leary, M. 2019. Eocene Bryozoans Preserved in Chert from the Wilson Bluff Limestone, Eucla Basin, Western Australia. *Australasian Palaeontological Memoirs* 52: 85–90.

Key, M.M., Jr., Wyse Jackson, P.N. & Felton, S.H. 2016. Intracolony Variation in Colony Morphology in Reassembled Fossil Ramose Stenolaemate Bryozoans from the Upper Ordovician (Katian) of the Cincinnati Arch region, USA. *Journal of Paleontology* 90: 400–412.

Key, M.M., Jr., Wyse Jackson, P.N. & Vitiello, L.J. 2011. Stream Channel Network Analysis Applied to Colony-wide Feeding Structures in a Permian Bryozoan from Greenland. *Paleobiology* 37: 287–302.

Key, M.M., Jr., Wyse Jackson, P.N., Patterson, W.P. & Moore, M.D. 2005. Stable Isotope Evidence for Diagenesis of the Ordovician Courtown and Tramore Limestones, South-eastern Ireland. *Irish Journal of Earth Sciences* 23: 25–38.

Koromyslova, A.V. & Pakhnevich, A.V. 2016. New Species of *Pachydermopora* Gordon, 2002 and *Beisselina* Canu, 1913 (Bryozoa: Cheilostomida) from a Campanian Erratic Block (Belarus) and Their MicroCT Investigation. *Paleontological Journal* 50: 41–53.

Limaye, A. 2012. Drishti: A Volume Exploration and Presentation Tool. *SPIE Proceedings* 8506: 1–9.

Maliva, R.G. & Siever, R. 1988. Mechanism and Controls of Silicification of Fossils in Limestones. *Journal of Geology* 96: 387–398.

Martha, S.O., Ruthensteiner, B., Taylor, P.D., Hillmer, G. & Matsuyama, K. 2019. Description of a New Cyclostome Species from the Middle Santonian of Germany Using Micro-computed Tomography. *Australasian Palaeontological Memoirs* 52: 91–99.

Matsuyama, K., Titschack, J., Baum, D. & Freiwald, A. 2015. Two New Species of Erect Bryozoa (Gymnolaemata: Cheilostomata) and the Application of Non-destructive Imaging Methods for Quantitative Taxonomy. *Zootaxa* 4020: 81–100.

Murray, R.W., Jones, D.L. & Buchholtz ten Brink, M.R. 1992. Diagenetic Formation of Bedded Chert: Evidence from Chemistry of the Chert-shale Couplet. *Geology* 20: 271–274.

Ngan-Tillard, D.J.M. & Huisman, D.J. 2017. Micro-CT Scanning. In C. Nicosia & G. Stoops (eds), *Archaeological Soil and Sediment Micromorphology*: 441–449. Oxford: Wiley.

Nye, O.B., Jr., Dean, D.A. & Hinds, R.W. 1972. Improved Thin Section Techniques for Fossil and Recent Organisms. *Journal of Paleontology* 46: 271–275.

O'Leary, M.J., Ward, I., Key, M.M., Jr., Burkhart, M. S., Rawson, C. & Evans, N. 2017. Challenging the 'Offshore Hypothesis' for Fossiliferous Chert Artefacts in Southwestern Australia and Consideration of Inland Trade Routes. *Quaternary Science Reviews* 156: 36–46.

Ryland, J.S. 2005. Bryozoa: An Introductory Overview. *Denisia* 16: 9–20.

Salomé, M., Peyrina, F., Cloetens, P., Odet, C., Laval-Jeantet, A.-M., Baruchel, J. & Spanne, P. 1999. A Synchrotron Radiation Microtomography System for the Analysis of Trabecular Bone Samples. *Medical Physics* 26: 2194–2204.

Sandberg, P.A. 1983. Ultrastructure and Skeletal Development in Cheilostomate Bryozoa. In R.A. Robinson (ed.), *Treatise on Invertebrate Paleontology, Part G, Bryozoa, Revised, Volume 1*: 238–286. Boulder & Lawrence: Geological Society of America & University of Kansas.

Schindelin, J., Arganda-Carreras, I., Frise, E., Kaynig, V., Longair, M., Pietzsch, T., Preibisch, S., Rueden, C., Saalfeld, S., Schmid, B., Tinevez, J.-Y., White, D. J., Hartenstein, V., Eliceiri, K., Tomancak, P. & Cardona, A. 2012. Fiji: An Open-source Platform for Biological-image Analysis. *Nature Methods* 9: 676–682.

Schmidt, R. 2013. High Resolution Non-destructive Imaging Techniques for Internal Fine Structure of Bryozoan Skeletons. In A. Ernst, P. Schäfer & J. Scholz (eds), *Bryozoan Studies 2010*: 321–326, Berlin: Springer-Verlag.

Schwaha, T.F., Handschuh, S., Ostrovsky, A.N. & Wanninger, A. 2018. Morphology of the Bryozoan *Cinctipora elegans* (Cyclostomata, Cinctiporidae) with First Data on Its Sexual Reproduction and the Cyclostome Neuro-Muscular System. *BMC Evolutionary Biology* 18: 92.

Snyder, E.M. 1991. Revised Taxonomic Procedures and Paleoecological Applications for Some North American Mississippian Fenestellidae and Polyporidae (Bryozoa). *Paleontographica Americana* 57: 1–275.

Spaggiari, C.V. & Smithies, R.H. 2015. *Eucla Basement Stratigraphic Drilling Results Release Workshop. Extended Abstracts.* Department of Mines and Petroleum, Government of Western Australia, 75 p.

Suárez Andrés, J.L. & Wyse Jackson, P.N. 2015. Feeding Currents: A Limiting Factor for Disparity of Palaeozoic Fenestrate Bryozoans. *Palaeogeography, Palaeoclimatology, Palaeoecology* 433: 219–232.

Sutton, M., Rahman, I. & Garwood, R. 2014. *Techniques for Virtual Palaeontology*. New York: Wiley.

Sutton, M.D. 2008. Tomographic Techniques for the Study of Exceptionally Preserved Fossils. *Proceedings of the Royal Society* B275: 1587–1593.

Sutton, M.D., Briggs, D.E.G., Siveter, D.J. & Siveter, D. J. 2001. Methodologies for the Visualization and Reconstruction of Three-dimensional Fossils from the Silurian Herefordshire Lagerstätte. *Palaeontologia Electronica* 4: 1–17.

Taylor, P.D., Howard, L. & Gundrum, B. 2008. Microcomputed Tomography X-ray Microscopy of Cyclostome Bryozoan Skeletons. In N. Vávra & A. Ostrovksy (eds), *Abstracts of the VIII Larwood Meeting 2008*: 23. Vienna: University of Vienna.

Viskova, L.A. & Pakhnevich, A.V. 2010. A New Boring Bryozoan from the Middle Jurassic of the Moscow Region and Its microCT Research. *Paleontological Journal* 44: 157–167.

Ward, I., Key, M.M., Jr., O'Leary, M., Carson, A., Shaw, J. & Maksimenko, A. 2019a. Synchrotron X-ray Tomographic Imaging of Embedded Fossil Invertebrates in Aboriginal Stone Artefacts from Southwestern Western Australia: Implications for Sourcing, Distribution and Chronostratigraphy. *Journal of Archaeological Science: Reports* 26: 101840.

Ward, I., Key, M.M., Jr., Riera, R., Carson, A. & O'Leary, M. 2019b. Insights into the Procurement and Distribution of Fossiliferous Chert Artefacts Across Southern Australia from the Archival Record. *Australian Archaeology* 85: 170–183.

Ward, I., O'Leary, M., Key, M. & Carson, A. 2021. Response to Comment on Ward *et al.*'s 'Insights into the Procurement and Distribution of Fossiliferous Chert Artefacts Across southern Australia from the Archival Record.' *Australian Archaeology* 87: 330–332.

Wilson, M.A. & Palmer, T.J. 1989. Preparation of Acetate Peels. *The Paleontological Society Special Publications* 4: 142–145.

Wyse Jackson, P.N. & Buttler, C.J. 2015. Preparation, Imaging, and Conservation of Paleozoic Bryozoans for Study. In *Treatise on Invertebrate Paleontology Online Number 63, Part G, Revised, Volume 2, Chapter 3*: 1–15. Lawrence: Paleontological Institute.

Wyse Jackson, P.N. & McKinney, F.K. 2013. A Micro-computed Tomography and Scanning Electron Microscopy Investigation of the Structure of *Polyfenestella* Bancroft, 1986 (Bryozoa, Fenestrata), from the Mississippian of Scotland; Revealing the Nature of its Heteromorphs. *Irish Journal of Earth Sciences* 31: 19–24.

Zhang, Z., Zhang, Z., Ma, J., Taylor, P.D., Strotz, L. C., Jacquet, S.M., Skovsted, C.B., Chen, F., Han, J. & Brock, G.A. 2021. Fossil Evidence Unveils an Early Cambrian Origin for Bryozoa. *Nature* 599: 251–255.

Bryozoan Studies 2022 – Key, Porter & Wyse Jackson (eds)

Revision of Japanese *Watersipora* based on microscopic morphology and molecular data

Chihiro Kubo*
School of Marine Biosciences, Kitasato University, Kitasato, Sagamihara-Minami, Kanagawa, Japan

Masato Hirose
School of Marine Biosciences, Kitasato University, Kitasato, Sagamihara-Minami, Kanagawa, Japan
ORCID ID: 0000-0001-9991-7238

ABSTRACT: *Watersipora* is one of the common bryozoan genera in Japanese shallow waters, but reports with detailed taxonomic descriptions of this group are limited. Recently, Japanese *Watersipora* has been classified into several species based on the morphology of their pseudopores and operculum. However, the distribution of each species and the microscopic differences in their morphology along the Japanese coast is still unknown. In this study, we reexamine the morphological characters of Japanese *Watersipora* collected from several different localities on the Japanese coast based on detailed observations of pseudopores and molecular phylogenetic analyses using the cytochrome c oxidase subunit 1 mitochondrial gene (COI), we identify the species and to verify the diagnosis of the species so that they reflect genetic differences within the genus. In the preliminary results, the number of pseudopores varies within the same genetic group. In contrast, the characters of their operculum better reflect the genetic distances between taxa.

1 INTRODUCTION

Several species of *Watersipora* have recently been widely reported as invasive in tropical and temperate port areas (Harmer 1957; Mackie *et al.* 2012; Reverter-Gil & Souto 2019; Ryland 1974; Ryland *et al.* 2009; Soule & Soule 1975; Vieira *et al.* 2014). *Watersipora* lack certain taxonomic characters such as avicularia, spines, and ovicells that are generally used as diagnostic characters for species additional identification in other bryozoan groups. Therefore, it is difficult to identify species of *Watersipora* without observations of detailed morphologies using the scanning electron microscope (SEM) (Reverter-Gil & Souto 2019; Vieira *et al.* 2014). Recently, detailed observations and comparisons of morphological characters have revealed that the opercular patterning and pseudopore diameters can be used as the diagnostic tools for some *Watersipora* species (Vieira *et al.* 2014). In addition to morphological characters, cytochrome c oxidase subunit I (COI) mitochondrial gene sequences have also been used for identification of some *Watersipora* species (Mackie *et al.* 2012). However, many of the COI gene sequences in GenBank are based on incorrect identifications, which may lead to confusion.

Watersipora is one of the common bryozoan groups in Japanese shallow waters. The genus is widely distributed from Hokkaido to Kyushu, where five species reported to date (Vieira *et al.* 2014). For many years, only a single species of *Watersipora* had been reported along the Japanese coast, though reports with detailed taxonomic descriptions are few. *Watersipora* is also known as an invasive species and has been reported to invade various marine areas (Ferrario *et al.* 2015; Kuhlenkamp & Kind 2013; Mackie *et al.* 2012; Porter *et al.* 2015; Porter *et al.* 2017; Reverter-Gil & Souto 2019; Ryland *et al.* 2009; Vieira *et al.* 2014). Some *Watersipora* are thought to be imported from Japanese waters with farmed oysters or by ship (Ryland *et al.* 2009). It is therefore necessary to reexamine *Watersipora* specimens collected along the Japanese coast to further understand their species diversity and the geographical distribution.

In this study, we re-examined the species taxonomy of Japanese *Watersipora* collected from several different localities on the Japanese coast, based on detailed observations of pseudopores and brief molecular phylogenetic analyses using the COI gene, and attempted to verify the diagnostic characters of the species that reflect genetic differences within the genus.

2 MATERIALS AND METHODS

2.1 *Specimen collection*

Specimens were collected from Kikonai and Muroran, Hokkaido; Asamushi, Aomori Prefecture; Otsuchi, Yoshihama, and Okirai Bays, Iwate Prefecture;

*Corresponding Author: mf21008@st.kitasato-u.ac.jp

DOI: 10.1201/9781003454915-8

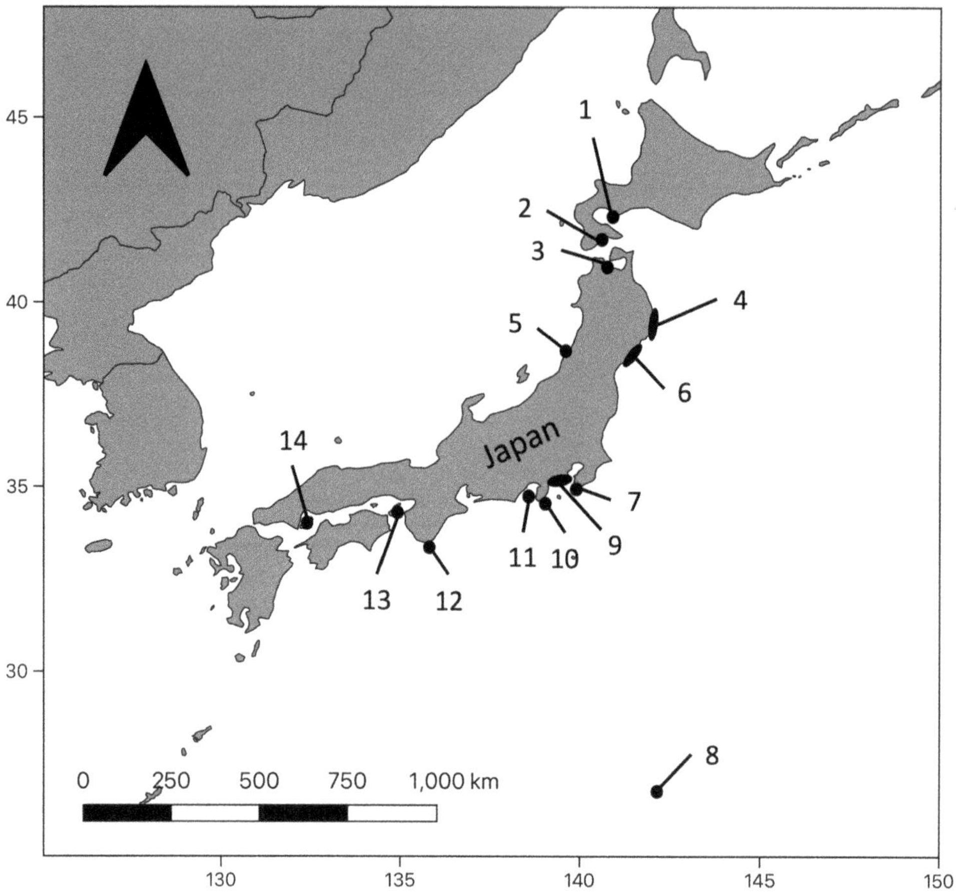

Figure 1. Map showing localities where examined specimens were collected. 1, Muroran, Hokkaido; 2, Kikonai, Hokkaido; 3, Asamushi, Aomori Pref.; 4, Otsuchi, Yoshihama and Okirai Bays, Iwate Pref.; 5, Kamo, Yamagata Pref.; 6, Kesennuma, Shizugawa, Onagawa and Matsushima Bays, Miyagi Pref.; 7, Okinoshima, Chiba Pref.; 8, Ogasawara Islands; 9, Kannonzaki (Tokyo Bay), Hayama and Manazuru (Sagami Bay); 10, Shimoda, Shizuoka Pref.; 11, Shirahama (Suruga Bay); 12, Shionomisaki, Wakayama Pref.; 13, Awaji Island; 14, Mukaishima Island, Hiroshima Pref.

Kesennuma, Shizugawa, Onagawa, and Matsushima Bays, Miyagi Prefecture; Kamo, Yamagata Prefecture; Okinoshima, Chiba Prefecture; Ogasawara Islands; Kannonzaki in Tokyo Bay, Hayama and Manazuru in Sagami Bay; Shirahama, Suruga Bay and Shimoda, Shizuoka Prefecture; Shionomisaki, Wakayama Prefecture; Awaji Island; and Mukaishima Island, Hiroshima Prefecture (Figure 1, Table 1). Some specimens were obtained from older collections of various other researchers completed in previous studies kept at Kitasato University. Newly collected samples were obtained from rocks in the intertidal zone or from settlement panels submerged in Otsuchi, Okirai, and Matsushima Bays. Subsequently, all samples were fixed with 99% ethanol.

2.2 *Preparation and observation of specimens*

The morphology of the collected colonies was observed using a stereomicroscope (OLYMPUS, SZX16). Approximately 1–5 mm pieces of a portion of each colony was removed, soaked in a 6% sodium hypochlorite solution to remove the soft tissue, rinsed in tap water, air dried, and mounted with double-sided adhesive carbon tape on an aluminum SEM stub. The dried specimens were observed using SEM (KEYENCE, VE-9800) without sputter coating.

Measurements were taken from SEM images by using ImageJ 1.37v software (Abramoff *et al.* 2004). All measurements are presented in micrometers (μm), and results are given as a range or average ± one (1) standard deviation with the following abbreviations; length of the zooid, ZL; width of the zooid, ZW; length of the orifice, OrL; width of the orifice, OrW; length of the sinus, SinL; width of the sinus, SinW; length of the pseudopore, PorL; width of the pseudopore, PorW; number of pseudopores, PorN.

2.3 *Molecular methods*

In this study, 15 specimens were prepare for molecular analysis. Total genomic DNA was

Table 1. Collection localities of specimens examined in this study.

Number on Figure 1	Latitude	Longitude	Water depth(m)	Location
1*	–	–	–	Muroran
2*	–	–	–	Kikonai
3*	–	–	–	Asamushi
4	39°20′06.1″N	141°55′41.9″E	10–20	Otsuchi
4	39°08′46.2″N	141°50′10.1″E	0–1	Yoshihama
4	39°05′27.2″N	141°52′08.4″E	10–20	Okirai
5	38°45′46.8″N	139°43′43.2″E	0–1	Kamo
6	–	–	–	Kesennuma
6	–	–	–	Shizugawa
6	38°26′36.4″N	141°27′44.8″E	0–1	Onagawa
6	38°20′34.4″N	141°05′38.4″E	10	Matsushima
7	34°59′27.8″N	139°49′31.8″E	3–5	Okinoshima
8	27°06′13.0″N	142°12′22.1″E	2–3	Ogasawara
9	–	–	0–1	Kannonzaki
9	35°15′55.4″N	139°34′22.5″E	0–1	Hayama
9	35°09′01.0″N	139°08′55.8″E	0–1	Manazuru
10	34°39′09.9″N	138°57′52.9″E	0–1	Shimoda
11	34°44′32.8″N	138°45′36.0″E	2	Shirahama
12	33°26′42.0″N	135°45′17.1″E	1	Shionomisaki
13	34°17′31.3″N	134°56′44.9″E	1–5	Awaji
14	–	–	–	Mukaishima

*Old collections kept at Kitasato University.

extracted from the lophophore of ethanol-fixed specimens using the DNeasy Blood & Tissue Kit (Qiagen) following the manufacture's protocol. The COI fragments were amplified with primers HCO2198 (5'-TAAACTTCAGGGTGACCAAA AAATCA-3') and LCO1490 (5'-GGTCAAC AAATCATAAAGAT ATTGG-3') (Folmer *et al.* 1994). PCR reactions were perfomed with *ExTaq* (TaKaRa). Conditions for hot-start thermal cycling were 4 min at 94°C; 40 cycles of 30 sec at 94°C, 30 sec at 43–54°C, and 90 sec at 72°C; and 7 min at 72°C. Subsequently, PCR products were purified with ExoSAP-IT (Affymetrix). All PCR products were sequenced in both directions by a commercial sequencing facility (Eurofin Genomics). The obtained sequences were aligned with those from other *Watersipora* species deposited in GenBank using Clustal W (Thompson *et al.* 1994) implemented in MEGAX (Kumar *et al.* 2018) in order to construct a maximum likelihood (ML) tree. The best-fit model was determined by the AICc implemented in MEGAX with T92 + G + I (Tamura 3-parameter model (Tamura 1992)) and gamma-distributed rates and invariant rates among sites. Optimal ML trees were found by a tree-bisection reconnection (TBR) search, starting with a tree topology generated by the BIONJ method (Gascuel 1997) using maximum composite likelihood (MCL) distances (Tamura *et al.* 2004). One-thousand bootstrap pseudoreplicates were analysed to obtain nodal support values. The COI dataset showed significant sequence saturation at the 3rd codon position as inferred by a larger value for the index of substitution saturation (Iss) than Iss.c (Xia & Lemey 2009; Xia *et al.* 2003). Therefore, we used only 1st + 2nd codon positions in our phylogenetic analyses. The tree was rooted with *Bugula neritina* as the outgroup taxa.

3 RESULTS

3.1 *Species diversity of examined Watersipora specimens*

Eight species of *Watersipora* were classified based on the observation of morphologies and the molecular phylogeny: *Watersipora bidentata*, *Watersipora* cf. *platypora*, *Watersipora arcuata*, *Watersipora* sp. 1, *Watersipora* sp. 2, *Watersipora* sp. 3, *Watersipora* sp. 4 and *Watersipora* sp. 5. Each species forms a clade in the results of our molecular phylogeny.

Although genomic DNA could not be obtained for the specimen from Okinoshima, Chiba Prefecture, we assigned it to *W. bidentata* according to the rectangular sinus. The specimens obtained from Sagami Bay and Awaji Island had a rectangular sinus that was concave toward the front, and the number of pseudopores was 113 ± 11 and 111 ± 9 in Sagami Bay and Awaji Island, respectively. The condyles

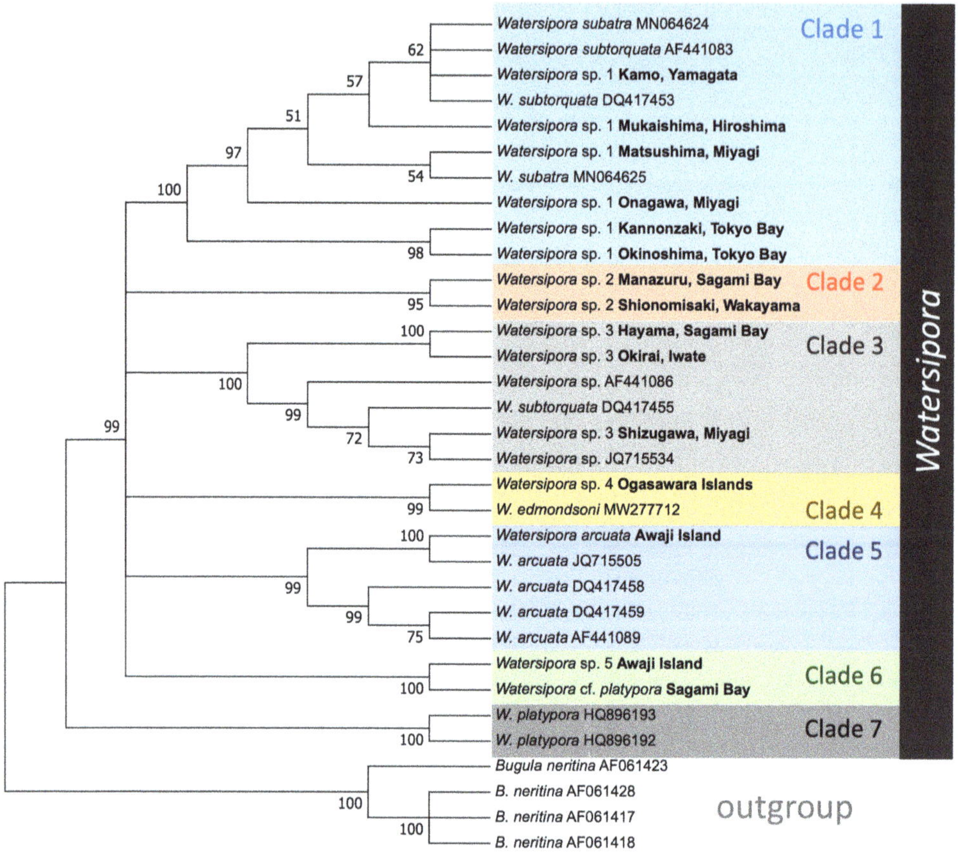

Figure 2. Maximum likelihood phylogeny reconstructed based on the mitochondrial cytochrome *c* oxidase subunit I (COI) sequences. The branch values indicate bootstrap values, and the phylogenetic tree is a consensus tree with bootstrap values greater than ≥ 50%. Bold letters indicate sequences obtained in this study.

formed small triangles with an orifice width of 333 ± 19 μm in Sagami Bay and 328 ± 9 μm on Awaji Island and were therefore assigned to *W.* cf. *platypora*. In addition to *W. platypora*, *W. arcuata* was found in the specimens from Awaji Island. This specimen had an orifice width was 232 ± 5 μm, and a pair of spots were present on the operculum. A rugby-ball-shaped cleft was present on the basal surface of the zooid, and the number of pseudopores was 132 ± 10, which was therefore assigned to *W. arcuata*. The specimen obtained from Ogasawara Islands, which was assigned to *Watersipora* sp. 4, has a small orifice and small pseudopores; the length and width of this orifice of the specimen was 164–184 μm and 195–225 μm, respectively. This specimen differs from *W. edmondsoni*, which is synonymized with *W. subtorquata*, in lacking white spots on the operculum and has interzooidal septula.

3.2 *Morphological characters*

3.2.1 *Pseudopore*

Here we describe and compare the detailed morphologies of unidentified species of *Watersipora* sp. 1, 2, 3 and 5 in this study (Table 2). Mean length of pseudopores for *Watersipora* sp. 1 ranged from 15–23 μm (15 ± 1 μm in Kamo, 16 ± 1 μm in

Hiroshima, 18 ± 2 μm in Matsushima, 16 ± 2 μm in Onagawa, 23 ± 3 μm in Kannonzaki, and 16 ± 1 μm in Okinoshima). The length of the pseudopores of *Watersipora* sp. 2 ranged from 13–19 μm (19 ± 3 μm in Manazuru and 13 ± 1 μm in Shionomisaki), while that of *Watersipora* sp. 3 was slightly larger and, ranged from 17–25 μm (17 ± 2 μm in Hayama, 25 ± 2 μm in Okirai Bay, and 19 ± 1 μm in Shizugawa). The length of the pseudopores of *Watersipora* sp. 5 was remarkably larger than these other species, ranged from 23–32 μm (30 ± 3 μm in Awaji Island).

The mean width of pseudopores of *Watersipora* sp. 1 ranged from 14–22 μm (14 ± 2 μm in Kamo, 16 ± 1 μm in Hiroshima, 20 ± 1 μm in Matsushima, 17 ± 1 μm in Onagawa, 22 ± 3 μm in Kannonzaki, and 16 ± 2 μm in Okinoshima). The width of the pseudopores of *Watersipora* sp. 2 ranged from 11 to 20 μm (20 ± 4 μm in Manazuru and 11 ± 1 μm in Shionomisaki), while that of

Table 2. Measurements of *Watersipora* sp. 1, 2, 3 and 5. All measurements are presented in micrometer (μm), and results are given as average ± standard deviation. Specimens for which genomic DNA was not extracted are also included.

Species	Site	ZL	ZW	OrL	OrW	SinL	SinW	N
sp.1	Kamo*	1071 ± 75	434 ± 30	247 ± 13	276 ± 18	59 ± 3	141 ± 12	10
	Kannonzaki*	981 ± 93	443 ± 43	250 ± 13	291 ± 16	56 ± 6	140 ± 8	10
	Kikonai	963 ± 58	563 ± 35	244 ± 11	288 ± 9	61 ± 6	145 ± 11	10
	Matsushima*	832 ± 56	478 ± 47	219 ± 8	268 ± 10	57 ± 2	135 ± 7	10
	Mukaishima*	946 ± 41	544 ± 37	247 ± 10	296 ± 14	55 ± 4	146 ± 9	10
	Okinoshima*	1025 ± 53	440 ± 13	233 ± 4	286 ± 5	61 ± 3	138 ± 3	10
	Onagawa*	1019 ± 125	471 ± 20	236 ± 16	279 ± 9	51 ± 6	140 ± 15	10
sp.2	Manazuru*	1023 ± 113	349 ± 59	231 ± 14	252 ± 15	49 ± 7	131 ± 10	10
	Shimoda	865 ± 43	479 ± 40	194 ± 10	245 ± 7	49 ± 3	123 ± 9	10
	Shionomisaki*	943 ± 72	444 ± 22	213 ± 6	272 ± 6	47 ± 4	129 ± 9	10
sp.3	Asamushi	864 ± 113	466 ± 41	212 ± 19	272 ± 23	52 ± 2	118 ± 13	10
	Hayama*	902 ± 56	481 ± 50	213 ± 7	258 ± 11	46 ± 4	101 ± 5	10
	Okirai*	1275 ± 55	483 ± 31	252 ± 12	299 ± 13	50 ± 3	136 ± 6	10
	Otsuchi	1447 ± 238	450 ± 66	271 ± 18	331 ± 37	53 ± 6	136 ± 6	10
	Shirahama	1098 ± 46	539 ± 33	242 ± 8	289 ± 11	53 ± 3	119 ± 9	10
	Shizugawa*	1064 ± 51	479 ± 21	232 ± 9	297 ± 10	42 ± 3	129 ± 6	10
sp. 4	Awaji*	1088 ± 48	565 ± 38	264 ± 13	324 ± 10	70 ± 4	161 ± 13	10
	Kesennuma	1055 ± 47	467 ± 16	236 ± 5	299 ± 5	55 ± 4	139 ± 7	10
	Muroran	1006 ± 65	485 ± 72	267 ± 17	297 ± 2	66 ± 14	139 ± 22	10
	Yoshihama	1284 ± 69	450 ± 29	262 ± 20	283 ± 11	58 ± 9	145 ± 6	10

*COI sequences obtained in this study

Watersipora sp. 3 was 18–27 μm (18 ± 1 μm in Hayama, 27 ± 2 μm in Okirai, and 19 ± 1 μm in Shizugawa). The width of the pseudopores of *Watersipora* sp. 5 was also slightly larger than the other species, ranged from 23–34 μm (29 ± 3 μm in Awaji Island).

Mean number of pseudopores of *Watersipora* sp. 1 ranged from 118 to 223 (170 ± 18 in Kamo, 150 ± 16 in Hiroshima, 223 ± 21 in Matsushima, 155 ± 12 in Onagawa, 162 ± 22 in Kannonzaki, and 118 ± 7 in Okinoshima). The number of pseudopores of *Watersipora* sp. 2 ranged 167–177 (177 ± 26 in Manazuru and 167 ± 9 in Shionomisaki), while that of *Watersipora* sp. 3 ranged 148–209 (175 ± 17 in Hayama, 209 ± 10 in Okirai Bay, and 148 ± 16 in Shizugawa). The number of pseudopores of *Watersipora* sp. 5 was 156 ± 14 in Awaji Island.

3.2.2 *Operculum pattern and ratio of the sinus width to orifice width*

Watersipora sp. 1, 2, 3 and 5 differed in the pattern and morphology of the operculum, and the ratio of the sinus width to the orifice width. The pattern and morphology of the operculum was divided into three types, A, B, and C (Figure 3). In type A, the operculum fulfilled the sinus area, and the line separating the central black band reached to both of the edges of the operculum. In type B, the operculum did not fill the sinus area, the proximal part was notably thin, and the line separating the central black band did not reach to both of the edges. Type C had an operculum that fully covered the sinus area, but the line separating the central black band did not reach to both of the edges of the operculum.

There were two types of ratios of sinus width to orifice width. The results showed that in *Watersipora* sp. 1, the operculum was type A and the ratio of sinus width to orifice width was 0.5, while in type B and 0.5 in *Watersipora* sp. 2 (Table 3). In *Watersipora* sp. 3, the operculum was type B and the ratio of sinus width to orifice width was 0.4, and in *Watersipora* sp. 5, the operculum was type C and the ratio was 0.5.

The specimens for which genomic DNA was not extracted were able to be classified into the species based on the pattern of the operculum and the ratio of the sinus width to orifice width (Table 3).

3.3 *Molecular results*

In this study, sequences were obtained from 15 specimens; the sequences have been deposited with GenBank under accession numbers LC746765–746779. The COI dataset comprised 696 aligned positions, with 264 variable sites, for

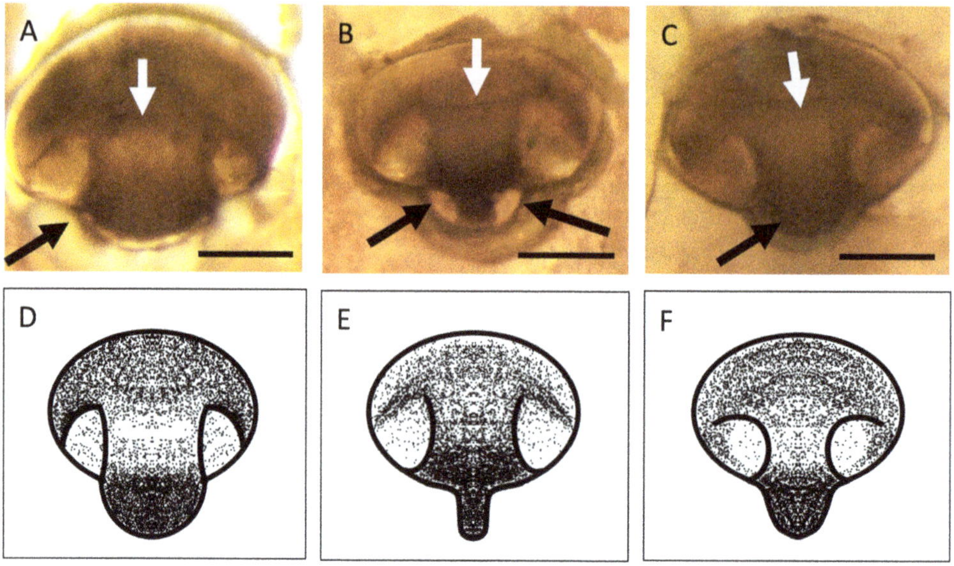

Figure 3. Pattern of operculum morphology observed in this study. A, D, operculum type A; B, E, operculum type B; C, F, operculum type C. Black arrows indicate whether or not the operculum fulfills sinus. White arrows indicate a black band. Scale bars = 100 μm.

Table 3. Operculum pattern (type) and ratio of the sinus width to orifice width of *Watersipora* sp. 1, 2, 3 and 5. Specimens for which genomic DNA was not extracted are also included.

Species	Site	Operculum Type	SinW/OrW	N
sp. 1	Kamo*	A	0.5 ± 0.03	10
	Kannonzaki*	A	0.5 ± 0.02	10
	Kikonai	A	0.5 ± 0.04	10
	Matsushima*	A	0.5 ± 0.02	10
	Mukaishima*	A	0.5 ± 0.02	10
	Okinoshima*	A	0.5 ± 0.01	10
	Onagawa*	A	0.5 ± 0.02	10
sp. 2	Manazuru*	B	0.5 ± 0.02	10
	Shimoda	B	0.5 ± 0.03	10
	Shionomisaki*	B	0.5 ± 0.03	10
sp. 3	Asamushi	B	0.4 ± 0.02	10
	Hayama*	B	0.4 ± 0.02	10
	Okirai*	B	0.4 ± 0.02	10
	Otsuchi	B	0.4 ± 0.03	10
	Shirahama	B	0.4 ± 0.02	10
	Shizugawa*	B	0.4 ± 0.03	10
sp. 4	Awaji*	C	0.5 ± 0.02	10
	Kesennuma*	C	0.5 ± 0.02	10
	Muroran	C	0.5 ± 0.02	10
	Yoshihama*	C	0.5 ± 0.03	10

*COI sequences obtained in this study.

29 ingroup taxa. In the resulting ML tree (Figure 2) (log L = −2245.99) not all nodes are resolved or well supported within the *Watersipora* clade, but seven clades were formed. Clade 1, represented by *Watersipora* sp. 1, was comprised of specimens from Kamo, Mukaishima, Matsushima, Onagawa, Kannonzaki, Okinoshima and several sequences of *W. subatra* and *Watersipora subtorquata* from GenBank. Clade 2, represented by *Watersipora* sp. 2, was comprised of only specimens from Manazuru and Shionomisaki, and none of the sequences obtained from GenBank were included in this clade. Clade 3, represented by *Watersipora* sp. 3, was comprised of specimens from Hayama, Okirai Bay, Shizugawa, and several sequences of *Watersipora* sp. and *W. subtorquata* from GenBank. Clade 4 was comprised of the *Watersipora* sp. 4 from Ogasawara and nominal "*W. edmondsoni*" (it is registered in Genbank as *W. edmondsoni*) from GenBank. Clade 5 included the *W. arcuata* from Awaji Island and several sequences of *W. arcuata* from GenBank. Clade 6 comprised the *W.* cf. *platypora* and *Watersipora* sp. 5 from Sagami Bay and Awaji Island respectively. Clade 7 was comprised of the only two sequences of *Watersipora platypora* from GenBank.

Pairwise distances among *Watersipora* species for 407 bp of all codon positions of COI ranged from 0.00 to 1.00. Within each clade, pairwise distances were 0.00–0.06 among *Watersipora* sp. 1 (clade 1), 0.07 between *Watersipora* sp. 2 from Manazuru and Shionomisaki (clade 2), 0.00–0.04 among *Watersipora* sp. 3 (clade 3), 0.03 between nominal "*W. edmondsoni*" and *Watersipora* sp. 4 from Ogasawara (clade 4), 0.00–0.04 among *W. arcuata* (clade 5), and 0.02 between *Watersipora platypora* and *Watersipora* sp. 5 from Sagami Bay and Awaji Island (clade 6). In contrast,

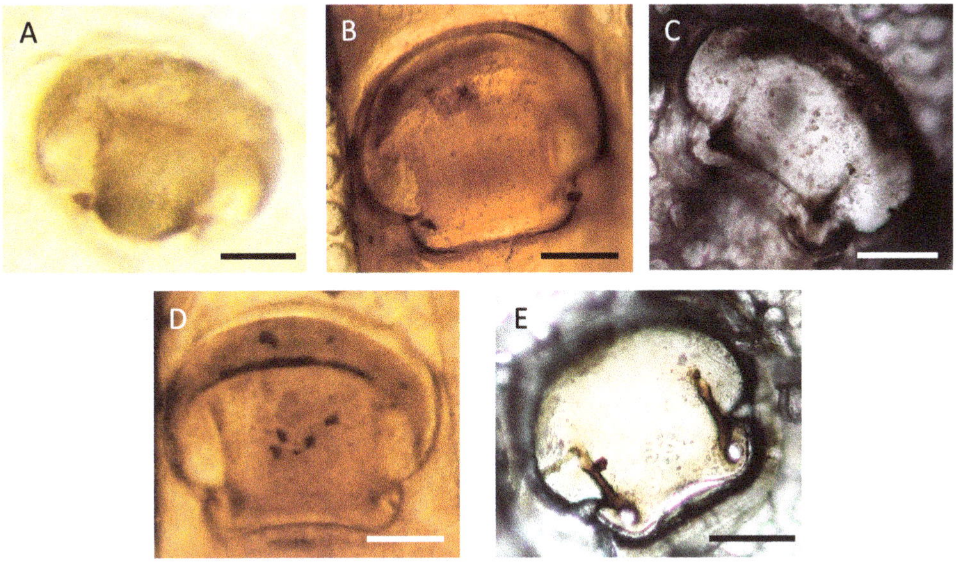

Figure 4. Operculum pattern of Japanese *Watersipora* species other than those shown in Figure 3. A, *Watersipora* sp. 4 from Ogasawara; B, *W. bidentata* from Okinoshima; C, *W. platypora* from Awaji Island; D, *W. platypora* from Sagami Bay; E, *W. arcuata* from Awaji Island. Scale bars = 100 μm.

Table 4. Pairwise distances among the clades in the result of the phylogeny of *Watersipora* for 407 bp of all codon positions of COI. Analyses were conducted using the Tamura 3-parameter model (Tamura 1992).

	Clade 1	Clade 2	Clade 3	Clade 4	Clade 5	Clade 6
Clade 1						
Clade 2	0.18–0.21					
Clade 3	0.27–0.32	0.23–0.27				
Clade 4	0.36–0.39	0.30–0.34	0.32–0.36			
Clade 5	0.26–0.27	0.21–0.23	0.23–0.25	0.25–0.26		
Clade 6	0.32–0.37	0.27–0.32	0.29–0.33	0.31–0.35	0.20–0.22	
Clade 7	0.79–0.82	0.74–0.77	0.76–0.78	0.79–0.80	0.67	0.64–0.66

comparisons of the pairwise distances between clades showed the values ranging from 0.18 to 0.82 (Table 4).

3 DISCUSSION

Recently, six species of *Watersipora* have been reported in Japanese waters: *Watersipora subatra*, *Watersipora aterrima*, *Watersipora typica*, *Watersipora mawatarii*, *Watersipora platypora*, and *Watersipora bidentata* (Hirose 2010; Vieira *et al.* 2014). They have also been reported to inhabit Hokkaido, Onagawa Bay, Sagami Bay, and Amakusa (Hirose 2010; Okada & Mawatari 1937; Ortmann 1890; Vieira *et al.* 2014). In this study, we found *Watersipora* sp. 4 and *Watersipora arcuata* from Ogasawara Islands and Awaji Island, respectively, based on molecular phylogenetic analysis and morphological characters, in addition to the six species mentioned above. *Watersipora* sp. 1, 2, 3 and 5 showed a high range of variation in pseudopore characters and condyle morphology within the same clade, but can be distinguished based on the operculum and orifice morphologies.

In this study, we examined whether pseudopore morphology reflects phylogenetic relationships. In the previous study, diameter of the pseudopores considered to vary among species (Vieira *et al.* 2014). Although we used only the COI dataset for the molecular phylogeny, the resulted ML tree showed seven clades within the examined genus *Watersipora*; the tree showed that the number, length, and width of pseudopores varies within each clade. Therefore, our results suggest that the morphology of pseudopores are not diagnostic characters that truly reflecting phylogenetic relationships.

Although condyle morphology and pseudopore diameter have also been used as taxonomic features to distinguish species, our phylogeny suggest that these characters also may not reflect phylogenetic relationships. In contrast, the pattern and morphology of the operculum and the ratio of the sinus width to the orifice width were clearly different in each clade. In recent years, the width of the sinus has been also used as a taxonomic character (Vieira *et al.* 2014); our results support the utility of this character to reflect phylogenetic relationships.

Bryozoan zooid size and orifice widths are known to decrease when the colony is grown at high water temperatures (Amui-Vedel *et al.* 2007; O'Dea & Okamura 1999). Furthermore, some studies have reported that the zooid size increases with depth in some species (Stępień *et al.* 2017). The difference in the width of the orifice is reported as a taxonomic character that distinguishes *W. subatra* from *W. aterrima* (Vieira *et al.* 2014), but this is also thought to vary depending on water temperature. Differences in the number, length, and width of pseudopores among the clades in this study may be due to the effects of different water temperatures and feeding environments. Therefore, it Is necessary to further examine intraspecific variation in these characters under different water temperatures in future experiments on *Watersipora.*

Whether fine morphological traits reflect genetic differences has not been studied. In this study, it was suggested that pseudopores may not reflect genetic differences, while the operculum pattern and the ratio of orifice width to sinus width may reflect genetic differences. Therefore, other taxonomic characters that are currently used should also be carefully examined to verify whether they are taxonomic traits that reflect the characteristics of the species.

ACKNOWLEDGEMENTS

This study was partly supported by a JSPS Grant-in-Aid for Scientific Research (A) Grant Number 21H04741 and JSPS Grant-in-Aid for Scientific Research (C) Grant Number 18K05791 and 22K05787.

REFERENCES

Abramoff, M.D., Magelhaes, P.J. & Ram, S.J. 2004. Image Processing with ImageJ. *Biophotonics International* 11(7): 36–42.

Amui-Vedel, A.-M., Hayward, P. & Porter, J.S. 2007. Zooid Size and Growth Rate of the Bryozoan *Cryptosula pallasiana* Moll in Relation to Temperature, in Culture and in Its Natural Environment. *Journal of Experimental Marine Biology and Ecology* 353: 1–12.

Ferrario, J., d'Hondt, J.L., Marchini, A. & Occhipinti-Ambrogi, A. 2015. From the Pacific Ocean to the Mediterranean Sea: *Watersipora arcuata*, a New Non-indigenous Bryozoan in Europe. *Marine Biology Reseach* 11: 909–919.

Folmer, O., Black, M., Hoeh, W., Lutz, R. & Vrijenhoek, R.C. 1994. DNA Primers for Amplification of Mitochondrial Cytochrome *c* Oxidase subunit I from Diverse Metazoan Invertebrates. *Molecular Marine Biology and Biotechnology* 3: 294–299.

Gascuel, O. 1997. BIONJ: An Improved Version of the NJ Algorithm Based on a Simple Model of Sequence Data. *Molecular Biology and Evolution* 14: 685–695.

Harmer, S.F. 1957. Polyzoa of the Siboga Expedition. Part 4. Cheilostomata Ascophora II. *Siboga-Expeditie*. 28d, xv, 641–1147.

Hirose, M. 2010. Cheilostomatous Bryozoa (Gymnolamata) from Sagami Bay, with Notes on Bryozoan Diversity and Faunal Changes over the past 130 Years. PhD Dissertation, Department of Natural History Science, Hokkaido University, Sapporo, Japan, 177 pp + 261 plates.

Kuhlenkamp, R. & Kind, B. 2013. Arrival of the Invasive *Watersipora subtorquata* (Bryo-Aoa) at Helgoland (Germany, North Sea) on Floating Macroalgae (Himanthalia). *Marine Biodiversity Records* 6: e73.

Kumar, S., Stecher, G., Li, M., Knyaz, C. & Tamura, K. 2018. MEGA X: Molecular Evolutionary Genetics Analysis Across Computing Platforms. *Molecular Biology and Evolution* 35: 1547–1549.

Mackie, J.A., Darling, J.A. & Geller, J.B. 2012. Ecology of Cryptic Invasions: Latitudinal Segregation Among *Watersipora* (Bryozoa) Species. *Scientific Reports* 2: 871.

O'Dea, A. & Okamura, B. 1999. Influence of Seasonal Variation in Temperature, Salinity and Food Availability on Module Size and Colony Growth of the Estuarine Bryozoan *Conopeum seurati. Marine Biology* 135: 581–588.

Okada, Y. & Mawatari, S. 1937. On the Collection of Bryozoa Along the Coast of Onagawa Bay and Its Vicinity, the Northern Part of Honshu, Japan. *The Science Report of Tohoku Imperial University, Biology* 11 (4): 433–445, pl. 1, textfigs, 1–5.

Ortmann, A. 1890. Die Japanische Bryozoenfauna. Bericht über die von Herrn Dr.L.Döderlein in Jahre 1880–81, Gemachten Sammlungen. *Archiv für Naturgeschichte* 56: 1–74.

Porter, J., Nunn, J., Ryland, J.S. Minchin, D. & Spencer Jones, M.E. 2017. The Status of Non-native Bryozoans on the North Coast of Ireland. *BioInvasions Records* 6: 321–330.

Porter, J., Spencer Jones, M., Kuklinski, P. & Rouse, S. 2015. First Records of Marine Invasive Non-native Bryozoa in Norwegian Coastal Waters from Bergen to Trondheim. *BioInvasions Records* 4: 157–169.

Reverter-Gil, O. & Souto, J. 2019. Watersiporidae (Bryozoa) in Iberian Waters: An Update on Alien and Native Species. *Marine Biodiversity* 49: 2735–2752.

Ryland, J.S. 1974. Bryozoa in the Great Barrier Reef province. In: Cameron, A.M. *et al.* (Eds.), *Proceedings of the Second International Reef Symposium. Vol. 2.* The Great Barrier Reef Committee, Brisbane, pp. 341–348.

Ryland, J.S., De Blauwe, H., Lord, R. & Mackie, J. 2009. Recent Discoveries of alien *Watersipora* (Bryozoa) in Western Europe, with Redescriptions of Species. *Zootaxa* 2093: 43–59.
Soule, D.F. & Soule, J.D. 1968. Bryozoan Fouling Organisms from Oahu, Hawaii, with a New Species of *Watersipora*. *Bulletin of the Southern California Academy of Science* 67: 203–218.
Soule, D.F. & Soule, J.D. 1975. Species Groups in Watersiporidae. *Bryozoa 1974. Documents des Laboratoires de Géologie de la Faculté des Science de Lyon, h.s.* 3(2): 299–309.
Stępień, A., Kukliński, P., Włodarska-Kowalczuk, M., Krzemińska, M. & Gudmundsson, G. 2017. Bryozoan Zooid size Variation Across a Bathymetric Gradient: A Case Study from the Icelandic Shelf and Continental Slope. *Marine Biology* 164(197): 1–13.
Tamura K, Nei M & Kumar S 2004. Prospects for Inferring Very Large Phylogenies by Using the Neighbor-joining Method. *Proceedings of the National Academy of Sciences* 101: 11030–11035.
Tamura, K. 1992. Estimation of the Number of Nucleotide Substitutions When There are Strong Transition-transversion and G + C-content Biases. *Molecular Biology and Evolution* 9: 678–687.
Thompson, J.D., Higgins, D.G. & Gibson, T.J. 1994. CLUSTAL W: Improving the Sensitivity of Progressive Multiple Sequence Alignment Through Sequence Weighting, Positions-specific Gap Penalties and Weight Matrix Choice. *Nucleic Acids Research* 22: 4673–4680.
Vieira, L.M., Spencer Jones, M.E. & Taylor, P.D. 2014. The Identity of the Invasive Fouling Bryozoan *Watersipora subtorquata* (d'Orbigny) and Some Other Congeneric Species. *Zootaxa* 3857(2): 151–182.
Xia, X. & Lemey, P. 2009. Assessing Substitution Saturation with Dambe: Practice. *Phylogenetic Handbook*. 624–630.
Xia, X., Xie, Z., Salemi, M., Chen, L & Wang, Y. 2003. An Index of Substitution Saturation and Its Application. *Molecular phylogenetics and evolution* 26: 1–7

Bryozoan Studies 2022 – Key, Porter & Wyse Jackson (eds)
© 2023 The Author(s), ISBN: 978-1-032-59487-3

A genome-skimmed phylogeny of California cheilostome bryozoans

Hannah E. Lee*
Department of Biological Sciences, California State Polytechnic University, Humboldt, Arcata, CA, USA
ORCID ID: 0000-0002-0064-3810

Russell J.S. Orr
Natural History Museum, University of Oslo, Oslo, Norway
ORCID ID: 0000-0002-1972-1321

Linda D. McCann
Smithsonian Environmental Research Centre, Tiburon, CA, USA
ORCID ID: 0000-0002-1571-1871

Sean F. Craig
Department of Biological Sciences, California State Polytechnic University, Humboldt, Arcata, CA, USA
ORCID ID: 0000-0002-0149-3455

ABSTRACT: The phylum Bryozoa is an incredibly diverse and widely distributed group of marine invertebrates whose phylogenetic relationships are still poorly understood, especially for species found at shallow subtidal to intertidal depths. This study aimed to expand the taxonomic sampling and assessment of the phylogenetic diversity of cheilostome bryozoan species present along the rocky outer coast and within sheltered harbors of California, USA. High-throughput genome skimming was used to produce multiple mitochondrial genes for samples collected from two rocky intertidal and two harbor sites. Phylogenetic analyses generated evolutionary hypotheses of species relationships and identified 15 distinct species that represent 10 different families to form the first comprehensive mitogenomic phylogeny for multiple bryozoan families sampled across approximately 973 km of the California coastline. These results support existing bryozoan phylogenies and provide significant additions to the available mitogenomic data on cheilostome bryozoans for both California and the North American west coast.

1 INTRODUCTION

The phylum Bryozoa is a taxonomically rich invertebrate group of colonial, benthic organisms commonly referred to as "moss animals". The phylum has over 6,000 described extant species that are globally distributed (Bock & Gordon 2013). Bryozoans have many roles in the community structure of benthic ecosystems, including the formation of habitat for other invertebrate taxa such as crustaceans and flatworms (Cocito 2004; Waeschenbach *et al.* 2012b; Wilson 2011) as well as food for predators, including certain species of nudibranchs and polychaetes (Lidgard 2008).

The order Cheilostomata includes the largest number of living bryozoan species, containing approximately 80% of extant bryozoan species (Taylor & Waeschenbach 2015). Cheilostome bryozoans are known for their calcified skeletal features which contribute to their extensive fossil record and allow the group to be a good target for understanding morphological character evolution. With the first cheilostome fossil appearing in the Late Jurassic (Pohowsky 1973), evidence suggests that cheilostome bryozoans originally had non-calcified frontal walls that were membranous and evolved the development of basally jointed spines around the frontal membrane later on (Dick *et al.* 2009; McKinney & Jackson 1989). Cheilostomes were estimated to have diverged from "ctenostome"-like bryozoans (which have non-calcified frontal walls) approximately 245–282 million years ago (MYA) during the Permian to the Early Triassic (Hao *et al.* 2005). This evolutionary hypothesis of cheilostomes having a Paleozoic origin has since been confirmed by recent research utilizing multi-gene sampling and fossil calibrations, indicating that there was a large evolutionary period between the initial divergence of cheilostomes from the ctenostomes and the first fossil records of their presence during the Late Jurassic

*Corresponding Author: hannahelee56@gmail.com

DOI: 10.1201/9781003454915-9

(Orr *et al.* 2022). Rates of morphological and molecular evolution often vary between and within different lineages, but genes that have experienced rapid evolution can be useful in assessments of phylogenetic divergence between different genera and in turn calculations of their evolutionary origins (Fuchs *et al.* 2009; Schwaninger 2008).

Despite advances in estimating the evolutionary relationships of bryozoans based on morphological character data, studies have shown that morphology-based phylogenies should be relied upon with caution and that these phylogenies should also be based upon other types of data (e.g. molecular data) to formulate more robust evolutionary hypotheses for the relationships between bryozoan species (Orr *et al.* 2019b). Many evolutionary studies use mitochondrial genes for species identification because mitochondrial genes are maternally inherited, often include protein-coding genes, evolve in relatively short time frames (generation time of a few years in invertebrate species), and are often easily recovered in DNA sequencing (Ge *et al.* 2021; Thomas *et al.* 2010).

Species level molecular phylogenies should include multiple genes because the molecular rate of evolution will vary for each gene within a species and the use of larger molecular datasets can provide a more robust phylogenetic inference of the evolutionary relationships between species (Duchêne *et al.* 2011). The use of complete or nearly complete mitogenomes (a.k.a. mitochondrial genomes) alone can produce phylogenies with much greater resolution and precision relative to those developed from only a few targeted genes (Duchêne *et al.* 2011; Trevisan *et al.* 2019). Studies that use multi-gene molecular phylogenies can both expand the available dataset of sequences available and better resolve the interrelationships on which taxonomic classifications should be based (Waeschenbach *et al.* 2012b).

Here we apply a genome skimming technique to cheilostome bryozoans sampled from the California coastline. High throughput sequencing (HTS) is used to "skim" the genome to quickly perform shallow sequencing of entire genomes of multiple species at once (Al-Haggar *et al.* 2013; Trevisan *et al.* 2019). This reduces the amount of time spent troubleshooting samples with low DNA yield or amplification processes with low specificity that may come with traditional polymerase chain reaction (PCR) methods while also increasing the relative abundance of total genomic DNA due to the capture of high copy genes including mitochondrial and housekeeping genes (Straub *et al.* 2012). It is often difficult to extract DNA from bryozoans and they generally have low DNA yields, but genome skimming has been proven to be successful for the recovery of genomic DNA from many cheilostome species (Orr *et al.* 2019a, 2021, 2022).

The California, USA coastline, which spans approximately 1351 km, is characterized by various geographical features including long stretches of sandy and rocky shores, estuaries, and headlands which provide numerous opportunities to study the evolutionary relationships of various marine communities (Blanchette *et al.* 2008; Kelly & Palumbi 2010). Since the last major taxonomic studies of bryozoans along the North American Pacific coastline were performed in the late 1990's and early 2000s (Soule *et al.* 1995, 2004, 2007), this study focuses its molecular sequencing effort on a subsample of cheilostome bryozoan species from both rocky outer coast and harbor habitats found in California due to its expansive coastline and the high likelihood that there may be several undiscovered species in this region whose phyletic inclusion in bryozoan taxonomies can be valuable to marine ecologists and conservationists studying these invertebrate community assemblages (Soule *et al.* 2007).

Many cheilostome bryozoan larvae are lecithotrophic and generally do not disperse far from their maternal source colony, so the lack of long-distance dispersal in most species may affect their overall geographic range. However, many species have been known to settle on the outside of ship hulls or be taken up in ballast water tanks of commercial vessels, thus allowing for larval dispersal outside of their natural region and the further promotion of gene flow and potentially marine invasions to new harbors (Holland 2000; Mackie *et al.* 2006). While certain species (e.g. *Watersipora subtorquata* complex) are known to invade different bays via anthropogenic sources, there is relatively little knowledge of the ability of bryozoan species to spread across different rocky intertidal areas or between the rocky outer coast and bays in California (Mackie *et al.* 2006).

The main objectives of this investigation were to utilize mitochondrial DNA as well as geographic distribution data to assess the evolutionary relationships of cheilostome bryozoans found in the rocky intertidal and bay sites along the California coastline. This study is based on collections of cheilostome bryozoans from two different habitats: 1. from small rocks from two rocky intertidal sites from Humboldt County, California, USA. and 2. from shallow subtidal sites within harbors in Morro Bay and Long Beach, California, which were previously collected by the Smithsonian Environmental Research Center (SERC). The latter were included in this study to expand the available diversity and geographic range of the dataset as well as examine the phylogenetic relationships of species found on the rocky outer coastline versus those found in protected bays and harbors. A genome-skimming approach was used to examine molecular data from the cheilostome bryozoan samples collected based on 15 mitochondrial genes. Their phylogenetic relationships were assessed using Maximum Likelihood and Bayesian inference methods to form a comprehensive molecular phylogeny of the cheilostome bryozoan species in this study. The aim of this assessment of the phylogenetic relationships of cheilostome bryozoans is to further the understanding of the speciation and spread of bryozoans throughout California.

2 METHODS

2.1 *Sample collection sites*

Samples of cheilostome bryozoans were collected at two rocky intertidal sites located on the outer coast of Humboldt County, California, USA during November and December in 2019 (Fig. 1 and Table 1, see Lee 2022 for full metadata from all collected samples). Baker Beach is a relatively secluded cove located approximately 2 km south of Trinidad Head. The intertidal region at Baker Beach is a semi-protected area characterized by large sea stacks at a peninsular-like point that is surrounded by a large boulder field. Sampling was restricted to the southern, more protected side of that boulder field. The second collecting site, Palmer's Point, is an exposed boulder field located approximately 10 km north of Baker Beach and is found within Sue-meg State Park (formerly known as Patrick's Point State Park). The intertidal region there is characterized by a large gently sloping field of small to medium-sized boulders with a few giant sedentary boulders. Sampling occurred at Baker Beach on three separate occasions (November 10, 23, and 24, 2019) while sampling at Palmer's Point occurred only once (December 9, 2019) due to weather restrictions.

Bryozoans were also collected from protected harbors or marinas by the Smithsonian Environmental Research Center (SERC: see Fig. 1 and Table 1) as part of a broader study of non-native marine invertebrate species in coastal ecosystems (Ruiz & Geller 2015; Ruiz unpubl.). These samples were collected from PVC settlement plates placed at marinas in four different localities within Morro Bay in 2013 and one locality within Long Beach in 2017. Harbor sites will henceforth be referred to by the city name (either Morro Bay or Long Beach) rather than the individual marina (locality).

2.2 *Sampling methods*

Outer coast samples were collected at low tides from small boulders found in the low intertidal zone (Ricketts & Calvin 1968). At each collecting site on a given date, approximately 10 different rocks or boulders were randomly sampled for bryozoan colonies. Rocks ranged in size from 12–75 cm in diameter. Upon collection, all bryozoan samples were catalogued individually in the field and were air dried or stored in 95% or 200 proof ethanol. Encrusting bryozoans attached to the undersides of rocks were chipped/cut off the rock using a portable Dremel tool and a hammer and chisel. All samples collected were kept at California State Polytechnic University, Humboldt and later shipped for genetic processing at the Bryozoan Lab for Ecology, Evolution and Development (a.k.a. BLEED) at the Natural History Museum at the University of Oslo in Oslo, Norway.

Bryozoans found at shallow subtidal depths in protected harbors or marinas were collected from settlement panels suspended one meter below randomly chosen floating docks in each marina by SERC (Table 1). Settlement panels consisted of bare, dark gray, lightly sanded PVC plates measuring 13.7 × 13.7 cm that were attached to bricks with the experimental surface facing downward, parallel to the seafloor, to mimic floating docks (Marraffini *et al.* 2017). Panels were suspended for approximately three months during the summer (June to September) to coincide with the period of high seasonal recruitment and provide sufficient time for mature communities to develop before collection (Jimenez *et al.* 2018; Marraffini *et al.* 2017). Sessile marine invertebrates on each panel were examined live under a dissecting microscope in the lab and each morphospecies was identified to the lowest possible taxonomic level. Cheilostome bryozoan samples were preserved in 95% ethanol at SERC and a subset were shipped to the BLEED laboratory for genetic processing.

Each colony was subsampled so that both DNA isolation and scanning electron microscopy (SEM) morphological examination could be completed. Subsamples for morphological vouchers from all sequenced colonies were dried, bleached in diluted household bleach to remove soft tissues, and prepped according to standard SEM protocols. All SEM images were taken at the University of Oslo using a Hitachi TM4000Plus tabletop scanning electron microscope so that skeletal features could be distinguished and measured for morphological taxonomic identification. Morphological identification to the lowest taxonomic level was performed for each specimen independently of the phylogenetic inference to avoid identification bias.

2.3 *DNA Isolation and sequencing*

Ethanol-preserved samples were dried and then rinsed in phosphate-buffered saline before isolating genomic DNA using the DNeasy Blood and Tissue kit following the manufacturer's instructions (QIAGEN, Germantown, MD, USA). Colonies were homogenized in lysis buffer, using a pestle, in the presence of proteinase K (50 µg/mL). DNA templates were sequenced directly by high-throughput sequencing (HTS) at the Norwegian Sequencing Centre (Oslo, Norway) using Illumina HiSeq4000 150 bp paired-end (PE) sequencing with a 350 bp insert size.

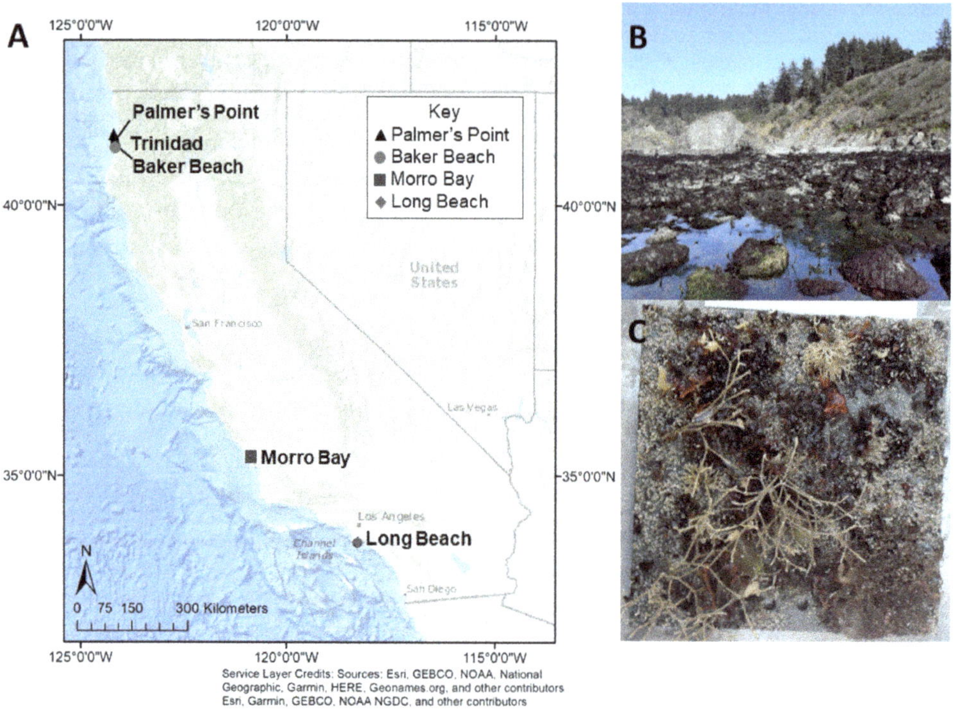

Figure 1. A. Map of sample collection sites along the California coastline. Two sites were sampled from the rocky outer coast near Trinidad. At the harbor sites, Morro Bay had samples collected from four different marina localities within the bay, whereas Long Beach had one marina sampled. Sites are indicated on the map with a corresponding shape (Palmer's Point = triangle, Baker Beach = circle, Morro Bay = square, and Long Beach = diamond). B. Example of a rocky outer coast boulder field (Palmer's Point, Trinidad, CA). C. Example of PVC settlement panels that were deployed in protected marinas.

Table 1. Location of specimen collection sites in California. The collection site or locality, city, and collection date are given as well as descriptions of latitude and longitude and substrate type from which a sample was collected. Samples collected in 2019 were retrieved from boulders within the rocky intertidal. Samples collected between 2013–2017 were retrieved from settlement panels within protected harbors by the Smithsonian Environmental Research Center (SERC).

Site	City	Collection Date	Latitude	Longitude	Substrate
Palmer's Point	Trinidad	Dec 9, 2019	41°07′51.6″N	124°09′50.4″W	Boulders
Baker Beach	Trinidad	Nov 10, 23, & 24, 2019	41°02′56.4″N	124°07′33.6″W	Boulders
City Harbor	Morro Bay	Aug 2013	35°22′15.6″N	120°51′28.8″W	Settlement panels
201 Main	Morro Bay	Aug 2013	35°21′21.6″N	120°50′49.2″W	Settlement panels
Launch Ramp	Morro Bay	Sept 2013	35°21′28.8″N	120°51′03.6″W	Settlement panels
Tidelands	Morro Bay	Sept 2013	35°21′36.0″N	120°51′07.2″W	Settlement panels
Cabrillo Marina	Long Beach	2017	33°43′04.8″N	118°16′40.8″W	Settlement panels

2.4 *Sequence assembly and alignment*

Illumina HiSeq reads were quality checked using FastQC v.0.11.9 (Andrews 2010), then quality and adapter trimmed using TrimGalore v.0.4.4 (Krueger 2015) with a quality score cut off of 35 and a minimum length of 100. Trimmed reads were assembled with MEGAHIT v.1.2.9 (Li *et al.* 2015) using k-mers of 21, 33, 55, 77, 99 and 127. Orthologous genomic sequences for the mitogenome were identified using blastn in NCBI BLAST (Altschul *et al.* 1990). A seed-extension based assembly was then performed with Novoplasty 4.1 (Dierckxsens *et al.* 2017).

Mitogenome sequences were annotated with Mitos 2 using a metazoan reference and invertebrate genetic code (Bernt *et al.* 2013) to identify 15 mitochondrial genes: two rRNA genes (rrnL and rrnS) and 13 protein coding genes (atp6, atp8, cox1, cox2, cox3, cob, nad1, nad2, nad3, nad4, nad4l, nad5, and nad6). Suitable orthologous sequences for each gene deposited in the NCBInr database were downloaded and aligned with MAFFT v.7 (Katoh & Standley 2013) using the respective default parameters for rRNA and protein-coding genes. All ingroup sequences were aligned with five ctenostome sequences (*Alcyonidioides mytili*, *Amathia citrina*, *Anguinella palmata*, *Flustrellidra hispida*, and *Paludicella* sp.) obtained from NCBI as the outgroup.

Each alignment was edited manually using MESQUITE v.3.61 (Maddison & Maddison 2018). Ambiguously aligned characters were removed from each alignment using Gblocks (Talavera & Castresana 2007) with the least stringent parameters. Each single-gene alignment was concatenated using catfasta2phyml perl script (Nylander 2010). A total of 31 cheilostome sequences obtained from this study were used in each alignment.

2.5 *Phylogenetic analysis*

Maximum likelihood (ML) phylogenetic analyses were performed for each single gene using the 'AUTO' parameter in RAxML v.8.2.12 (Stamatakis 2014) to establish the evolutionary model with the best fit. The general time reversible (GTR) was the preferred model for the two rRNA genes (rrnL and rrnS), and the MtZoa was the preferred model for the thirteen protein-coding genes. The concatenated dataset, divided into separate partitions for each gene for which each will have a separate gamma distribution, was analyzed using RAxML. The topology with the highest likelihood score of 100 heuristic searches was chosen as the best fit model. Bootstrap values were calculated from 500 pseudoreplicates. Taxa with unstable phylogenetic affinities were identified and removed (following previously outlined sampling rules) using MESQUITE v.3.61 based on evaluation of a 70% majority rule (MR) consensus tree.

Bayesian inference (BI) was performed using MrBayes v.3.2.7a incorporating the mtREV evolutionary model (Huelsenbeck & Ronquist 2001; Ronquist *et al.* 2012). The mtREV model is a general reversible Markov model in which each substitution probability is represented by an adjustable parameter for a total of 189 adjustable parameters (Adachi & Hasegawa 1996; Dimmic *et al.* 2000). The dataset was executed, as before, with a separate partition for each gene under which each will have a separate gamma distribution. Two independent runs, each with three heated and one cold Markov Chain Monte Carlo (MCMC) chain, were initiated from a random starting tree. The MCMC chains were run for 20,000,000 generations, with trees sampled every 1000th generation. The posterior probabilities and mean marginal likelihood values of these trees were calculated after the burn-in phase, which is determined from the marginal likelihood scores of the initially sampled trees. If the average standard deviation of split frequencies of the two runs is <0.01, then it indicates the convergence of the MCMC chains.

3 RESULTS

This study successfully sequenced and assembled 20 cheilostome colonies from four different sites (two rocky intertidal sites in northern California, one harbor site in central California, and one harbor site in southern California) over approximately 973 km of coastline between the northernmost site (Palmer's Point) and the southernmost site (Long Beach). A complete or circularized mitogenome (which included all 15 mitochondrial genes) was produced for 13 of the 20 collected samples (Table 2). Another six samples also produced a nearly complete mitogenome with 14 genes, but the bioinformatics pipeline was unable to recover the atp8 gene in these six samples. The phylogenetic analysis elucidated 15 lineages that correspond to 10 families and 12 genera (Table 2) in total, with the resulting alignment constituting 1,213 nucleotide and 2,794 amino acid characters (for a combined total of 4,007 characters). A summary table with individual gene availability for each sequence is available in Lee (2022). The phylogenetic placement of each colony, as well as its identity based on morphology, was confirmed to the species level by comparison to the results of Orr *et al.* (2020) and by examination by SEM of vouchers for each sequenced colony (except two colonies which were confirmed to the genus level). The SEM digital vouchers of all samples included in phylogenetic analysis can be found in Orr *et al.* 2022.

An additional 19 cheilostome sequences from NCBI were included in the ingroup phylogenetic analysis of the concatenated dataset of all newly sequenced California bryozoans from Orr *et al.* (2022) to confirm their phylogenetic placement, and these were aligned with an outgroup of five ctenostome bryozoan species (See Lee 2022 for available metadata for taxa from NCBI). Taxa from NCBI were chosen based on their broad phylogenetic placement throughout the order Cheilostomata as well as confirmation of the species identification from previous studies by Orr *et al.* (2020). NCBI sequences were also chosen based on their inclusion of multiple mitochondrial genes in the sequence. Statistical support for nodes based on bootstrap values (BS) in the ML analysis and posterior probability (PP) in the Bayesian analysis were defined as either full support (100 BS/1 PP), high support (>90 BS/0.99 PP), moderate support (>65 BS/0.95 PP),

Table 2. Taxa analyzed in this study. BLEED stands for Bryozoan Lab for Ecology, Evolution and Development, Natural History Museum, University of Oslo, Norway, and BLEED numbers are numerical tags for the specimens. The location given for outer coast samples is the local site name and the location for harbor samples is the city name. The mitogenome (Mt) size, in base pairs (bp), are only shown if it is complete/circularized, with NO indicating that a complete mitogenome was not formed. Genes, represents the number of genes, a maximum of 15, recovered and used in the alignments for each taxon. Accession nr. refer to those California bryozoan species (from Orr *et al.* 2022) or outgroup sequences retrieved from NCBI. Sequences were aligned with a five ctenostome bryozoan outgroup (*Alcyonidioides mytili*, *Amathia citrina*, *Anguinella palmata*, *Flustrellidra hispida*, and *Paludicella* sp.).

Family	Genus	Species	BLEED	Location	Mt size (bp)	Genes	Accession nbr
Calloporidae	*Tegella*	*horrida*	1894	Palmer's Point	NO	15	OK245189, OK245190, OK245191
Hippothoidae	*Celleporella*	*hyalina*	1895	Palmer's Point	NO	15	OK244797, OK244798
Calloporidae	*Cauloramphus*	*californiensis*	1896	Palmer's Point	14,573	15	OK244778
Microporellidae	*Microporella*	sp.	1897	Palmer's Point	13,563	14	OK245155
Phidoloporidae	*Rhynchozoon*	cf. *tumulo-sum*	1899	Palmer's Point	14,102	14	OK244996
Microporellidae	*Microporella*	*umbonata*	1900	Palmer's Point	13,931	14	OK245161
Candidae	*Pomocellaria*	*californica*	1901	Palmer's Point	15,733	15	OK244975
Fenestrulinidae	*Fenestrulina*	*umbonata*	1903	Baker Beach	14,780	14	OK244859
Chaperiidae	*Chaperiopsis*	*patula*	1906	Baker Beach	NO	15	OK244804
Smittinidae	*Parasmittina*	*collifera*	1907	Baker Beach	14,299	15	OK244944
Calloporidae	*Tegella*	*horrida*	1909	Baker Beach	NO	12	OK245192, OK245193, OK245194
Calloporidae	*Copidozoum*	*adamantum*	1910	Baker Beach	NO	13	OK244819
Candidae	*Pomocellaria*	*californica*	1912	Baker Beach	15,733	15	OK244976
Calloporidae	*Tegella*	*horrida*	1914	Baker Beach	NO	15	OK245195, OK245196, OK245197
Thalamoporellidae	*Thalamoporella*	*californica*	1211	Long Beach	13,928	14	OK245202
Pacificincolidae	*Primavelans*	*insculpta*	1214	Morro Bay	17,268	14	OK244984
Smittinidae	*Parasmittina*	*collifera*	1216	Morro Bay	14,305	15	OK244952
Smittinidae	*Parasmittina*	sp.	1219	Morro Bay	14,302	15	OK244953
Fenestrulinidae	*Fenestrulina*	*delicia*	1244	Morro Bay	NO	15	OK244853
Candidae	*Pomocellaria*	*californica*	1246	Morro Bay	15,733	15	OK244974
Alcyonidiidae	*Alcyonidioides*	*mytili*		Unavailable		5	JN681069, JN681102, AEV21493, AEV21531
Vesiculariidae	*Amathia*	*citrina*		Ferrol, San Felipe, Spain		4	KM373503, JN681121
Nolellidae	*Anguinella*	*palmata*		Parana, Ilha do Mel, Mor-ro do Sabao, Brazil		4	JN681101, AJB84768 AEV21530
Flustrellidridae	*Flustrellidra*	*hispida*		Mumbles and Aberystwyth, Wales, UK		13	NC_008192
Paludicellidae	*Paludicella*	sp.		Unavailable		5	JN681070, JN681103, AEV21494, AEV21532

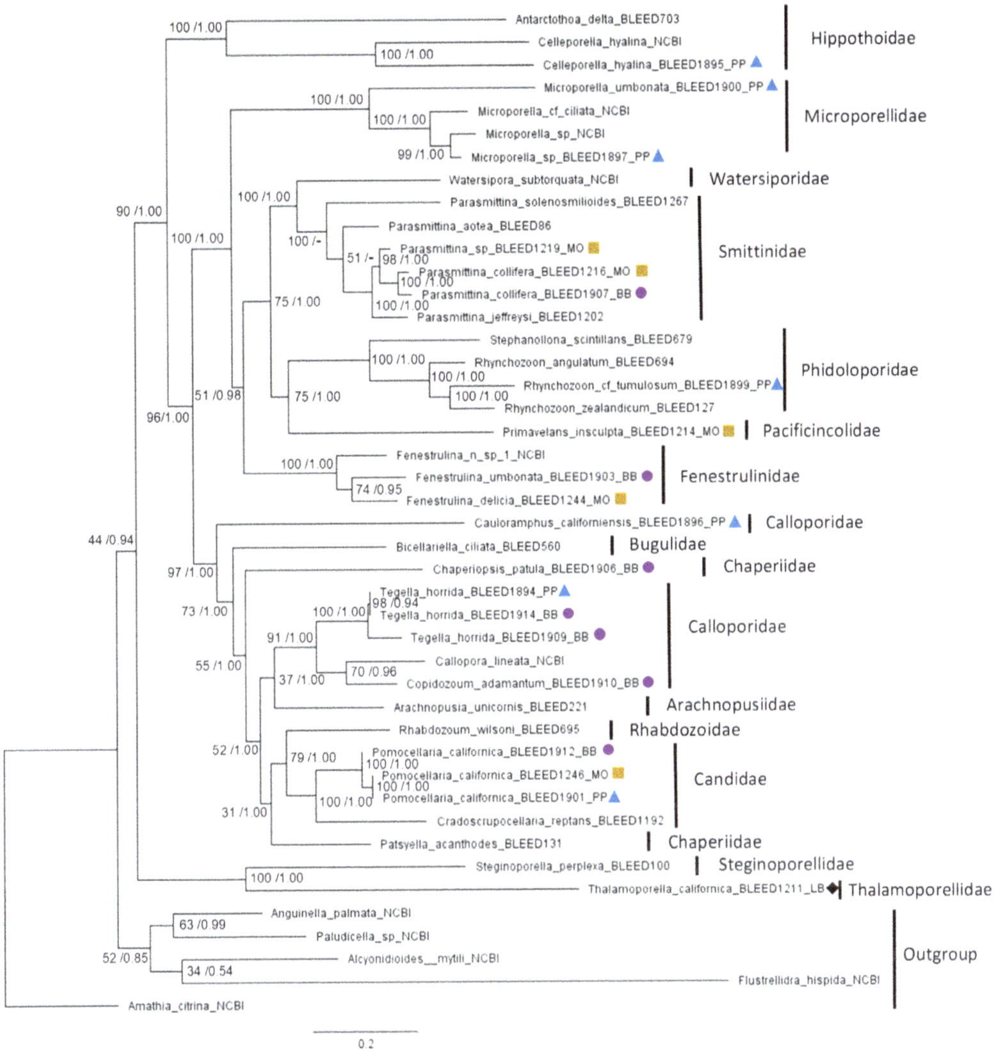

Figure 2. The phylogeny of cheilostome bryozoans collected along the California coastline between Trinidad, CA and Long Beach, CA based on 15 mitochondrial genes. Maximum likelihood topology of 39 ingroup taxa (20 California taxa from Orr *et al.* 2022) and 5 ctenostome outgroup taxa with 4,007 nucleotide and amino acid characters inferred using RAxML (100 heuristic searches and bootstrap values from 500 pseudoreplicates). The numbers on the internal nodes are ML bootstrap values (BS from RAxML) followed by posterior probabilities (PP from MrBayes). Black dashes indicate topological difference between the ML and Bayesian trees. Branches with samples sequenced in this study are labeled with the site abbreviation from which they were collected (PP = Palmer's Point, BB = Baker Beach, MO = Morro Bay and LB = Long Beach) and a corresponding shape (PP = triangle, BB = circle, MO = square, and LB = diamond). Scale bar indicates number of substitutions per site.

and low support (>50 BS/0.90 PP) (Enevoldsen 2016). Most of the nodes in the phylogeny received either high support or full support by both the ML and the Bayesian inferences (Fig. 2).

Out of the 20 California bryozoan sequences included in the phylogenetic analysis, 14 cheilostome colonies were collected from the two rocky intertidal sites at Baker Beach and Palmer's Point (seven samples from Palmer's Point and seven samples from Baker Beach) and represent 11 species from 10 genera and eight families. A complete mitogenome was produced for eight of the 14 samples. The remaining six cheilostome colonies that were successfully sequenced were collected from harbor sites in Morro Bay and Long Beach. These samples represent six species from five genera and five families (five samples from Morro Bay and one sample from Long Beach), and a complete mitogenome was produced for all of these species except *Fenestrulina delicia.* A summary of the species, genera, and families found to be present at each site is in Table 3.

Table 3. A summary of the total number of representative individuals sequenced for each cheilostome species (if applicable) at each site. Samples collected at Palmer's Point and Baker Beach were collected from boulders on the rocky outer coast, while those collected at Morro Bay and Long Beach were collected off settlement panels deployed within harbors. The total number of species per family is listed in the last column.

Family	Genus	Species	Palmer's Point	Baker Beach	Morro Bay	Long Beach	Total # Sp/Family
Calloporidae	*Cauloramphus*	*californiensis*	1	–	–	–	3
	Copidozoum	*adamantum*	–	1	–	–	
	Tegella	*horrida*	1	2	–	–	
Candidae	*Pomocellaria*	*californica*	1	1	1	–	1
Chaperiidae	*Chaperiopsis*	*patula*	–	1	–	–	1
Fenestrulinidae	*Fenestrulina*	*delicia*	–	–	1	–	2
	umbonata	–	1	–	–		
Hippothoidae	*Celleporella*	*hyalina*	1	–	–	–	1
Microporellidae	*Microporella*	sp.	1	–	–	–	2
		umbonata	1	–	–	–	
Pacificincolidae	*Primavelans*	*insculpta*	–	–	1	–	1
Phidoloporidae	*Rhynchozoon*	*tumulosum*	1	–	–	–	1
Smittinidae	*Parasmittina*	sp.	–	–	1	–	2
		collifera	–	1	1	–	
Thalamoporellidae	*Thalamoporella*	*californica*	–	–	–	1	1
	Total # of Samples Per Site		7	7	5	1	

The ML and Bayesian phylogenies were both found to have high or full support for the monophyly of each family present in the trees with the exception of two families (Fig. 2). The family Calloporidae included *Tegella horrida* (BLEED 1894, 1909, and 1914), *Copidozoum adamantum* (BLEED 1910) and *Callopora lineata* (NCBI) which together formed a highly supported monophyletic group (91 BS/1 PP) while *Cauloramphus californiensis* (BLEED 1896) was polyphyletic to the rest of the family (97 BS/1 PP). The family Chaperiidae was also polyphyletic, with *Chaperiopsis patula* (BLEED 1906) separated by several families from *Patsyella acanthodes* (NCBI) with moderate ML and full Bayesian support (73 BS/1 PP). For genera with multiple species, the ML and Bayesian phylogenies inferred monophyletic groupings for 7 genera with full support (100 BS/1 PP): *Celleporella, Microporella, Parasmittina, Rhynchozoon, Fenestrulina, Tegella, and Pomocellaria.*

The species *T. horrida*, *Pomocellaria californica*, and *Parasmittina collifera* each had multiple samples in these phylogenies which were all monophyletic for each species, showing strong support for this method in reflecting species identity. The species identity was also confirmed by morphological analysis from the SEMs for each of the three species. Full ML and Bayesian support was given to *P. californica* and *P. collifera*, while *T. horrida* had high to full support between the multiple samples of this same species with the exception of the posterior probability value for the relationship between samples BLEED 1894 and BLEED 1914 which had a low support value of 0.94 in the Bayesian analysis. Each of these three species were collected from multiple sites. *Tegella horrida* was found at both of the rocky outer coast locations in northern California. Representatives of *P. collifera* and *P. californica* were both observed in Morro Bay and Baker Beach, with *P. californica* also being observed at Palmer's Point.

The inferred topology for the ML and Bayesian trees did not agree with the nodal placement of NCBI taxa *Parasmittina solenosmilioides* and *Parasmittina aotea*. Both species were supported to be a part of the clade *Parasmittina* within the family Smittinidae, however the ML phylogeny found *P. solenosmilioides* (100 BS) to be basal to *P. aotea* (which had low support of 51 BS) while the reverse relationship in which *P. aotea* is basal to *P. solenosmilioides* was found to be fully supported (1 PP) in the Bayesian topology (See Lee 2022 for Bayesian topology).

4 DISCUSSION

The primary goals of this study were to increase the taxonomic sampling and assessment of the phylogenetic diversity of cheilostome bryozoans found within and between rocky outer coast sites and/or sheltered harbors along the California coastline. The 20 cheilostome colonies that were sequenced resulted in the identification of 15 distinct species from 10 different families and 12 genera in total, and this effort produced the first multi-family phylogenetic tree for marine bryozoans sampled from the California coastline based on phylogenomic data (Fig. 2). The diversity of the Californian bryozoans found in this study showed that there is much variation in species distribution both within and between the rocky intertidal and harbor sites. Eleven distinct species were sampled between the two rocky outer coast sites, with seven species at Palmer's Point and six species at Baker Beach. A total of six distinct species were sampled at the two harbor sites, with five species in Morro Bay and one species in Long Beach. Three species identified in the study (*T. horrida*, *P. collifera*, and *P. californica*) were found to have overlapping geographic ranges amongst the rocky intertidal and harbor sites along the California coastline, each with representatives at two to three (out of the four) different sites sampled.

The presence of *T. horrida*, *P. collifera*, and *P. californica* at multiple sites suggests that dispersal events between these sites occurred at some point in time and allowed for gene flow between different geographical populations despite their large physical separation by headlands as well as rocky and sandy shores. The phylogenetic analyses found that a highly to fully supported group was formed for each of these three species in both the ML and Bayesian phylogenies. These species have been previously known to be found in both intertidal and subtidal regions but are more commonly found at shallower depths (including the intertidal) in northern latitudes (Soule *et al.* 2007).

It has been previously hypothesized that gene flow between populations separated by large stretches of coastline is limited as a result of isolation by distance (Wright 1943). Given that Palmer's Point and Baker Beach are separated by a large headland as well as approximately 10 km of rocky and sandy shoreline, it is likely that different intertidal areas located between the two sites allowed for a stepping-stone pattern of gene flow to occur due to multiple dispersal events along the coastline (Kimura & Weiss 1964; Slatkin & Maddison 1990). However, transportation on the hulls or in ballast tanks of ships or rafting on plastic debris could also account for the larger distance (approximately 692 to 702 km) that *P. collifera* and *P. californica* were able to travel between the rocky intertidal sites in northern California and the harbor site in Morro Bay (García-Gómez *et al.* 2021; Wisely 1958, 1963).

With the exception of *F. delicia*, all of the species identified from the sequenced samples from California have historically been found to be distributed throughout the range between British Columbia to southern California (Soule *et al.* 2007). *F. delicia* has previously been reported in the Ocean Biodiversity Information System and the Global Biodiversity Information Facility databases to only be present in areas of western Europe and the northeastern Atlantic, but recent studies have found *F. delicia* to be present in various bays along the California coastline, suggesting that the species has undergone long-distance dispersal, likely being introduced to eastern Pacific waters anthropogenically via ships (Ruiz & Gellar 2018). All of the species identified in this study are known to possess lecithotrophic coronate larvae and are therefore reliant on settling on mobile substrates to be transported between different sites since these species spend very little time in a planktotrophic state (Orr *et al.* 2022). While there are few studies comparing the difference in bryozoan diversity between rocky intertidal regions and more protected bays and harbors, species that are more ecologically tolerant of variations in local climate as well as capable of long-distance dispersal are more likely to see a higher frequency of dispersal and subsequent gene flow and therefore a wider geographic distribution (Mackie *et al.* 2012; McKinney & Jackson 1989; Zabin *et al.* 2018).

Of the 20 California bryozoan samples that were sequenced in Orr *et al.* 2022, potentially one new species of *Microporella* (BLEED 1897; See Orr *et al.* 2022 and Lee 2022 for SEM) from Palmer's Point and one new species of *Parasmittina* (BLEED 1219; See Orr *et al.* 2022 and Lee 2022 for SEM) from Morro Bay were identified that are yet to be fully taxonomically described. These samples were found to belong to the genera *Microporella* and *Parasmittina* respectively but are not the same genetic species as other species previously identified within each genus in the phylogenies in this study and did not match the previous taxonomic descriptions for other species within each genus. This demonstrates that there are evolutionarily unique species present along the California coast that remain to be discovered.

Another important species that was identified within the dataset was sample BLEED 1895 (named *Celleporella hyalina* in the results of this study). Although only one genetic sequence of *C. hyalina* was included in the phylogenetic analysis of this study, this sample had two potential morphological variations identified (described as *Celleporella hyalina* sp. A and *Celleporella hyalina* sp. B in the SEM images available in Orr *et al.* 2022 and Lee 2022) whose only differences noted were a more developed umbo on the autozooids and a different distribution pattern of pseudopores on the ovicells in sp. B in comparison to sp. A. Although sample BLEED 1895 (whose genetic sequence corresponds with the morphology of *C. hyalina* sp. A) formed a fully supported monophyly with the *C. hyalina* sample from NCBI (100 BS/1 PP) in the genetic analysis, other studies suggest the possibility that BLEED 1895 could be a different morphotype if not subspecies of *C. hyalina* than the sequence included from NCBI (Hoare *et al.* 2001; Navarrete *et al.* 2005; Waeschenbach *et al.* 2012a). While *C. hyalina* is known to have a cosmopolitan distribution, studies have found that samples from different geographic regions are generally morphologically similar but have produced genetically distinct lineages that are reproductively incompatible with each other, suggesting that *C. hyalina* may be producing cryptic subspecies (Hoare *et al.* 2001; Waeschenbach *et al.* 2012a). Since this study included only one sample of *C. hyalina*, more research would need to be performed to examine the prevalence of the morphological variation seen in our collected sample of *C. hyalina* in northern California and how genetic diversity within this species compares to other global samples of *C. hyalina* (Navarrete *et al.* 2005; Waeschenbach *et al.* 2012a).

Due to the limited number of samples and sampling sites in this study, it is estimated that an additional 100 bryozoan species at minimum are missing from this phylogenetic analysis for the entire California coastline (Soule *et al.* 2007; Ruiz & Geller 2018). It is recognized that specimens obtained from the rocky outer coast could only be sequenced from two sites north of Cape Mendocino, a biogeographic transition zone for many invertebrate species, while the specimens obtained from harbor sites were all obtained from sites well south of another biogeographic break in San Francisco Bay (Connolly *et al.* 2001). Since there were no harbor sites from northern California, or any rocky intertidal sites from south of Cape Mendocino included in this study, a fully comprehensive survey of bryozoan diversity across the entire California coastline still remains to be completed, and such a study would be highly valuable give the rapid rate of introduction of exotic bryozoan species along the California coast.

5 CONCLUSIONS

This study utilized a dataset of 20 recently produced bryozoan mitogenomic sequences and identified 15 distinct species from 10 different families using high-throughput sequencing to form the first comprehensive phylogenomic analysis for multiple bryozoan families along the California coastline. A complete mitogenome including all 15 mitochondrial genes was also produced for 13 of the 20 bryozoans sequenced which significantly improves the availability of genetic data for this lesser studied phylum. Future studies would benefit from increased genetic and taxonomic investigation into the bryozoan biodiversity all along the California coastline. The new findings and insights gained from the phylogenetic analyses performed in this study are useful in providing a better understanding of the diversity and distribution of bryozoans throughout the California coastline and can help efforts to infer the presence of non-indigenous bryozoan species in bays and rocky intertidal sites in California. By continuing to increase the availability of combined molecular and morphological datasets to help describe local bryozoan diversity for intertidal and subtidal depths in California, further data can be integrated with the phylogenies developed in this study to examine more robust evolutionary hypotheses on bryozoan diversity and how their distribution impacts marine invertebrate communities throughout California and the eastern Pacific.

ACKNOWLEDGEMENTS

The authors thank Lee Hsiang Liow from the Natural History Museum at the University of Oslo, Norway for her guidance in the project aims and funding of the HTS sequencing and all collectors named in the SEM cards available in Orr *et al.* (2022) and Lee (2022). We would especially like to thank Ismael Chowdhury from California State Polytechnic University, Humboldt for assistance with the rocky intertidal sample collections. Specimens collected from California harbor sites were provided by Gregory Ruiz and his research team from the Smithsonian Environmental Research Center funded by the California Dept. of Fish and Wildlife Marine Invasive Species Program. We thank Emanuela Di Martino and Mali Ramsfjell for sorting and identifying samples under the SEM at the Natural History Museum at the University of Oslo, Norway as well as the Norwegian Sequencing Centre for assistance with the genomic sequencing. The HTS sequencing is funded by the European Research Council (ERC) under the European Union's Horizon 2020 research and innovation programme (grant agreement No 724324 to L.H. Liow). We also thank the International Bryozoology Association for funding to support travel to the 19th IBA Conference in Dublin, Ireland to give a talk on the research presented here.

REFERENCES

Adachi, J. & Hasegawa, M. 1996. Model of Amino Acid Substitution in Proteins Encoded by Mitochondrial DNA. *Journal of Molecular Evolution* 42(4): 459–468.

Al-Haggar, M.M., Khair-Allaha, B.A., Islam, M.M. & Mohamed, A.S. 2013. Bioinformatics in High Throughput Sequencing: Application in Evolving Genetic Diseases. *Data Mining in Genomics & Proteomics* 4(3): 1–5.

Altschul, S.F., Gish, W., Miller, W., Myers, E.W. & Lipman, D.J. 1990. Basic local Alignment Search Tool. *Journal of Molecular Biology* 215: 403–410.

Andrews, S. 2010. *FastQC: A Quality Control Tool for High Throughput Sequence Data (Version 0.11.9).* https://www.bioinformatics.babraham.ac.uk/projects/fastqc/

Bernt, M., Donath, A., Jühling, F., Externbrink, F., Florentz, C., Fritzsch, G., Pütz, J., Middendorf, M. & Stadler, P.F. 2013. MITOS: Improved De Novo Metazoan Mitochondrial Genome Annotation. *Molecular Phylogenetics and Evolution* 69(2): 313–319.

Blanchette, C.A., Miner, C.M., Raimondi, P.T., Lohse, D., Heady, K.E.K. & Broitman, B.R. 2008. Biogeographical Patterns of Rocky Intertidal Communities Along the Pacific Coast of North America. *Journal of Biogeography* 35: 1593–1607.

Bock, P.E. & Gordon, D.P. 2013. Phylum Bryozoa Ehrenberg, 1831. *Zootaxa* 3703(1): 67–74.

Cocito, S. 2004. Bioconstruction and Biodiversity: Their Mutual Influence. *Scientia Marina* 68: 137–144.

Connolly, S.R., Menge, B.A. & Roughgarden, J. 2001. A latitudinal Gradient in Recruitment of Intertidal Invertebrates in the Northeast Pacific Ocean. *Ecology* 82(7): 1799–1813.

Dick, M.H., Lidgard, S., Gordon, D.P. & Mawatari, S. F. 2009. The Origin of Ascophoran Bryozoans was Historically Contingent But Likely. *Proceedings of the Royal Society B* 276(1670): 3141–3148.

Dierckxsens, N., Mardulyn, P. & Smits, G. 2017. NOVOPlasty: De Novo Assembly of Organelle Genomes from Whole Genome Data. *Nucleic Acids Research* 45(4): e18.

Dimmic, M.W., Mindell, D.P. & Goldstein, R.A. 2000. Modeling Evolution at the Protein Level Using an Adjustable Amino Acid Fitness Model. *Pacific Symposium on Biocomputing* 5: 18–29.

Duchêne, S., Archer, F.I., Vilstrup, J., Caballero, S. & Morin, P.A. 2011. Mitogenome Phylogenetics: The Impact of Using Single Regions and Partitioning Schemes on Topology, Substitution Rate and Divergence Time Estimation. *PLoS One* 6(11): e27138.

Enevoldsen, E.L.G. 2016. *Microporellidae Phylogeny and Evolution* [Master's thesis, University of Oslo]. DUO Research Archive. http://urn.nb.no/URN:NBN:no-55791

Fuchs, J., Obst, M. & Sundberg, P. 2009. The First Comprehensive Molecular Phylogeny of Bryozoa (Ectoprocta) Based on Combined Analyses of Nuclear and Mitochondrial Genes. *Molecular Phylogenetics and Evolution* 52(1): 225–233.

García-Gómez, J. C., Garrigós, M. & Garrigós, J. 2021. Plastic as a Vector of Dispersion for Marine Species with Invasive Potential. A Review. *Frontiers in Ecology and Evolution* 9: e629756.

Ge, Y., Xia, C., Wang, J., Zhang, X., Ma, X. & Zhou, Q. 2021. The Efficacy of DNA Barcoding in the Classification, Genetic Differentiation, and Biodiversity Assessment of Benthic Macroinvertebrates. *Ecology and Evolution* 11(10): 5669–5681.

Hao, J., Li, C., Sun, X. & Yang, Q. 2005. Phylogeny and Divergence Time Estimation of Cheilostome Bryozoans Based on Mitochondrial 16S rRNA Sequences. *Chinese Science Bulletin* 50(12): 1205–1211.

Hoare, K., Goldson, A.J., Giannasi, N. & Hughes, R.N. 2001. Molecular Phylogeography of the Cosmopolitan Bryozoan *Celleporella hyalina*: Cryptic Speciation? *Molecular Phylogenetics and Evolution* 18(3): 488–92.

Holland, B.S. 2000. Genetics of Marine Bioinvasions. *Hydrobiologia* 420: 63–71.

Huelsenbeck, J.P. & Ronquist, F. 2001. MRBAYES: Bayesian Inference of Phylogenetic Trees. *Bioinformatics* 17(8): 754–755.

Jimenez, H., Keppel, E., Chang, A.L. & Ruiz, G.M. 2018. Invasions in Marine Communities: Contrasting Species Richness and Community Composition Across Habitats and Salinity. *Estuaries and Coasts* 41 (1): 484–494.

Katoh, K. & Standley, D.M. 2013. MAFFT Multiple Sequence Alignment Software Version 7: Improvements in Performance and Usability. *Molecular Biology and Evolution* 30(4): 772–780.

Kelly, R.P. & Palumbi, S.R. 2010. Genetic Structure Among 50 Species of the Northeastern Pacific Rocky Intertidal Community. *PLoS One* 5(1): e8594.

Kimura, M. & Weiss, G.H. 1964. The Stepping Stone Model of Population Structure and the Decrease of Genetic Correlation with Distance. *Genetics* 49(4): 561–576.

Krueger, F. 2015. *TrimGalore* (Version 0.6.5). https://www.bioinformatics.babraham.ac.uk/projects/trim_galore/

Lee, H.E. 2022. *Phylogenetic and Geographic Relationships of Cheilostome Bryozoans in the Eastern Pacific.* [Master's thesis, California State Polytechnic University, Humboldt]. Cal Poly Humboldt Digital Commons. https://digitalcommons.humboldt.edu/etd/552

Li, D., Liu, C.-M., Luo, R., Sadakane, K. & Lam, T.-W. 2015. MEGAHIT: An Ultra-fast Single-node Solution for Large and Complex Metagenomics Assembly Via Succinct de Bruijn Graph. *Bioinformatics* 31(10): 1674–1676.

Lidgard, S. 2008. Predation on Marine Bryozoan Colonies: Taxa, Traits and Trophic Groups. *Marine Ecology Progress Series* 359: 117–131.

Mackie, J.A., Darling, J.A. & Geller, J.B. 2012. Ecology of Cryptic Invasions: Latitudinal Segregation Among *Watersipora* (Bryozoa) Species. *Scientific Reports* 2 (871): 1–10.

Mackie, J.A., Keough, M.J. & Christidis, L. 2006. Invasion Patterns Inferred from Cytochrome Oxidase I Sequences in Three Bryozoans, *Bugula neritina*, *Watersipora subtorquata*, and *Watersipora arcuata*. *Marine Biology* 149: 285–295.

Maddison, W.P. & Maddison, D.R. 2018. *Mesquite: A Modular System for Evolutionary Analysis (Version 3.51).* http://www.mesquiteproject.org

Marraffini, M.L., Ashton, G.V., Brown, C.W., Chang, A.L. & Ruiz, G.M. 2017. Settlement Plates as Monitoring Devices for Non-indigenous Species in Marine Fouling Communities. *Management of Biological Invasions* 8(4): 559–566.

McKinney, F.K. & Jackson, J.B.C. 1989. *Bryozoan Evolution.* Chicago: The University of Chicago Press.

Navarrete, Z.A., Cancino, J.M., Moyano, H.I. & Hughes, R.N. 2005. Morphological Differentiation in the *Celleporella hyalina* (Linnaeus, 1767) Complex (Bryozoa, Cheilostomata) Along the Chilean Coast. In H.I. Moyano, J.M. Cancino & P.N. Wyse Jackson (eds), *Bryozoan Studies 2004*: 207–213. London: Taylor & Francis Group.

Nylander, J.A.A. 2010. *catfasta2phyml.* https://github.com/nylander/catfasta2phyml

Orr, R.J.S., Di Martino, E., Gordon, D.P., Ramsfjell, M.H., Mello, H.L., Smith, A.M. & Liow, L.H. 2021. A Broadly Resolved Molecular Phylogeny of New Zealand Cheilostome Bryozoans as a Framework for Hypotheses of Morphological Evolution. *Molecular Phylogenetics and Evolution* 161(5996): 107172.

Orr, R.J.S., Di Martino, E., Ramsfjell, M.H., Gordon, D. P., Berning, B., Chowdhury, I., Craig, S., Cumming, R.

L., Figuerola, B., Florence, W., Harmelin, J.G., Hirose, M., Huang, D., Jain, S.S., Jenkins, H.L., Kotenko, O. N., Kuklinski, P., Lee, H.E., Madurell, T., McCann, L., Mello, H.L., Obst, M., Ostrovsky, A.N., Paulay, G., Porter, J.S., Shunatova, N.N., Smith, A.M., Souto-Derungs, J., Vieira, L.M., Voje, K.L., Waeschenbach, A., Zágoršek, K., Warnock, R.C.M. & Liow, L.H. 2022. Paleozoic Origins of Cheilostome Bryozoans and Their Parental Care Inferred by a New Genome-skimmed Phylogeny. *Science Advances* 8(13): eabm7452.

Orr, R.J.S., Haugen, M.N., Berning, B., Bock, P., Cumming, R.L., Florence, W.K., Hirose, M., Di Martino, E., Ramsfjell, M.H., Sannum, M.M., Smith, A.M., Vieria, L.M., Waeschenbach, A. & Liow, L.H. 2019a. A Genome-skimmed Phylogeny of a Widespread Bryozoan Family, Adeonidae. *BMC Evolutionary Biology* 19(1): 235.

Orr, R.J.S., Sannum, M.M., Boessenkool, S., Di Martino, E., Gordon, D.P., Mello, H.L., Obst, M., Ramsfjell, M.H., Smith, A.M. & Liow, L.H. 2020. A Molecular Phylogeny of Historical and Contemporary Specimens of an Under-studied Microinvertebrate Group. *Ecology and Evolution* 11(1): 309–320.

Orr, R.J.S., Waeschenbach, A., Enevoldsen, E.L.G., Boeve, J.P., Haugen, M.N., Voje, K.L., Porter, J., Zagorsek, K., Smith, A.M., Gordon, D.P. & Liow, L. H. 2019b. Bryozoan Genera *Fenestrulina* and *Microporella* no Longer Confamilial; Multi-gene Phylogeny Supports Separation. *Zoological Journal of the Linnean Society* 186(1): 190–199.

Pohowsky, R.A. 1973. A Jurassic Cheilostome from England. In G.P. Larwood (ed.), *Living and Fossil Bryozoa*: 447–461. London: Academic Press.

Ricketts, E.F. & Calvin, J. 1968. *Between Pacific Tides* (4th ed.). Stanford: Stanford University Press.

Ronquist, F., Teslenko, M., van der Mark, P., Ayres, D., Darling, A., Höhna, S., Larget, B., Liu, L., Suchard, M.A. & Huelsenbeck, J.P. 2012. MrBayes 3.2: Efficient Bayesian Phylogenetic Inference and Model Choice Across a Large Model Space. *Systematic Biology* 61: 539–542.

Ruiz, G.M. Unpublished. *Spatial and Temporal Analysis of Marine Invasions in California.*

Ruiz, G.M. & Geller, J. 2015. *Spatial and Temporal Analysis of Marine Invasions in California: Morphological and Molecular Comparisons Across Habitats.* Smithsonian Environmental Research Center & Moss Landing Laboratories, Edgewater MD, Moss Landing CA.

Ruiz, G.M. & Geller, J. 2018. *Spatial and Temporal Analysis of Marine Invasions in California, Part II: Humboldt Bay, Marina del Rey, Port Hueneme, and San Francisco Bay.* Smithsonian Environmental Research Center & Moss Landing Laboratories, Edgewater MD, Moss Landing CA.

Schwaninger, H.R. 2008. Global Mitochondrial DNA Phylogeography and Biogeographic History of the Antitropically and Longitudinally Disjunct Marine Bryozoan *Membranipora membranacea* L. (Cheilostomata): Another Cryptic Marine Sibling Species Complex? *Molecular Phylogenetics and Evolution* 49: 893–908.

Slatkin, M. & Maddison, W.P. 1990. Detecting Isolation by Distance Using Phylogenies of Genes. *Genetics* 126 (1): 249–260.

Soule, D.F., Chaney, H.W. & Morris, P.A. 2004. Additional New Species of *Microporelloides* from Southern California and American Samoa. In *Irene McCulloch Foundation Monograph Series 6A*: 1–14. Los Angeles: Hancock Institute for Marine Studies University of Southern California.

Soule, D.F., Soule, J.D. & Chaney, H.W. 1995. Taxonomic Atlas of the Benthic Fauna of the Santa Maria Basin and Western Santa Barbara Channel. The Bryozoa. In *Irene McCulloch Foundation Monograph Series 2*. Santa Barbara: Santa Barbara Museum of Natural History.

Soule, D.F, Soule, J.D, Morris, P.A. & Chaney, H.W. 2007. Bryozoa. In S.F. Light & J. Carlton (eds.), *The Light & Smith Manual: Intertidal Invertebrates from Central California to Oregon, Fourth Edition*: 866–904. Berkeley: University of California Press.

Stamatakis, A. 2014. RAxML Version 8: A Tool for Phylogenetic Analysis and Post-analysis of Large Phylogenies. *Bioinformatics* 30(9): 1312–1313.

Straub, S.C.K., Parks, M., Weitemier, K., Fishbein, M., Cronn, R.C. & Liston, A. 2012. Navigating the Tip of the Genomic Iceberg: Next-generation Sequencing for Plant Systematics. *American Journal of Botany* 99(2): 349–364.

Talavera, G. & Castresana, J. 2007. Improvement of Phylogenies After Removing Divergent and Ambiguously Aligned Blocks from Protein Sequence Alignments. *Systematic Biology* 56(4): 564–577.

Taylor, P.D. & Waeschenbach, A. 2015. Phylogeny and Diversification of Bryozoans. *Palaeontology* 58(4): 585–599.

Thomas, J.A., Welch, J.J., Lanfear, R. & Bromham, L. 2010. A Generation Time Effect on the Rate of Molecular Evolution in Invertebrates. *Molecular Biology and Evolution* 27(5): 1173–1180.

Trevisan, B., Alcantara, D.M.C., Machado, D.J., Marques, F.P.L. & Lahr, D.J.G. 2019. Genome Skimming is a Low-cost and Robust Strategy to Assemble Complete Mitochondrial Genomes from Ethanol Preserved Specimens in Biodiversity Studies. *PeerJ* 7: e7543.

Waeschenbach, A., Porter, J.S. & Hughes, R.N. 2012a. Molecular Variability in the *Celleporella hyalina* (Bryozoa; Cheilostomata) Species Complex: Evidence for Cryptic Speciation from Complete Mitochondrial Genomes. *Molecular Biology Reports* 39(9): 8601–8614.

Waeschenbach, A., Taylor, P.D. & Littlewood, D.T.J. 2012b. A Molecular Phylogeny of Bryozoans. *Molecular Phylogenetics and Evolution* 62(2): 718–735.

Wilson, E.E. 2011. *The Facilitative Role of an Introduced Bryozoan (Watersipora Spp.): Structuring Fouling Community Assemblages within Humboldt Bay* [Master's thesis, Humboldt State University]. ScholarWorks. http://hdl.handle.net/2148/856

Wisely, B. 1958. The Settling and Some Experimental Reactions of a Bryozoan Larva, *Watersipora cucullata* (Busk). *Australian Journal of Marine and Freshwater Research* 9: 362–371.

Wisely, B. 1963. Effects of Antifouling Paints on Settling Larvae of the Bryozoan *Bugula neritina* L. *Australian Journal of Marine and Freshwater Research* 14(1): 44–59.

Wright, S. 1943. Isolation by Distance. *Genetics* 16: 97–159.

Zabin, C.J., Maraffini, M., Lonhart, S.I., McCann, L., Ceballos, L., King, C., Watanabe, J., Pearse, J.S. & Ruiz, G.M. 2018. Non-native Species Colonization of Highly Diverse, Wave Swept Outer Coast Habitats in Central California. *Marine Biology* 165: 31.

Bryozoan Studies 2022 – Key, Porter & Wyse Jackson (eds)
© 2023 The Author(s), ISBN: 978-1-032-59487-3

Early proliferation of avicularia in the Cretaceous cheilostome bryozoan *Wilbertopora*: a diversification event guided by ecological exploration

Sarah Leventhal*
Department of Geological Sciences, University of Colorado Museum of Natural History University of Colorado, Boulder, CO, USA
ORCID ID: 0000-0002-6537-1553

Kayli Stowe
Department of Geological Sciences, University of Colorado Museum of Natural History University of Colorado, Boulder, CO, USA

Carl Simpson
Department of Geological Sciences, University of Colorado Museum of Natural History University of Colorado, Boulder, CO, USA
ORCID ID: 0000-0003-0719-4437

ABSTRACT: Cheilostome bryozoans are a diverse clade of colonial animals that first appeared during the Jurassic period. For the first 100 Myr of their existence, colonies of cheilostomes were monomorphic, composed entirely of autozooids. Divergent body types, termed avicularia, first appeared in colonies in the early Cretaceous period, in the genus *Wilbertopora*. Over the course of the Cretaceous, *Wilbertopora* diversified into 27 species, spanning the early Albian through Maastrichtian stages. In this study, we quantify autozooid and avicularia shape and size to evaluate how the morphological disparity of zooid types in colonies changes over the course of *Wilbertopora*'s diversification. We find that taxonomic diversity and morphological disparity are largely decoupled, with disparity outpacing diversity for much of *Wilbertopora*'s evolutionary history. Increases in disparity are primarily driven by evolution of avicularian morphology, indicating that *Wilbertopora*'s avicularia may have served an array of purposes in different lineages.

1 INTRODUCTION

Extant cheilostome bryozoans possess a wide range of polymorphs that have proved to be an important innovation for the success of the clade (Jablonski *et al.* 1997; Taylor 2020). Polymorphs, in the form of avicularia, first evolved in the late Albian cheilostome *Wilbertopora* (Cheetham 1954; Ostrovsky *et al.* 2008) (Figure 1), which underwent a brief but explosive diversification event until the end of the Maastrichtian (Cheetham *et al.* 2006; Taylor 2020). Over the course of this period, *Wilbertopora* diversified into a total of 27 species.

While *Wilbertopora* is the first genus of cheilostomes to have avicularia, the evolution of its avicularian morphology has never been subject to quantitative study. Here we evaluate the taxonomic diversification of *Wilbertopora* (Cheetham 1954; Ostrovsky *et al.* 2008; Cheetham *et al.* 2006) and quantify the morphological disparity of zooid types in colonies over the course of this diversification. We present an analysis of the trajectory of morphological disparity and taxonomic diversity in 25 of the 27 species of *Wilbertopora* from the genus's first appearance in the Albian to its last appearance in Maastrichtian. To better understand the evolving morphologies of different zooid types in *Wilbertopora*, we also directly compare disparity of autozooids and avicularia over time.

2 METHODS

2.1 *Data aggregation and analytical methodology*

To estimate of the number of *Wilbertopora* species that existed over the genus' life span, we compiled publications containing references to the genus from Bock (2020). Then, we filtered out publications that mention the genus but do not describe a specific occurrence of its species. This process left us with publications detailing 27 distinct species. Fossils of distinct *Wilbertopora* species have been found in the Albian, Cenomanian, Campanian, and Maastrichtian stages of Texas (Cheetham *et al.* 2006); the Campanian and Maastrichtian of

*Corresponding Author: sarah.leventhal@colorado.edu

DOI: 10.1201/9781003454915-10

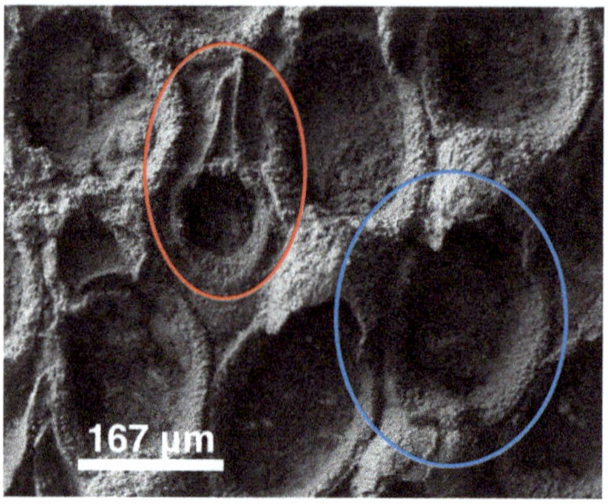

Figure 1. Photograph of *Wilbertopora acuminata*, a species of *Wilbertopora* from the late Albian/early Cenomanian. Avicularium is circled in red, and autozooid is circled in blue. Specimen is from NHM London, NHM PI BZ 1479.

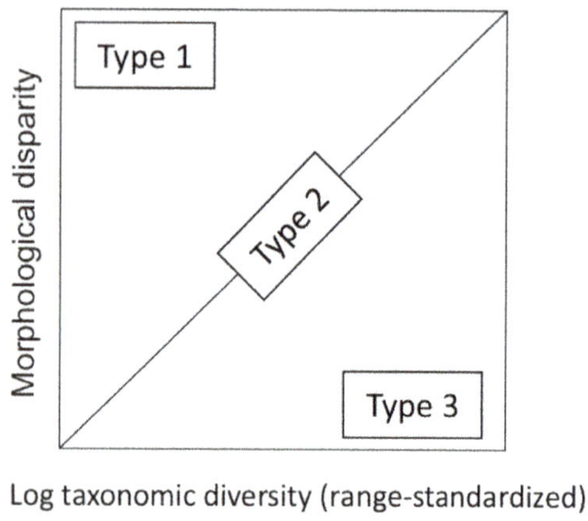

Figure 2. Diversity-disparity analytical framework applied in this study as adapted from Jablonski (2017, fig. 4). There are three main patterns of diversification. Type 1 refers to diversification events where increases in disparity outpace increases in taxonomic diversity, or "ecological exploration." Type 2 refers to events with concordant increases of disparity and taxonomic diversity, which follows a line with a slope of 1. Type 3 refers to events with increases in taxonomic diversity that outpace increases in disparity, or "ecological exploitation."

southern California, Baja California (Taylor 2008), Kazakhstan (Koromyslova *et al.* 2018), and the southeastern United States (Taylor & McKinney 2006); the Cenomanian, Turonian, and Campanian of England (Smith and Batten 2002; Taylor & Martha 2017); the late Cretaceous of the United Arab Emirates (Di Martino & Taylor 2013); the mid-late Cretaceous of western and central Europe (Martha & Taylor 2016; Voigt 1981); the Cenomanian of India (Guha 1989; Taylor & Badve 1994); and the Maastrichtian of Madagascar (Di Martino *et al.* 2018).

We recorded occurrence information from each of these publications to get an estimate of first and last appearances for each species. We removed two species for which we could not access adequate SEM images or measurement information in monographs: *Wilbertopora stoliczki* (Guha 1989), and *Wilbertopora*? *bohemica* from the Czech Republic (Žítt *et al.* 2006). We therefore consider 25 species of *Wilbertopora* in our analysis.

To assess the macroevolutionary patterns in early avicularia evolution, we use Jablonski's (2017) diversity-disparity analysis. This type of analysis allows for us to make broad ecological interpretations about patterns of morphological disparity and taxonomic diversity (Figure 2). There are three general diversification patterns observable using this type of analysis (Jablonski 2017). Trajectories that fall below the 1:1 line (Type 3) exhibit ecological exploitation, where diversification occurs without major increase in disparity. The conservation of morphology through taxonomic diversification suggests that niche exploitation guides diversification. Conversely, trajectories falling above the 1:1 line (Type 1) exhibit what we term ecological exploration. In these types of diversification events, increases in morphological disparity outpace increases in taxonomic diversity. This pattern would suggest that different lineages of the subject taxon may be pursuing different ecological niches or experiencing genetic drift. Trajectories tracking with the 1:1 line (Type 2) follow the expected pattern of concordant increases in disparity and diversity (Jablonski 2017).

2.2 *Estimating diversity*

To assess diversity over time, we first compiled all occurrence information from monographs describing the 25 *Wilbertopora* species. Due to the incompleteness of the fossil record, we generated a tip-dated phylogeny to better estimate fossil ranges for each species. Since this is one of the few morphology-based phylogenies of more than a handful of bryozoan species, we first compiled a list of applicable characters based on previous smaller phylogenetic studies (Cheetham *et al.* 2006), but this did not provide us with enough characters to construct a phylogeny for our 25 species. Only 15 of the 20 significant characters included in Cheetham *et al.*'s (2006) analysis could be determined from monographs or available specimen SEMs. Characters derived from the aperture measurements of autozooids and measurements of zooids distal to ovicells, while taxonomically significant, were not included in our character matrix because monographs typically report opesia measurements instead of aperture measurements for autozooids, and few specify measurements of autozooids distal to ovicells.

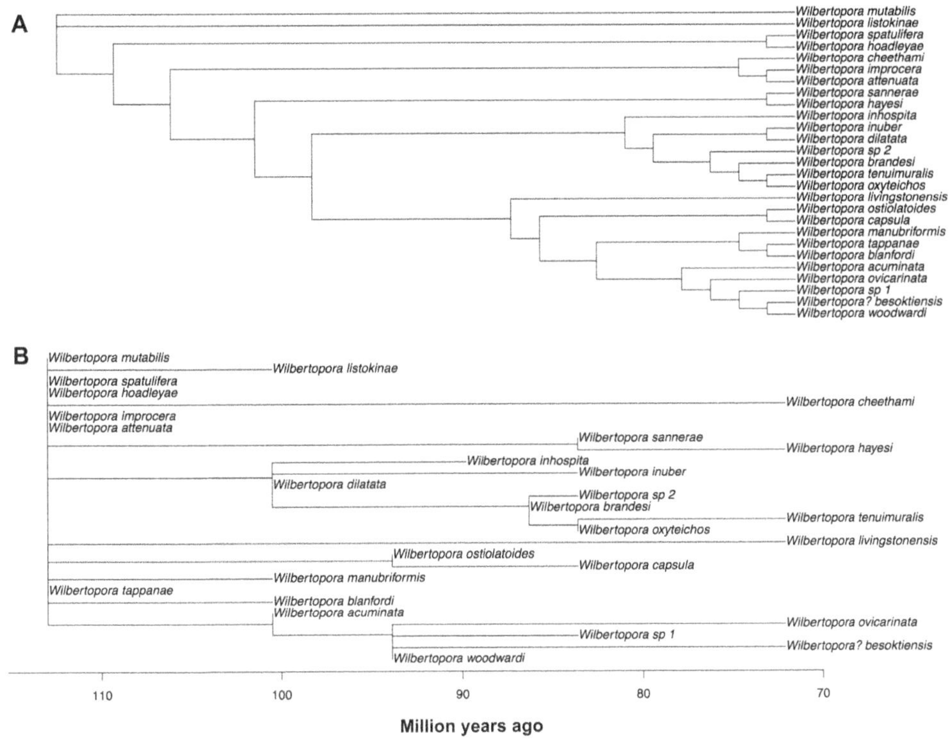

Figure 3. Phylogenetic trees used in this study. A, maximum parsimony tree for *Wilbertopora* completed in PAUP* (Swofford 2003). B, maximum-likelihood tip-dated phylogeny generated using Paleotree (Bapst 2012). Tip dates clump according to geologic stage, with species lineages terminating at the start of the geologic stage of their extinction. Note: due to the small size of our character matrix, the topology of this phylogeny contains several polytomies. We primarily used this phylogeny for the estimation of species richness over time.

Thus, we use phylogenetic studies on other groups of animals that identify novel and phylogenetically significant characters, like shape ratios of features and presence/absence of certain ecologically significant characters (Atteberry & Eberle 2021; Cole 2019), as guides for developing an additional five characters (11, 15, 16, 19, 20 in our character matrix, Table 1). Since zooid shape characteristics and colony-level characters (e.g., the ratio of autozooid opesia to autozooid length, or the presence/absence of spines, respectively) both contain information about the soft tissue features or potential ecological adaptations of zooids and colony ecology, we use these features as additional characters in our character matrix.

Using our character matrix, we generated a set of 10 maximum parsimony phylogenetic trees using a heuristic search over 10,000 iterations in PAUP* (Swofford 2003) and selected the one with the fewest polytomies for tip dating (Figure 3A). Then, using the package Paleotree in Rstudio (Bapst 2012), we used the first and last appearances of each species to generate a maximum-likelihood tip-dated phylogeny (Figure 3B). The purpose of this phylogeny was to supplement our occurrence information due to the low species sampling during the Santonian and Coniacian stages. The small number of characters in the character matrix led to the generation of several polytomies in our phylogeny, but since this phylogeny is for the purpose of diversity estimation and not topology analysis, we accept this preliminary phylogeny as a measure of species diversity over the course of the Cretaceous.

2.3 *Estimating disparity*

There are two fundamental ways in which morphologies can vary: shape and size. To get an accurate estimate of size disparity in *Wilbertopora*, we use species mean linear measurements either taken from monographs or from mean values from available SEM images in museum databases. The characters we use for our size estimates of disparity are autozooid length, autozooid width, avicularium length, and avicularium width. While these size metrics do pertain to the functional biology of bryozoans, zooid size may also be influenced by temperature and seasonality (O'Dea & Okamura 1999; Okamura 1992). However, since climate

Table 1. Character matrix for *Wilbertopora* phylogeny. Enumerated characters and their corresponding state values are specified in Appendix 1.

	Characters																			
Species	1	2	3	4	5	6	7	8	9	10	11	12	13	14	15	16	17	18	19	20
W. acuminata	3	0	0	0	1	0	1	1	1	1	1	2	4	2	0	0	1	1	3	3
W. attenuata	2	1	1	0	2	1	0	0	1	1	0	2	4	2	1	0	3	1	3	2
W. blanfordi	5	1	0	0	1	2	1	1	1	1	0	0	1	2	0	1	2	2	4	5
W. brandesi	1	4	1	1	4	4	1	0	0	1	0	1	5	3	1	2	5	4	3	3
W. capsula	4	3	2	2	1	2	2	1	1	1	1	0	0	2	3	2	0	2	3	2
W. cheethami	2	3	1	0	0	0	0	0	1	1	0	1	1	2	1	1	0	0	3	2
W. dilatata	3	4	0	0	4	4	0	1	0	1	0	0	5	1	0	1	5	4	3	1
W. hayesi	2	0	1	0	2	1	0	1	1	1	1	2	3	1	3	2	3	0	4	5
W. hoadleyae	1	1	1	0	0	1	1	1	1	1	0	2	1	2	1	1	0	0	3	3
W. improcera	2	1	1	0	1	1	0	0	1	1	1	2	2	2	1	2	2	1	3	2
W. inhospita	6	4	1	1	4	4	0	1	0	1	0	0	5	0	1	1	5	4	2	2
W. inuber	2	4	3	1	4	4	0	2	0	1	0	2	5	0	0	0	5	4	0	0
W. listokinae	0	2	1	0	2	2	1	1	1	1	1	2	1	3	1	1	3	3	3	1
W. livingstonensis	5	1	1	1	2	2	0	1	1	1	1	0	1	2	1	1	2	1	3	1
W. manubriformis	4	1	0	0	1	2	1	1	1	1	1	0	1	2	0	0	2	1	3	2
W. mutabilis	2	2	1	0	2	2	1	1	1	1	0	2	1	2	1	1	2	3	3	1
W. ostiolatoides	3	3	1	1	0	1	1	1	1	1	1	0	0	1	1	2	0	0	2	3
W. ovicarinata	3	1	0	0	1	1	0	0	1	1	1	1	1	1	0	0	1	1	2	2
W. oxyteichos	1	4	2	2	4	4	1	2	0	1	0	0	5	1	4	3	5	4	4	3
W. sannerae	0	0	1	0	2	1	0	1	1	1	0	2	4	1	1	1	1	1	2	2
W. sp 1	6	1	0	0	0	0	4	4	1	0	0	1	2	4	0	0	0	0	2	2
W. sp 2	4	4	1	1	4	4	3	2	0	1	0	0	5	2	2	2	5	4	3	2
W. spatulifera	1	1	1	0	2	1	1	1	1	1	0	3	3	2	1	1	3	2	3	2
W. tappanae	2	2	0	0	1	2	1	0	1	1	1	0	1	3	0	0	2	3	2	1
W. tenuimuralis	0	4	1	1	4	4	1	1	0	1	0	0	5	1	2	2	5	4	3	3
W. woodwardi	5	1	4	3	3	3	2	3	1	1	0	1	2	1	4	4	4	3	3	2
W.? besoktiensis	2	1	3	2	3	2	3	3	1	1	0	1	2	1	2	2	1	1	2	1

zones of the Cretaceous were relatively stable at mid-latitudes (Hay & Floegel 2012), and *Wilbertopora* is primarily found in mid-latitudinal regions (excepting *W. tenuimuralis* (Di Martino *et al.* 2018)), the impact of temperature and seasonality on the differences in zooid sizes across species is likely minimal.

To get an estimate of shape disparity, we used Elliptical Fourier Analysis (EFA) to quantify autozooid and avicularium operculum shape (Crampton 1995). We chose to measure the outlines of opercula because that is the portion of the zooid exoskeleton most divergent between autozooids and avicularia (Carter 2008; Kauffman 1971). The divergence in opercular shape corresponds to the different functionalities of autozooids and avicularia, as the operculum indirectly reflects polypide morphology. We used Fiji (Schindelin *et al.* 2012) to manually collect outlines of the opercula of autozooids and avicularia in all 25 species. On each measured zooid, we marked the starting measurement point as the most distal end of the zooid operculum and captured outlines in a clockwise fashion. We collected operculum outlines of two zooids belonging to each zooid type in each species and averaged their EFA coefficients to get an estimate of shape disparity of autozooids and avicularia for each species. We ran a principal components analysis of the EFA coefficients across all species and took resulting first principal components of autozooids and avicularia as our shape metrics.

To calculate disparity using these values, we calculated centroid size for each geologic stage from the Albian to the Maastrichtian (Foote 1993). We computed centroid size by taking species means for each character, and then summing the square distances of each species mean from the pooled mean for each geologic stage. We scaled these summed square distances by the number of species in each geologic stage, so that stages with different levels of species richness could be directly compared.

We computed three disparity measures for each geologic stage: total disparity, avicularian disparity, and autozooidal disparity. We can therefore assess how each zooid type contributes to the signature of total disparity across each geologic stage and assess whether different zooid types have different evolutionary trajectories as *Wilbertopora* diversifies.

3 RESULTS AND DISCUSSION

3.1 *Disparity and diversity over time*

Our results show that disparity and diversity are largely decoupled (Figure 4). Disparity increases monotonically from the Albian (beginning 113 Ma) to the Turonian (ending 89.6 Ma), where it reaches an absolute maximum. For the rest of the Cretaceous, disparity declines monotonically.

The diversity curve of *Wilbertopora* follows a different pattern from disparity (Figure 4). Taxonomic diversity begins at a local maximum in the Albian, and then it alternates between local maxima and local minima until it reaches an absolute peak in the Campanian (beginning 83.6

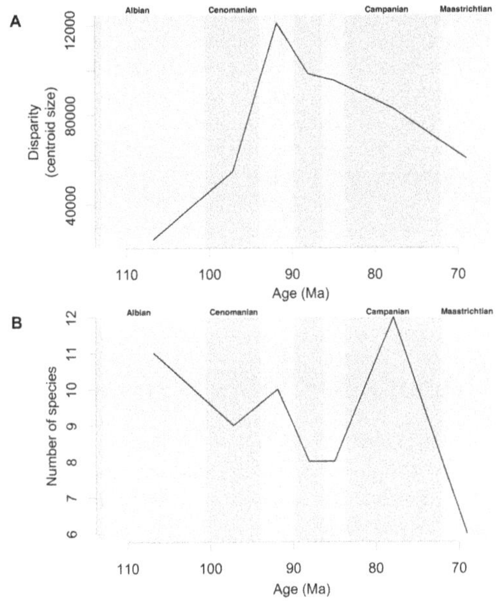

Figure 4. Disparity (A) and diversity (B) of *Wilbertopora*. Disparity reaches a single peak during the Turonian (93.9 Ma – 89.8 Ma), whereas diversity exhibits a more varied trajectory.

Ma). The Maastrichtian sees the largest drop in taxonomic diversity in the Cretaceous, with species richness dropping from 12 to 6 species.

3.2 *Diversity-disparity analysis*

Directly comparing diversity and total disparity using Jablonski's (2017) diversity-disparity analysis shows the extent of the decoupled nature of disparity and diversity during *Wilbertopora*'s genus lifespan (Figure 5A). Comparing *Wilbertopora*'s pattern of diversity and disparity with the three types of diversification trajectories explicated by Jablonski (2017) reveals surprising results. The diversity-disparity patterns of *Wilbertopora* have an unusual signature: the changes in diversity and disparity do not follow a single trajectory. Instead, the genus appears, across all levels of analysis (Figure 5), to begin its diversification according to ecological exploitation, then shifting into ecological exploration in most of the latter half of its stratigraphic range. Most of the genus's diversification occurs above the 1:1 line, indicating that different species of *Wilbertopora* are exhibiting disparate zooid morphologies.

3.3 *Avicularian versus autozooidal disparity*

Comparison of the avicularian and autozooidal diversity-disparity analyses shows that the two body types generally follow the same trajectory shape (Figure 5 B-C). However, diversity-disparity plots are scaled by value ranges, and are not suitable for direct comparisons of

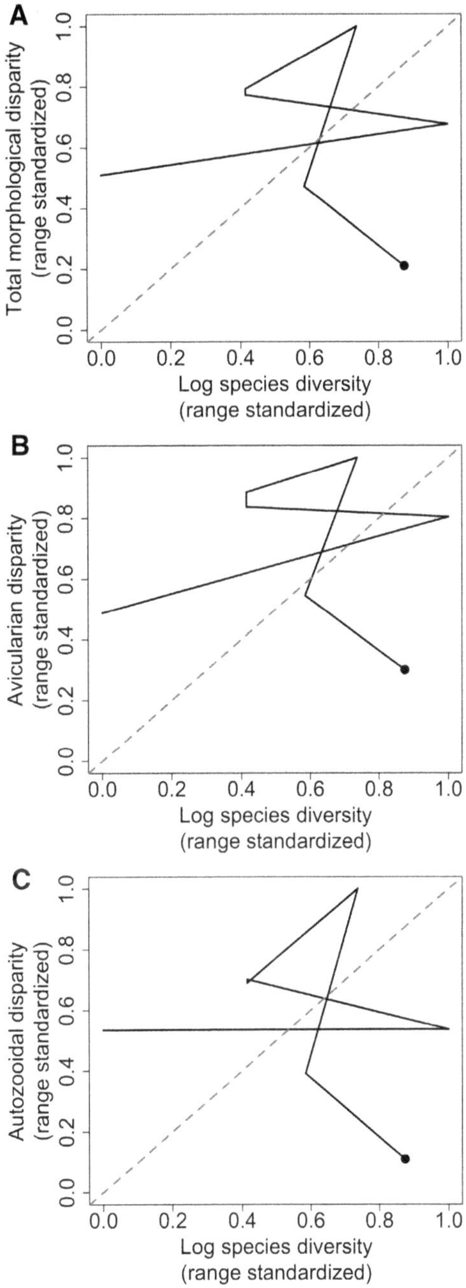

Figure 5. Disparity-diversity plots of *Wilbertopora*. A, total morphologic disparity versus species diversity of *Wilbertopora* over the genus' life span. B, disparity of avicularian characters in *Wilbertopora* versus diversity over the genus life span. C, autozooidal disparity of *Wilbertopora* versus diversity over the genus' life span. In all panels, the dashed red line shows a line with a slope of 1, which is the expected pattern in a diversification event where disparity and the log of diversity increase proportionately (i.e., Jablonski's (2017) Type 2). These plots show time series, beginning with the circle in the bottom right portion of each graph. Each subsequent change in slope corresponds to the onset of a new geologic stage.

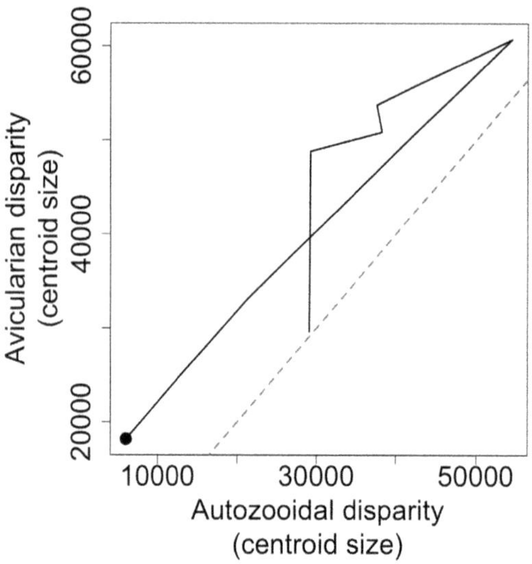

Figure 6. Plot of avicularian disparity versus autozooidal disparity. This represents a time series going from the Albian (circle) to the Maastrichtian (line ends). The raw values of disparity of each zooid type are directly compared to assess how their relationship changes over the course of *Wilbertopora* diversification. Dashed red line (i.e., slope = 1) indicates Jablonski's (2017) Type 2 event.

different disparity measures. Comparing raw avicularian disparity values and raw autozooidal disparity values directly is a more direct comparison of the two measures (Figure 6). For *Wilbertopora*'s genus life span, avicularia have a higher disparity during each geologic stage than autozooids. This result indicates that avicularia contribute more heavily to the signature of "ecological exploration" that *Wilbertopora* exhibits during its diversification than autozooids.

Overall, our results indicate that both avicularia and autozooids experienced increases in disparity relative to diversity as *Wilbertopora* diversified. *Wilbertopora* may have experienced selective pressure to generate autozooids of differing sizes during this time, as a response to substrate competition. Encrusting bryozoans must maximize occupation of substrate space to ensure survival, and many modern bryozoans exhibit wide ranges of autozooid shapes and sizes (McKinney & Jackson 1991). Perhaps expansion of autozooidal phenotypes in *Wilbertopora* is evidence of this selective pressure guiding the genus's diversification

Avicularia show the strongest pattern of ecological exploration during *Wilbertopora* diversification. The divergent morphology of avicularia is not unique to *Wilbertopora*. Avicularia in extant taxa have a variety of functions, including colony defense, hygiene, and the maintenance of feeding currents (Kaufmann 1971; McKinney & Jackson 1991; Winston 1986). The divergent morphologies of the early avicularia in *Wilbertopora* suggest that avicularia may have evolved to serve a variety of functions from its origination. Since there is no conserved morphological form for avicularia, even among the closely related members of this single genus, there is likely a range of avicularian function for each lineage of *Wilbertopora*. This suggests that from the first appearance of avicularia, their functional morphology has been widely divergent, even among different species of the same genus.

4 FUTURE CONSIDERATIONS

Since this study is preliminary by nature, there is a limit to the conclusions that can be drawn from these results. A lineage-specific analysis of *Wilbertopora* avicularian evolution is a natural next step but requires a resolved phylogeny of the genus. To make this possible, a much larger character matrix (with 100+ characters) is required. Furthermore, analysis of partial disparity of *Wilbertopora* species using whole-colony specimens would reveal whether there are species-level constraints on avicularian morphology in *Wilbertopora*, or whether each species can express a large range of avicularian morphology.

ACKNOWLEDGEMENTS

We would like to thank the Natural History Museum in London and the National Museum of Natural History in Washington, D.C. for making SEMs of their bryozoan specimens accessible via their online databases. This research was supported by the Bruce and Marcy Benson Graduate Student Fellowship in the Department of Geological Sciences at the University of Colorado, Boulder.

REFERENCES

Atteberry, M.R. & Eberle, J.J. 2021. New Earliest Paleocene (Puercan) Periptychid 'Condylarths' from the Great Divide Basin, Wyoming, USA. *Journal of Systematic Palaeontology* 19(8): 565–593.

Bapst, D.W. 2012. Paleotree: An R Package for Paleontological and Phylogenetic Analyses of Evolution. *Methods in Ecology and Evolution* 3: 803–807.

Bock, P. 2020. *Recent and Fossil Bryozoa*. The Bryozoa Home Page. Retrieved August 14, 2022, from http://bryozoa.net/.

Carter, M.C. 2008. *The Functional Morphology of Avicularia in Cheilostome Bryozoans*. Ph.D. thesis. Victoria University of Wellingon, 223 p.

Cheetham, A.H. 1954. A New Early Cretaceous Cheilostome Bryozoan from Texas. *Journal of Paleontology* 28(2): 177–184.

Cheetham, A.H., Sanner, J., Taylor, P.D. & Ostrovsky, A.N. 2006. Morphological Differentiation of Avicularia and the Proliferation of Species in Mid-Cretaceous *Wilbertopora* Cheetham, 1954 (Bryozoa: Cheilostomata). *Journal of Paleontology* 80(1): 49–71.
Cole, S.R. 2019. Phylogeny and Evolutionary History of Diplobathrid Crinoids (Echinodermata). *Palaeontology* 62(3): 357–373.
Crampton, J.S. 1995. Elliptic Fourier Shape Analysis of Fossil Bivalves: Some Practical Considerations. *Lethaia* 28(2):179–186.
Di Martino, E. & Taylor, P.D. 2013. First Bryozoan Fauna from a Tropical Cretaceous Carbonate: Simsima Formation, United Arab Emirates–Oman Border Region. *Cretaceous Research* 43: 80–96.
Di Martino, E., Martha, S.O. & Taylor, P.D. 2018. The Madagascan Maastrichtian bryozoans of Ferdinand Canu – Systematic Revision and Scanning Electron Microscopic Study. *Annales de Paléontologie* 104(2): 101–128.
Foote, M. 1993. Contributions of Individual Taxa to Overall Morphological Disparity. *Paleobiology* 19(4): 403–419.
Guha, A.K. 1989. Upper Cretaceous Cheilostomes (Bryozoa) from the Bagh Sediments of Madhya Pradesh. In P. Kalia (ed), *Micropalaeontology of the Shelf Sequences of India*: 118–131. New Delhi: Papyrus Publishing House.
Hay, W.W. & Floegel, S. 2012. New Thoughts About the Cretaceous Climate and Oceans. *Earth Science Reviews* 115(4): 262–272.
Jablonski, D. 2017. Approaches to Macroevolution: 2. Sorting of Variation, Some Overarching Issues, and General Conclusions. *Evolutionary Biology* 44(4): 451–475.
Jablonski, D., Lidgard, S. & Taylor, P.D. 1997. Comparative Ecology of Bryozoan Radiations; Origin of Novelties in Cyclostomes and Cheilostomes. *PALAIOS* 12(6): 505–523.
Kaufmann, K.W. 1971. *The Form and Functions of the Avicularia of Bugula [phylum Ectoprocta]*. New Haven: Peabody Museum of Natural History.
Koromyslova, A.V., Baraboshkin, E.Y. & Martha, S.O. 2018. Late Campanian to Late Maastrichtian Bryozoans Encrusting on Belemnite Rostra from the Aktolagay Plateau in Western Kazakhstan. *Geobios* 51(4): 307–333.
Martha, S.O. & Taylor, P.D. 2016. A New Western European Cretaceous Bryozoan Genus from the Early Cenomanian Radiation of Neocheilostomes. *Papers in Palaeontology* 2(2): 311–321.
McKinney, F.K. & Jackson, J.B.C. 1991. *Bryozoan Evolution*. Chicago: University of Chicago Press.
O'Dea, A. & Okamura, B. 1999. Influence of Seasonal Variation in Temperature, Salinity and Food Availability on Module Size and Colony Growth of the Estuarine Bryozoan *Conopeum seurati*. *Marine Biology*, 135(4): 581–588.
Okamura, B. 1992. Microhabitat Variation and Patterns of Colony Growth and Feeding in a Marine Bryozoan. *Ecology*, 73(4): 1502–1513.
Ostrovsky, A.N., Taylor, P D., Dick, M.H. & Mawatari, S. F. 2008. Pre-Cenomanian Cheilostome Bryozoa: Current State of Knowledge. In H. Okada, S.F. Mawatari, N. Suzuki & P. Gautam (eds), *Origin and Evolution of Natural Diversity: Proceedings of the International Symposium, The Origin and Evolution of Natural Diversity, held from 1–5 October 2007 in Sapporo, Japan*: 69–74. Sapporo: Hokkaido University.
Schindelin, J., Arganda-Carreras, I., Frise, E. *et al.* 2012. Fiji: An Open-source Platform for Biological-image Analysis. *Nature Methods* 9(7): 676–682.
Smith, A.B. & Batten, D.J. 2002. *Fossils of the Chalk*. London: Palaeontological Association.
Swofford, D.L. 2003. *PAUP*. Phylogenetic Analysis Using Parsimony (*and Other Methods). Version 4.* Sunderland, Massachusetts: Sinauer Associates.
Taylor, P.D. & Badve, R.M. 1994. The Mid-Cretaceous Bryozoan Fauna from the Bagh Beds of Central India: Composition and Evolutionary Significance. In P.J. Hayward, J.S. Ryland & P.D. Taylor (eds), *Biology and Palaeobiology of Bryozoans*: 181–186. Fredensborg: Olsen & Olsen.
Taylor, P.D. & Martha, S.O. 2017. Cenomanian Cheilostome Bryozoans from Devon, England. *Annales de Paléontologie* 103(1): 19–31.
Taylor, P.D. & McKinney, F.K. 2006. Cretaceous Bryozoa from the Campanian and Maastrichtian of the Atlantic and Gulf coastal plains, United States. *Scripta Geologica* 132: 1–346.
Taylor, P.D. 1988. Major Radiation of Cheilostome Bryozoans: Triggered by the Evolution of a New Larval Type? *Historical Biology* 1(1): 45–64.
Taylor, P.D. 2008. Late Cretaceous Cheilostome Bryozoans from California and Baja California. *Journal of Paleontology* 82(4): 823–834.
Taylor, P.D. 2020. *Bryozoan Paleobiology*. Hoboken, NJ: Wiley.
Voigt, E. 1981. Répartition et Utilisation Stratigraphique des Bryozoaires du Crétacé Moyen (Aptien-Coniacien). *Cretaceous Research* 2(3): 439–462.
Winston, J.E. 1986. Victims of Avicularia. *Marine Ecology* 7(2): 193–199.
Žítt, J., Vodrážka, R., Hradecká, L., Svobodová, M. & Zágoršek, K. 2006. Late Cretaceous Environments and Communities as Recorded at Chrtníky (Bohemian Cretaceous Basin, Czech Republic). *Bulletin of Geosciences* 81(1): 43–79.

APPENDIX 1

Characters are coded as follows:

1: first principal component (PC1) of outline of autozooid operculum. 0 = –0.051 to 0; 1 = 0.001 to 0.050; 2 = 0.051 to 0.100; 3 = 0.101 to 0.150; 4 = 0.151 to 0.200; 5 = 0.201 to 0.250; 6 = 0.251 to 0.300.
2: PC1 of outline of avicularium operculum. 0 = –0.49 to –0.30; 1 = –0.29 to –0.10; 2 = –0.09 to 0.10; 3 = 0.11 to 0.30; 4 = N/A.
3: autozooid length. 0 = 301 to 450 μm; 1 = 451 to 600 μm; 2 = 601 to 750 μm; 3 = 751 to 900 μm; 4 = 901 to 1050 μm.
4: autozooid width. 0 = 201 to 350 μm; 1 = 351 to 500 μm; 2 = 501 to 650 μm; 3 = 651 to 800 μm; 4 = N/A.
5: avicularium length. 0 = 201 to 350 μm; 1 = 351 to 500 μm; 2 = 501 to 650 μm; 3 = 651 to 800; 4 = N/A.
6: avicularium width. 0 = 51 to 150 μm; 1 = 151 to 250 μm; 2 = 251 to 350 μm; 3 = 351 to 450 μm; 4 = N/A.

7: ovicell length. 0 = 101 to 150 µm, 1 = 151 to 200 µm; 2 = 201 to 250 µm; 3 = 251 to 300 µm; 4 = N/A.
8: ovicell width. 0 = 101 to 170 µm; 1 = 171 to 240 µm; 2 = 241 to 310 µm; 3 = 311 to 380 µm; 4 = N/A.
9: presence/absence of avicularia. 0 = absent; 1 = present.
10: presence/absence of ovicells. 0 = absent; 1 = present.
11: presence/absence of spines. 0 = absent; 1 = present.
12: autozooid length / autozooid width ratio. 0 = 1.21 to 1.40; 1 = 1.41 to 1.60; 2 = 1.61 to 1.80; 3 = 1.81 to 2.00; 4 = N/A.
13: avicularia length / avicularia width ratio. 0 = 1.01 to 1.50; 1 = 1.51 to 1.90; 2 = 1.91 to 2.30; 3 = 2.31 to 2.70; 4 = 2.71 to 3.10; 5 = N/A.
14: ovicell length / ovicell width ratio. 0 = 0.501 to 0.650; 1 = 0.651 to 0.800; 2 = 0.801 to 0.950; 3 = 0.951 to 1.100; 4 = N/A.
15: autozooid opesia length. 0 = 191 to 290 µm; 1 = 291 to 390 µm; 2 = 391 to 490 µm; 3 = 491 to 590 µm; 4 = 591 to 690 µm.
16: autozooid opesia width. 0 = 101 to 180 µm; 1 = 181 to 260 µm; 2 = 261 to 340 µm; 3 = 341 to 420 µm; 4 = 421 to 500 µm.
17: avicularium aperture length. 0 = 81 to 200 µm; 1 = 201 to 320 µm; 2 = 321 to 440 µm; 3 = 441 to 560 µm; 4 = 561 to 680 µm.
18: avicularium aperture width. 0 = 51 to 100 µm; 1 = 101 to 150 µm; 2 = 151 to 200 µm; 3 = 201–250 µm; 4 = N/A.
19: autozooid opesia length / autozooid length ratio. 0 = 0.201 to 0.350; 1 = 0.351 to 0.500; 2 = 0.501 to 0.650; 3 = 0.651 to 0.800; 4 = 0.801 to 0.950.
20: autozooid opesia width / autozooid width ratio. 0 = <0.51; 1 = 0.51 to 0.60; 2 = 0.61 to 0.70; 3 = 0.71 to 0.80; 4 = 0.81 to 0.90, 5 = 0.91 to 1.00.

Bryozoan Studies 2022 – Key, Porter & Wyse Jackson (eds)

A review of the species belonging in *Stenopora* Lonsdale, 1844

Catherine M. Reid*
School of Earth and Environment, University of Canterbury, Christchurch, New Zealand
ORCID ID: 0000-0002-7204-2037

ABSTRACT: Following the recent redescription of the genus *Stenopora* Lonsdale, 1844, the species currently included in *Stenopora* are reviewed. Earlier concepts of the genus, that included diaphragms and two sizes of acanthostyles, have led many authors to erroneously include new species in *Stenopora*. While not all original descriptions or figured materials were sufficient to determine key zoarial characters, the number of species within the genus is now significantly reduced. Characters are tabulated but species not retained in *Stenopora* are not formally assigned to other genera. The primary reasons for recommendation for removal from the genus are the presence of diaphragms, two sizes of acanthostyles, or non-monilate walls. Following this review species considered to belong to *Stenopora* remain widespread in the Carboniferous and Permian in both hemispheres and exhibit a generally non-tropical, bipolar distribution, with distinctly fewer occurrences on the Panthalassan margin of North America.

1 INTRODUCTION

The genus *Stenopora* was established in 1844 by William Lonsdale with the formal description of the type species *Stenopora tasmaniensis* from collections from Tasmania, Australia. The genus was then described in detail by Lonsdale (1845a) alongside other species of the genus collected in Tasmania and Illawarra, New South Wales. At that time trepostome bryozoans were considered to belong to the corals, and subsequent workers describing species of the genus often interpreted artifacts of thin-sectioning to represent the presence of diaphragms, where in fact none existed. This issue of diaphragms was resolved almost a century later by re-description of the genus *Stenopora* (and others) in Bassler (1941). While the absence of diaphragms was formally resolved, Bassler had accessed the only material of *Stenopora tasmaniensis* available to him that was collected on the Wilkes Expedition and described by Dana (1849), as Lonsdale's type material had been lost (Nicholson & Etheridge 1879). This material described by Dana came from Harper's Hill, New South Wales, not from the type locality in Hobart, Tasmania (Reid 2021). While other workers had recorded *S. tasmaniensis* from the locality (Etheridge 1891; Nicholson & Etheridge 1886), and probably correctly, the material assigned by Dana that Bassler later re-described, was not in fact *Stenopora tasmaniensis* Lonsdale, 1844 (Reid 2021). The Harper Hill material clearly possesses two sizes of acanthostyles (Bassler 1941, figure 5 p. 175), where Lonsdale (1844, 1845a) neither figured nor described more than one size of acanthostyle. Bassler's (1941) description and figures were repeated in the *Treatise on Invertebrate Paleontology* record of *Stenopora* in Bassler (1953). Review of all *Stenopora* species described by Lonsdale from Australia (Lonsdale 1844, 1845a) alongside his establishment of the genus has confirmed that these species only have one size of acanthostyle (Reid 2021). In erecting a neotype for *S. tasmaniensis* Lonsdale, 1844, Reid (2021) also re-described the genus *Stenopora* returning it to the concept of the genus described in Lonsdale (1844, 1845a). This descripton follows.

Genus *Stenopora* Lonsdale, 1844

Type species. *Stenopora tasmaniensis* Lonsdale, 1844; Sakmarian of Tasmania, Australia.

Diagnosis. Zoarium ramose, frondescent, encrusting, massive or discoidal. One or more morphological forms may be exhibited in one zoarium or species. Endozonal walls thin and tubes free of diaphragms. Exozonal walls thickened by irregularly or regularly spaced monilae, diaphragms absent. Autozooecial apertures oval to rounded, exilazooecia present, rare to common. Apertures surrounded by acanthostyles of one size, either restricted to autozooecial wall junctions or regularly spaced about apertures. Maculae may be developed. (from Reid 2021, p. 1150).

*Corresponding Author:
catherine.reid@canterbury.ac.nz

DOI: 10.1201/9781003454915-11

This study aims to review the currently identified species of *Stenopora* to determine if they belong in the re-established concept of the genus, or if they should be considered for placement elsewhere. Any results will remain recommendations only.

2 METHODS

Data were compiled from review of the available literature and no physical material was examined. Currently known species of *Stenopora* were compiled primarily from the Bryozoa Home Page (Bock 2022) and augmented with additional species from McKinney (1974) that had not been subsequently assigned elsewhere. Lastly the literature was searched for described or undescribed species of *Stenopora* not included in Bock (2022) or McKinney (1974) or subsequently described. Data was compiled only from new species descriptions, not from subsequent records of existing species. Where original descriptions could not be accessed the named species was not included in this revision. Where original descriptions were in a language other than English, figures, and any accompanying data tables, from original descriptions, were used to record key features. Descriptions of some species listed on the Bryozoa Home Page (Bock 2022) and McKinney (1974) were unable to be sourced and are not included in the final data set.

Parameters compiled from the literature were the presence or absence of:- monilate walls in the exozone, diaphragms of any sort, and one or two sets of acanthostyles, as these parameters are the best indicator of inclusion or exclusion from the genus. Age and location of the holotype were also recorded. The age given by Bock (2022) from original sources was used here, recognizing that the age given in older original descriptions may require review.

Species were recorded as included or likely included in *Stenopora* or excluded from or query exclusion from *Stenopora.* No new genera are described for excluded species as this is outside the scope of this review.

3 RESULTS

Data from a total of 102 species of *Stenopora* were able to be compiled. Of the 109 species listed on the Bryozoa Home Page (Bock 2022) data for 94 were able to be sourced (Tables 1 & 2). Data for the remaining 15 species from the Bryozoa Home Page could not be found (Table 3) either because the original citation was not able to be confirmed or the reference obtained, or the reference did not contain sufficient descriptive or figured information to be useful. An additional 8 species were included in this data set from species listed in McKinney (1974), where references could be sourced, and from unknown species referred to *Stenopora* by Girty (1908), Lee (1912), Ross & Ross (1962), Ernst (2000) and Ernst *et al.* (2005, 2008). Species listed in McKinney (1974) that do not appear to be listed in other genera by Bock (2022), and where a reference could not be obtained are also listed in Table 3. There may be other described or unidentified species that were unintentionally omitted.

Of the 102 species compiled 37 exhibited monilate walls in the exozone, did not have diaphragms described or figured, only had one set of acanthostyles described or figured and exhibited other features that are consistent with *Stenopora* as redescribed in Reid (2021). These 37 species span the Early Carboniferous (Tournasian) to Late Permian (Lopingian) and have a distribution that includes North America, Russia, Svalbard, Greenland, Mongolia, China, Timor, Australia and India (Table 1).

The remaining 65 of 102 species assessed exhibit either one or more of non-monilate walls, diaphragms (whole or perforated) and either two sizes of acanthostyles or in a few cases no apparent acanthostyles (Table 2). Of these 65 species 34 have two sizes of acanthostyles either described or figured. Where acanthostyle size was assessed from figures only, acanthostyles with the clear appearance of a bimodal size distribution were included in this count, see Bassler (1941, fig 5, p 175) or Bassler (1953, fig 65 1a & b, G101-102) for the typical appearance of the acanthostyles when there are two sizes. A continuous range of acanthostyle sizes from smaller to larger were considered to represent one size of acanthostyles only. Three species appeared to have no acanthostyles at all. Of the same 65 species 29 had diaphragms recorded or figured, which were either common or rare. Those with rare diaphragms also had either two sizes of acanthostyles, or lacked monilate walls, supporting their inclusion in this group of 65. Five species with common diaphragms also possess two sizes of acanthostyles. Of this group of 65 species 36 are either described as not having monilate walls, or figures indicate the exozone walls do not have monilae. Only nine of these species also lack diaphragms and only have one size of acanthostyles.

4 DISCUSSION

Not all species currently assigned to *Stenopora* were able to be examined in this study, as not all original descriptions could be obtained. Likewise, there may be additional currently described species that have been unintentionally omitted. Of those that were obtained, however, only 37 of 102 or 36% were considered to be correctly assigned to *Stenopora* as re-described in Reid (2021). Of the 102 species assessed 65, or 64% do not align with

Table 1. List of species considered consistent with the revised description of *Stenopora*. Details of author, age and location, along with characters of monilae, diaphragms and number of acanthostyles are given. Except where noted all data has been gathered from original descriptions – sources for exceptions [1]Reid (2021). Species where descriptions could not be accessed, or interpreted, have not been included.

Species Key features consistent with *Stenopora*	Author	Age	Location	Monilae	Diaphr.	Acanth.
S. aequalis	Reid, 2003	Perm. Art.-Kungurian	Australia Tasmania	yes	no	1
S. angularis	Ulrich, 1890	Carb. Viséan	USA Illinois	yes	no?	1
S. brucei	Lee, 1908	C-P. Mosc.-Kung?	Norway Svalbard?	yes	no	1
S. clara	Kiseleva, 1970	Perm. Wordian	Russia Primorye	yes	no	1
S. contigua	Crockford, 1945	Perm. Cisuralian	Australia NSW	yes	no	1
S. crassa	Lonsdale, 1845b	Perm. Guadalupian	Russia Vaga River	yes	no	?
S. crinita	Lonsdale, 1845a	Perm. Cisuralian	Australia NSW	yes	no	1
S. frondescens	Crockford, 1943	Perm. Artinskian	Australia NSW	yes	no	1
S. grandis	Morozova, 1970a	Perm. Wordian	Russian Platform	yes	no	1
S. granulosa	Girty, 1908	Perm. Guadalupian	USA Texas	yes	no	1
S. inclinis	Liu, 1980	Perm. Cisuralian	NE China	yes	no?	1?
S. inflata	Liu, 1980	Perm. Cisuralian	NE China	yes	no	1?
S. informis	Lonsdale, 1845a	Perm. Cisuralian	Australia Tasmania	yes	no	?
S. jackii	Nich. & Etheridge, 1879	Permian	Australia NSW	yes	no	?
S. jalaidensis	Liu, 1980	Perm. Cisuralian	NE China	yes	no?	?
S. jungersenensis	Ross & Ross, 1962	Perm. Sakmarian	Greenland	yes	no?	1
S.? kashmirensis	Sakagami, 1981	Perm. Guadalupian	India	yes	no	1?
S. labrosa	Yang & Lu, 1962	Carb. Mississippian	(South?) China	yes	no	1
S. labrosa florida	Yang & Lu, 1962	Carb. Mississippian	(South?) China	yes	no	1
S. leichhardti	Nich. & Etheridge, 1886	Carb. Pennsylvanian	Australia NSW	yes	no?	1
S. libita	Liu, 1976	Perm. Cisuralian	North China	yes?	no	1?
S. nicholsoni	Waagen & Wentzel, 1886	Perm. Lopingian	India	yes	no	1
S. nigris	Crockford, 1943	Perm. Artinskian	Australia NSW	yes	no	1
S. normalis	Liu, 1980	Perm. Cisuralian	NE China	yes?	no	1?
S. obesa	Morozova, 1970b	Perm. Guadalupian	Mongolia	yes	no	1
S. ovata	Lonsdale, 1844	Permian	Australia Tasmania	yes	no	1
S. parvaexozona	Gilmour *et al.*, 1997	Perm. Wordian	USA Nevada	yes	no	1
S.? pertusa	Trizna, 1958	Carb. Tournaisian	Russia Kuznetsk Bsn	?	no	1
S. punctata	Crockford, 1957	Perm. Artinskian	Australia also China	yes	no	1
S. rugosa	Crockford, 1945	Perm. Artinskian	Australia NSW	yes	no	1
S. spicata	(Bassler 1929)	Perm. Sak.-Artinsk.	Timor	yes	no	1
S. spiculata	Crockford, 1945	Perm. Sakmarian	Australia NSW	yes	no	1
S. subzonata	Liu, 1976	Perm. Cisuralian	North China	yes	no	1
S. tasmaniensis	Lonsdale, 1844[1]	Perm. Sak.-Artinsk.	Australia Tasmania	yes	no	1
S. thula	Ross & Ross, 1962	Perm. Sakmarian	Greenland	yes	no	1
S. toriyamai	Sugimura & Ota, 1980	Permian	Japan	yes	no	1
S. zabaycalica	Nekhoroshev, 1956	Carb. Tournaisian	Russia Transbaikal	yes	no	1

Table 2. List of species recommended for exclusion from *Stenopora*. Details of author, age and location, along with characters of monilae, diaphragms and number of acanthostyles are given. Except where noted all data has been gathered from original descriptions – sources for exceptions [2]A. Ernst pers. comm. (July 2022), [3]Crockford (1943), [4]Sakagami (1981). Species where descriptions could not be accessed, or interpreted, have not been included.

Species Key features inconsistent with *Stenopora*	Author	Age	Location	Monilae	Diaphr.	Acanth.
S. akagoensis	Sugimura & Ota, 1971	Carb. Visean	Japan	yes	yes	1
S. americana	Ulrich, 1890	Carb. Visean	USA Illinois	yes	yes	1
S. americana varsoviensis	Ulrich, 1890	Carb. Visean	USA Illinois	yes	yes	1
S. amplopora	Nikiforova, 1939	Perm. Sak.-Artinsk.	Russia Urals	no?	yes?	1
S. armata	Morozova, 1970a	Perm. Lopingian	Russian Platform	no	no	2
S. armatiformis	Xia, 1986	Perm. Guadalupian	Tibet	no	no	2
S. australis	Etheridge *in* Eth. & Jk, 1892	Permian	Australia NSW	yes	yes?	absent
S. berriedalensis	Reid, 2003	Perm. Art.-Kungurian	Australia Tasmania	no	no	1
S. castletonensis	Lee, 1912	Carb. Viséan	England	yes	yes	1
S. cidariformis	Lee, 1908	Perm.? Art.-Kung.?	Norway Svalbard?	yes	no	2
S. circina	Coryell *in* Morgan, 1924	Carb. Bashkirian	USA Oklahoma	yes	yes	1
S. columnaris incrustans	Geinitz, 1861[2]	Perm. Guadalupian	Germany	no	no	1
S. columnaris ramosa	Geinitz, 1861[2]	Perm. Guadalupian	Germany	no	no	1
S. confusionensis	Karklins, 1986	Carb. Serpukhovian	USA Utah	yes	no	2
S. daheshenensis	Liu, 1980	Perm. Cisuralian	North China	yes	yes?	?
S. decimana	Gorjunova, 1983	Carb. Serpukhovian	Mongolia	no	yes	1
S. distans	Condra 1902	?	Nebraska	no?	yes	2
S. etheridgei	Crockford, 1945	Perm. Sakmarian	Australia NSW	no?	no	2
S. gimpiensis	Etheridge *in* Eth. & Jk, 1892	Perm. Sak.-Artinsk.	Australia Qslnd.	yes	yes	1
S. gracilis	(Dana 1849)[3]	Perm. Artinskian	Australia NSW	yes	rare	2?
S. hemispherica	Waagen & Wentzel, 1886	Perm. Cisuralian	India	yes	yes?	1?
S. hirsuta	Crockford, 1945	Perm. Artinskian	Australia	yes	no	2
S. incondita	Trizna, 1958	Carb. Viséan	Russia Kuznetsk Bsn	no	no	2?
S. intercalaris	Ulrich, 1890	Carb. Viséan	USA Illinois	yes	yes	1
S. lineata	Crockford, 1957	Perm. Artinskian	Australia WA.	no	no	2
S. lineata (homonym)	Yang & Lu, 1962	Perm. Lopingian	South? China	no?	no	2
S. miseri	Girty, 1911	Carb. Visean-Serp.	USA Arkansas	yes	yes	1
S. miseri tubulata	Girty, 1911	Carb. Visean-Serp.	USA Arkansas	?	yes	1
S. multifida	Yang & Lu, 1962	Carb. Mississippian	South? China	no?	no	?
S. nishiyamensis	Sakagami, 1962	Carb. Bashkirian	Japan	yes	rare	2
S. parallela	Crockford, 1945	Perm. Cisuralian	Australia Tasmania	yes	no	2

(*continued*)

Table 2. Continued

Species Key features inconsistent with *Stenopora*	Author	Age	Location	Monilae	Diaphr.	Acanth.
S. perattenuata	Girty, 1911	Carb. Visean-Serp.	USA Arkansas	?	yes	absent
S. permiana	Yang, 1958 non (Bassler 1929)	Perm. Cisuralian	South? China	no	rare	1
S.? perpolita	Trizna & Klaucan, 1961	Perm. Artinskian	Russia Urals	no	no	?
S.? polyspinosa	Condra 1902	?	Nebraska	no	yes	2
S. polyspinosa richardsoni	Girty, 1908	Perm. Guadalupian	USA Texas	no	rare	1
S. prolifica	(Fritz 1932)	Perm. Artinskian	Canada BC	no?	no	2
S. pusilimonila	Sakagami, 1964[4]	Carb. Serpukhovian	Japan	yes	yes	1
S. qaganqulutensis	Liu, 1980	Perm. Cisuralian	NE China	no?	no	1
S. quasitimanensis	Morozova, 1986 *in* Moro & Krutchinina 1986	Perm. Wordian	Russia Nov. Zem-lya	no	no	2
S. quinghaiensis	Yang & Lu, 1962	Perm. Lopingian	South? China	yes	no	2?
S. redesalensis	Lee, 1912	Carb. Viséan	England	yes?	yes	2
S. saranaensis	Trizna & Klaucan, 1961	Perm. Artinskian	Russia Urals	yes	no	2
S. seriatensis	Reid, 2003	Perm. Kungurian	Australia NSW	no	no	1
S.? signata	Ulrich, 1890	Carb. Pennsylvanian	USA Illinois	no	no	2
S. simplex	Liu, 1980	Perm. Cisuralian	NE China	no?	no	1
S. simulans	Girty, 1911	Carb. Visean-Serp.	USA Arkansas	no?	yes	1
S. spicata obtusa	Crockford, 1957	Perm. Artinskian	Australia WA	yes	rare	2
S. spinigera	Lonsdale, 1845b	Perm. Guadalupian	Europe Urals	no?	no	1
S. spinulosa	Rogers, 1900	Carb. Pennsylvanian	USA Kansas	no?	no	2
S. spissa	Rogers, 1900	Carb. Pennsylvanian	USA Kansas	no	no	1
S. stereotoechusi formis	Yang, Hu & Xia, 1988	Carb. Tournaisian	South China	no	no	1
S. stereotoech. pristina	Yang, Hu & Xia,1988	Carb. Tournaisian	South China	no	no	absent?
S. tenuipora	Lee, 1912	Carb. Viséan	England	no	yes	1
S. thuliformis	Morozova, 1970a	Perm. Wordian	Russian Platform	yes	no	2?
S. timanensis	Morozova, 1970a	Perm. Wordian	Russian Platform	no	no	2
S. veriparallela	Xia, 1986	Perm. Guadalupian	Tibet	no	rare?	2?
S. veriparallela munilimuralis	Xia, 1986	Perm. Guadalupian	Tibet	no	no	2?
S. yunzhugensis	Xia, 1986	Perm. Cisuralian	Tibet	no	no	1?
Stenopora sp.	Girty 1908	Perm. Guadalupian	Guadalupe Mtns	no	no	2
Stenopora sp.	Ernst *et al.*, 2008	Low.?-Mid. Perm.	Oman	yes	rare	2
Stenopora sp.	Lee, 1908	Perm? Art.-Kung.?	Norway Svalbard?	yes	yes	2
Stenopora sp.	Ross & Ross, 1962	Perm. Sakmarian	Greenland	yes	no	2
Stenopora sp.	Ernst 2000	Upper Permian	Tunisia	yes	yes	2
Stenopora sp.	Ernst *et al.*, 2005	Carb. Pennsylvanian	Spain	no	no	2

the revised criteria for inclusion in *Stenopora*, and their placement in this genus is questioned.

The original description of *Stenopora* by Lonsdale (1845a) made it clear that walls in the exozone are monilate and the zooecial tubes did not possess diaphragms. He did not specify the size of acanthostyles, however only one size of acanthostyles were figured in Lonsdale (1844 and 1845a). At that time *Stenopora* was considered to be a coral and subsequent authors attributed diaphragms to *Stenopora* (Etheridge 1891; Nicholson & Etheridge 1886) although they were never figured, even in sections where they should be expected to appear if they existed (Reid 2021). The absence of diaphragms was recorded by Bassler (1941) in his re-description of the genotype. Prior to this however, numerous authors had already assigned new species, with diaphragms, to *Stenopora.* Of the species where assignment to *Stenopora* is questioned on the basis of diaphragms most were described prior to 1941 (Table 2).

The material re-described by Bassler, from Harper Hill NSW, not the Hobart type locality, clearly shows two sizes of acanthostyles (Bassler 1941, figure 5, page 175) and as a result the description of the genus also specifies two sizes of acanthostyles. This description and figure was repeated in Bassler (1953) in the *Treatise of Invertebrate Paleontology*. Re-examination of *S. tasmaniensis* from the type locality in Tasmania and consideration of other species described by Lonsdale in 1844 and 1845a has shown that the material examined by Lonsdale bore only one size of acanthostyle (Reid 2021). The 34 species (Table 2) in which two sizes of acanthostyles are described or figured do not align with the revised description of *Stenopora* and should be assigned elsewhere. Many species exhibit occasional large acanthostyles at the junctions of exozonal walls, and smaller acanthostyles between. The contrast in size between the two sets of acanthostyles is variable. Some species, such as *S. timanensis* Morozova, 1970a show very large macro-acanthostyles at every exozonal wall junction and a clear row of very small numerous micro-acanthostyles in the exozonal wall between apertures. Others tend to have macro-acanthostyles present at some, but not all, exozonal wall junctions, and the difference in size between micro- and macro-acanthostyles is not as marked. Further to this variation in acanthostyle style and arrangement, the species that show or appear to show two sizes of acanthostyles, differ from each other in other characters, such as the presence or absence of diaphragms and monilate walls. This suggests that there may be more than one taxon represented in this group with two sizes of acanthostyles.

Lonsdale (1844, 1845a) was very clear in describing monilate walls or "successive narrowing in each tube" (Lonsdale 1844, p. 161). He repeated this in his descrption of *Stenopora* species from Europe and Russia (Lonsdale 1845b). Examination of material from the type locality in Hobart Tasmania by Reid (2021) show that monilate walls are present in *Stenopora tasmaniensis*, but also that monilae may be crowded and almost confluent. In other species described by Lonsdale (1844, 1845a) monilae are clearly separated. In reconsidering which species may not belong to *Stenopora*, those with clearly evenly thickened exozonal walls are questioned on the basis of monilae alone. Of those species where descriptions or figures of monilae were ambiguous, the nature of acanthostyles or the occurrence of diaphragms was required before their assignment to *Stenopora* was questioned.

4.1 *Problematic taxa*

Not withstanding taxa with characters that no longer align with the revised concept of *Stenopora*, some species remain problematic. An example is *Stenopora columnaris* (Schotheim 1813), originally described as *Coralliolites columnaris* but insufficiently described to determine true placement (Duncan 1949). Following establishment of the species subsequent workers assigned material to *columnaris* as *Stenopora* and included subspecies (ie Geinetz 1861) but descriptions are insufficient to determine relevant characters. It appears that some records should be attributed to anthozoans (Ernst pers comm. 2022; Weyer 1984). It would seem appropriate to suspend the use of the name *S. columnaris* and subspecies, and all material so assigned, until original material can be examined and correct taxonomic placement at phylum level can be determined.

Lee (1908) described three species from the Permo-Carboniferous of Prins Karls Forland in Norway (Table 1 & 2). As noted by Lee some of the materials were not in-situ, and rocks of Permo-Carboniferous age are not known from Prins Karls Forland (Dallmann 2015), and the bryozoans may be from glacial erratics. As a result these species are not included on Figures 1 or 2.

4.2 *Potential for new genera*

Numerous species fail to meet the criteria for inclusion in *Stenopora* based on the presence of diaphragms or two sizes of acanthostyles, or absence of monilate exozonal walls (Table 1). Several taxa, that have no diaphragms and only one size of acanthostyles, but lack monilate exozonal walls, likely can be assigned to *Dyscritella* Girty, 1911. Taxa that have monilate walls, one size of acanthostyle and only fail inclusion in *Stenopora* on the presence of diaphragms, or ring septa, in the exozone likely can be assigned, respectively, to *Stenodiscus* Crockford, 1945, or *Tabulipora* Young, 1883.

Only published records have been evaluated, and no physical material has been directly

Table 3. Names, authors, age and location of species from Bock (2022) and McKinney (1974) that were not included in this study. Citations are not included in the reference list where it was not possible to confirm or source the citation. Note Kiseleva (1970) is not the same as the citation shown in Table 1.

Species	Author	Age	Location
S. approximata	Eichwald, 1855	Carb. Viséan	Russian Platform
S. buragoi	Kiseleva, 1970?	Perm. Guadalupian	Russia Primorye
S. columnaris	(von Schlotheim 1813)	Perm. Cisuralian	Europe
S. columnaris buchiana	(King 1850)	Perm. Guadalupian	England
S. columnaris tuberosasparsigemmata	(Grabau 1931)	Permian	Mongolia
S. derilangmaensis	Liu & Wang, 1987	Carb.-Perm.	Tibet
S. dubia	(McCoy 1844)	Carb. Mississippian	Ireland
S. gorskji	Baranova, 1960a	Permian	Russia
S. guangdongensis	Li, 1977	Perm. Lopingian	China
S.? kocki	Vinassa de Regny, 1901	Triassic Carnian	Hungary
S. kojimensis	Baranova, 1960a	Permian	Russia
S. lahuseni	Romanowsky, 1891	Carb.-Perm.	Russia
S. mazhongshanensis	Lu, 1983	Perm. Cisuralian	China
S. meekana	(Girty 1907) non Ulrich, 1890	Permian	China
S. mirabilis	Yang, Lu & Zhou, 1979	Perm. Cisuralian	China
S. patula	Billings, 1859	Ord. Darriwilian	Canada Quebec
S. uralica	Lebedev, 1924?	Carb. Pennsylvanian	Russia Urals

examined. Examination of original material, whether it be type or subsequently assigned material, is required before confirmation of exclusion from *Stenopora*, or assignation to new genera can be confirmed. It is likely that new genera should be described to accommodate the taxa that do not appear to meet the criteria for inclusion in *Stenopora*, however no attempt is made here to do so as original materials have not been examined.

4.3 *Palaeobiogeography*

The distribution of all species currently assigned to *Stenopora* is global, with occurrences appearing in high southern latitudes, mid and low latitude Tethys Ocean and mid-latitude northern hemisphere, and including Panthalassa Ocean locations (western North America and Japan) (Figures 1–2), although there is a tendency to bipolarity (Reid & James 2010).

The overall distribution of Permian taxa that, in this assessment, are likely to be correctly assigned to *Stenopora* is the same, however there are fewer low latitude taxa, and few taxa from western North America (Panthalassan margin). The general bipolarity of *Stenopora* (Reid & James 2010) in the Permian is apparently strengthened in this reassessment.

Carboniferous taxa that likely are correctly assigned to *Stenopora* do have a lower latitude distribution. The Late Palaeozoic Ice Age was well-developed in the Carboniferous and did not begin to wane until the Permian (Crowell 1999) and low and mid-latitude locatons may have had cooler ocean temperatures. If the bipolarity of *Stenopora* reflects a preference for cooler water, this may have influenced the occurrence of the genus in lower latitudes during the Carboniferous.

5 CONCLUSIONS

Of the 102 species able to be evaluated in this study 37 are likely to be correctly assigned to *Stenopora* as re-defined by Reid (2021). The remaining 65 taxa possess or appear to possess diaphragms in the endozone or exozone, posess both mega/macroacanthostyles and microacanthostyles, or lack monilae in the exozone.

It is likely that new genera should be erected to accommodate at least some of these 65 taxa. This is not done here as original material was not examined.

Permian taxa that are likely correctly assigned to the revised concept of *Stenopora* exhibit a distinct mid to high-latitude bipolarity, with only Carboniferous taxa appearing at lower latitudes, during the cooler waters of the Late Palaeozoic Ice Age.

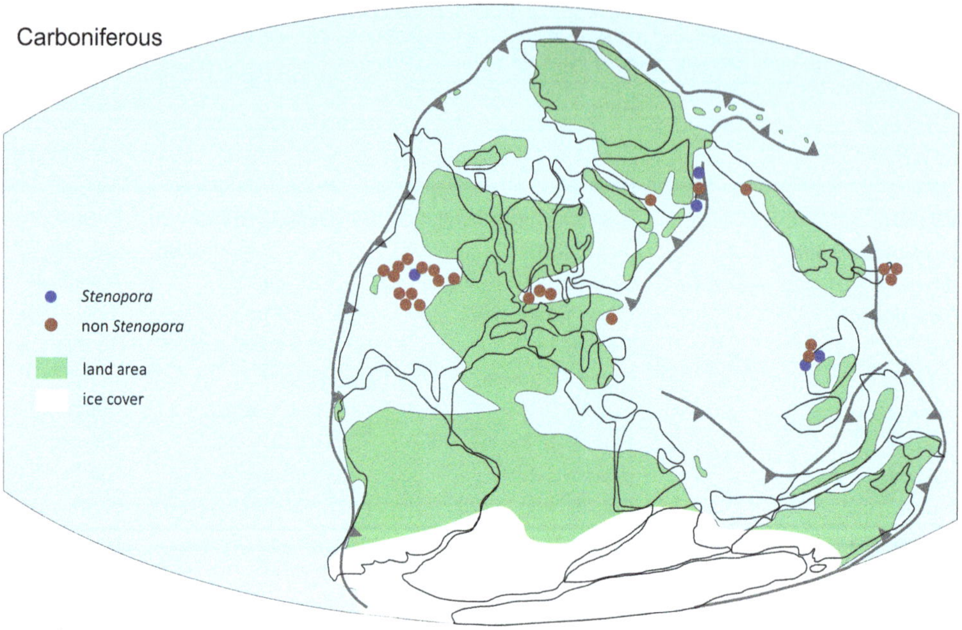

Figure 1. Distribution of *Stenopora* in the Carboniferous. Blue markers indicate species that are likely true *Stenopora*, and brown colours are those that are recommended for revision. Base-map after Scotese (2002) Early and Late Carboniferous maps. Locations of taxa are based on the original description and the relevant location of landmasses in Scotese (2002).

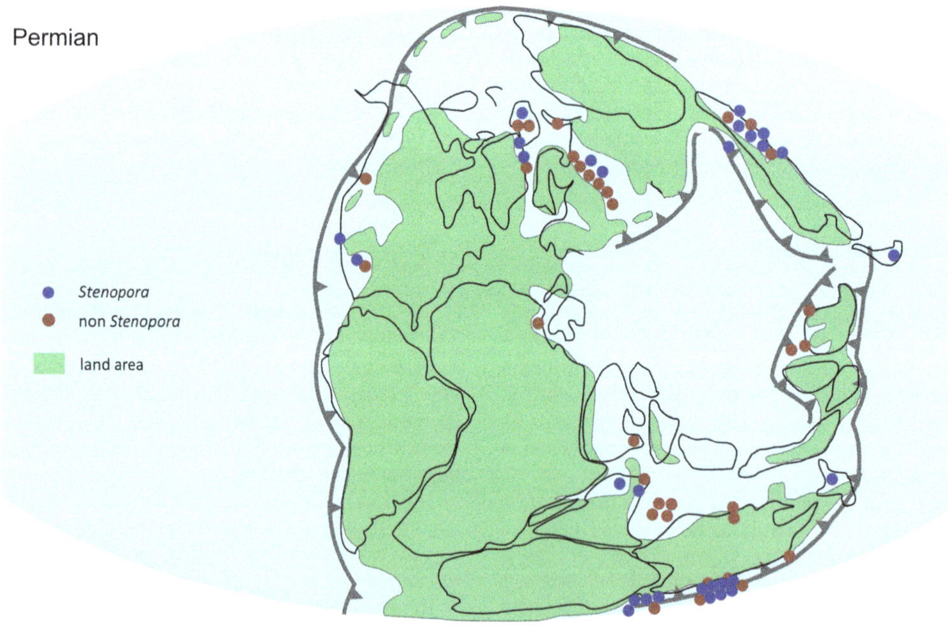

Figure 2. Distribution of *Stenopora* in the Permian. Blue markers indicate species that are likely true *Stenopora*, and brown colours are those that are recommended for revision. Base-map redrawn from Scotese (2002) Late Permian map. Locations of taxa are based on the original description and the relevant location of landmasses in Scotese (2002).

ACKNOWLEDGEMENTS

The author is grateful to Patrick Wyse Jackson, Andrej Ernst, Hans Arne Nakrem and Paul Taylor for providing copies of difficult to source references. Additional thanks are extended to Andrej Ernst for his comments on some taxa. Thanks to Caroline Buttler, Hans Arne Nakrem and Patrick Wyse Jackson for reviews of the original manuscript.

REFERENCES

Baranova, G.L. 1960. To the Question of the Kungurian Stage of the Basin of the River Pechory. *Vses Scientific Research Geological Institute Sbornik* 24: 27–35.

Bassler, R.S. 1929. The Permian Bryozoa of Timor. *Paläeontologie von Timor*, XVI Lieferung 28: 36–89.

Bassler, R.S. 1941. Generic Descriptions of Upper Paleozoic Bryozoa. *Journal of the Washington Academy of Sciences* 31(5): 173–179.

Bassler, R.S. 1953. *Treatise on Invertebrate Paleontology, Part G, Bryozoa.* Lawrence, Kansas: Geological Society of America and University of Kansas Press.

Bock, P.E. 2022. Bryozoa Home Page. Indexes to Taxa, *Stenopora* Lonsdale, 1844. http://www.bryozoa.net/trepostomatida/stenoporidae/stenopora.html (downloaded April 2022).

Condra, G. 1902. New Bryozoa from the Coal Measures of Nebraska. *The American Geologist* 30(6): 337–358.

Crockford, J.M. 1943. Permian Bryozoa of Eastern Australia. Part III. Batostomellidae and Fenestrellinidae from Queensland, New South Wales, and Tasmania. *Journal and Proceedings of the Royal Society of New South Wales* 78: 260–266.

Crockford, J.M. 1945. Stenoporids from the Permian of New South Wales and Tasmania. *Proceedings of the Linnean Society of New South Wales* 70: 9–24.

Crockford, J.M. 1957. Permian Bryozoa from the Fitzroy Basin, Western Australia. *Bureau of Mineral Resources, Geology and Geophysics, Bulletin* 34: 132 p.

Crowell, J.C. 1999. Pre-Mesozoic Ice Ages: Their Bearing on Understanding the Climate System. *Geological Society of America Memoir* 192: 106p.

Dallmann, W.K. 2015. Geoscience Atlas of Svalbard. *Norsk Polarinstitutt Rapportserie*, 148 (2015), pp. 1–292

Dana, J.D. 1849. United States Exploring Expedition: During the Years 1838, 1839, 1840, 1841, 1842 Under the Command of Charles Wilkes. *Atlas of Geology*. New York: Geo P. Putnam.

Duncan, H. 1949. Genotypes of Some Paleozoic Bryozoa. *Journal of the Washington Academy of Sciences* 39(4): 122–136.

Ernst, A. 2000. Permian Bryozoans of the NW-Tethys. *Facies* 43(1): 79–102.

Ernst, A., Schäfer, P. & Reijmer, J.J. 2005. Stenolaemate Bryozoa from the Upper Carboniferous of the Cantabrian Basin, Northern Spain. *Senckenbergiana Lethaea* 85(2): 301–317.

Ernst, A., Weidlich, O. & Schäfer, P. 2008. Stenolaemate Bryozoa from the Permian of Oman (Aseelah Unit, Batain Coast). *Journal of Paleontology* 82(4): 676–716.

Etheridge, R. 1891. Carboniferous and Permo-Carboniferous Invertebrata of New South Wales, Part 1, Coelenterata. *Memoir of the Geological Survey New South Wales, Palaeontology* 5: 1–64.

Etheridge, R. & Jack. R. L. 1892. *The Geology and Paleontology of Queensland and New Guinea.* Brisbane: Government Printer.

Fritz, M.A. 1932. Permian Bryozoa from Vancouver Island. *Transactions of the Royal Society of Canada* 26: 93–109.

Geinitz, H.B. 1861. *Dyas; Oder, Die Zechsteinformation und das Rothliegende: (Permische Formation zum Theil)*. Leipzig: W. Engelmann.

Gilmour, E.H., McColloch, M.E. & Wardlaw, B.R. 1997. Bryozoa of the Murdock Mountain Formation (Wordian, Permian), Leach Mountains, northeastern Nevada. *Journal of Paleontology* 71(2): 214–236.

Girty, G.H. 1908. *The Guadalupian Fauna.* United States Geological Survey, Professional Paper 58: 657p.

Girty, G.H. 1911. New Genera and Species of Carboniferous Fossils from the Fayetteville Shale of Arkansas. *Annals of the New York Academy of Sciences* 20: 189–238.

Gorjunova, R.V. 1983. First find of Carboniferous Trepostomids in Mongolia. In A. Y. Rozanov (ed.) *New Species of Fossil Invertebrates of Mongolia*: 62–66. Moscow. Trudi Sovmestnaya. Sovetsko-Mongol'skaya Paleontogicheskaya Ekspeditsiya. [In Russian]

Karklins, O.L. 1986. Chesterian (Late Mississippian) bryozoans from the upper Chainman Shale and the Lowermost Ely Limestone of western Utah. *Journal of Paleontology* 60 (S17): 1–48.

Kiseleva, A.V. 1970. Novyye Pozdne Permskiye mshanki Primorskogo Kraya (New Late Permian bryozoans from Primorsky Krai). In A.V. Kiseleva (ed.) *Novye Vidy Paleozoiskih Msanok I Korallov*: 76–80. Moscow: Nauka. [In Russian]

Lee, G.W. 1908. Notes on Fossils from Prince Charles Foreland: Brought Home by Dr William S. Bruce in 1906 and 1907. *Proceedings of the Royal Physical Society of Edinburgh* 17: 149–166.

Lee, G.W. 1912. The British Carboniferous Trepostomata. *Memoirs of the Geological Survey of Great Britain. Palaeontology* 1: 135–195.

Liu, X.L. 1976. Phylum Bryozoa. In *Palaeontological Atlas of North China. Inner Mongolia*: 131–155. Peking: Geological Publishing House. [In Chinese].

Liu, X.L. 1980. Phylum Bryozoa. In *Paleontological Atlas of Northeast China*: 189–254. Peking: Geological Publishing House. [In Chinese].

Lonsdale, W. 1844. Description of Six Species of Corals from the Palaeozoic Formation of Van Diemen's Land. In C. Darwin (ed.), *Geological Observations on the Volcanic Islands Visited During the Voyage of the HMS Beagle*: 161–169. London: Smith, Elder.

Lonsdale, W. 1845a. Polyparia. In P.E. Strzelecki de, (ed.), *Physical Description of New South Wales and Van Diemen's Land*: 262–270. London: Longman, Brown, Green and Longmans.

Lonsdale, W. 1845b. Description of Some Characteristic Palaeozoic Corals of Russia. In R.I. Murchison, d.V. E. Verneuil & A. von Keyserling (eds), *The Geology of Russia in Europe and the Ural Mountains*: 591–634. London: John Murray.

McKinney, F.K. 1974. Bibliography and Catalogue (1900–1969) of the Trepostomata (Phylum Ectoprocta). *Southeastern Geology Special Publication* 4: 147p.

Morgan, G.D. 1924. Geologic Map of the Stonewall Quadrangle, Oklahoma. *Bureau of Geology Bulletin* 2: 75–247.

Morozova, I. 1970a. Late Permian Bryozoa. *Transactions Paleontological Institute, Akademii Nauk, SSSR* 122:1–347. [In Russian]

Morozova, I.P. 1970b. Some New Late Permian Bryozoans from Southeastern Mongolia. In G.G. Astrova & I.I. Chudinova (eds). *New Species of Paleozoic Bryozoans and Corals*: 69–75. Moscow: Nauka. [In Russian]

Morozova, I.P. & Krutchinina, O.N. 1986. *Permian Bryozoa of the Arctic Region; Western sector.* Moscow: Nauka. [In Russian]

Nekhoroshev, V.P. 1956. Lower Carboniferous Bryozoa of Altai and Siberia. *Trudy Vsesoyuznogo Nauchno-issledovatelskogo Instituta (VSEGEI)* 13: 1–420. [In Russian]

Nicholson, H.A. & Etheridge, R. 1879. XXV—Descriptions of Palæozoic Corals from Northern Queensland, with Observations on the Genus *Stenopora. Annals and Magazine of Natural History* 4 (21): 216–226.

Nicholson, H.A. & Etheridge, R. 1886. XVI.—On the Tasmanian and Australian Species of the Genus *Stenopora*, Lonsdale. *Annals and Magazine of Natural History* 17(99):173–187.

Nikiforova, A.I. 1939. New Species of the Upper Paleozoic Bryozoa from the Bashkirian Urals. *Transaction of the Geological Oil Institute* 115: 70–102. [In Russian]

Reid, C.M. 2003. Permian Bryozoa of Tasmania and New South Wales: Systematics and their Use in Tasmanian Biostratigraphy. *Association of Australasian Palaeontologists*, Memoir 28: 133p.

Reid, C.M. 2021. A Redescription of *Stenopora* and the Type Species *Stenopora tasmaniensis* Lonsdale, 1844 (Trepostomata, Bryozoa). *Journal of Paleontology* 95 (6): 1147–1157.

Reid, C. M. & James, N. P. 2010. Permian Higher Latitude Bryozoan Biogeography. *Palaeogeography, Palaeoclimatology, Palaeoecology* 298: 31–41.

Rogers, A. 1900. New Bryozoans from the Coal Measures of Kansas and Missouri. *Kansas University Quarterly* 9: 1–12.

Ross, J.R.P. & Ross, C.A. 1962. Faunas and Correlation of the Late Paleozoic Rocks of North-east Greenland, pt 4: Bryozoa. *Meddelelser Grønland* 167(7): 1–65.

Sakagami, S. 1962. Lower Carboniferous Bryozoa from the Omi Limestone Japan part 1: Discovery of the *Profusulinella* Zone and Descriptions of *Profusulinella*, Cyclostomata, Trepostomata and *Fenestella. Transactions and Proceedings of the Paleontological Society of Japan. New series* 1962(48): 321–330.

Sakagami, S. 1964. Bryozoa of Akiyoshi. Part 2. Lower Carboniferous Bryozoa from the Uzura Quarry. *Transactions and Proceedings of the Palaeontological Society of Japan* 56: 295–308.

Sakagami, S. 1981. Upper Permian Bryozoa from Guryul Ravine and the Spur Three kilometres North of Barus. *The Upper Permian and Lower Triassic faunas of Kashmir, Palaeontologia Indica New Series* 46: 47–62.

Schlotheim, E.F. 1813. Beitrage zur Naturgeschichte der Versteinerungen in Geognostischer Hinsicht. *Taschenbuch Gesammte Mineralogie* 7(1): 3–134.

Scotese, C.R. 2002. *Earth History: Late Carboniferous and Permian Maps.* http://www.scotese.com (PALEOMAP website). Downloaded July 2022.

Sugimura, A. & Ota, M. 1971. A Bryozoan Reef Found Near the Lowest Part of the Akiyoshi Lime-stone Group, Japan. *Bulletin of the Akiyoshi-dai Science Museum* 7: 57–63.

Sugimura, A. & Ota, M. 1980. Some Lower Permian Bryozoa from the Akiyoshi Limestone Group, Southwest Japan. *Bulletin of the Akiyoshi-dai Museum of Natural History* 15: 49–58, pls 3–5.

Trizna, V. B. 1958. Early Carboniferous Bryozoans of the Kuznetsk Basin. *Trudy Vsesojuznolgo Neftjanogo Nauchno-issledovatelyskogo Geologo-razvedochnogo Instituta (VNIGRI)* 122: 436p. [In Russian]

Trizna, V.B. & Klaucan, R.A. 1961. Bryozoans of the Artinskian Stage of the Ufimian Plate and Their Role in the Stratigraphy of this Stage in Urals. *Trudy VNIGRI, 179th ed., Microfauna of the USSR* 13: 331–453. [In Russian]

Ulrich, E.O. 1890. Palaeozoic Bryozoa. In J. Lindahl (ed.) *Geology and Palaeontology Volume III*: 283–678. Springfield: Geological Survey of Illinois.

Waagen, W. & Wentzel, I. 1886. Salt Range Fossils. Pt. Coelenterata. *Memoir of the Geological Survey of India, Palaeontologica Indica Series* 13: 835–924.

Weyer, D. 1984. Korallen im Paläozoikum von Thüringen. *Hallesches Jahrbuch fur Geowissenschaften* 9: 5–33.

Xia, F. 1986. Carboniferous and early Permian bryozoans from Xainza, northern Xizang. *Bulletin of the Nanjing Institution of Geology and Palaeontology, Academia Sinica* 10(10): 201–254.

Yang, J. & Lu, L. 1962. Paleozoic Bryozoans of Gilianshan. *Geology of Gilianshan* 4: 1–114 [In Chinese]

Yang, K.C. 1958. Stenoporidae from Upper Palaeozoic of China. *Acta Palaeontologica Sinica* 6(2): 122–139.

Yang, K. C., Hu, Z. X. & Xia, F. 1988. Bryozoans from Late Devonian and Early Carboniferous of Central Hunan. *Palaeontologia Sinica* 174: 197p.

Young, J. 1883. On Ure's "Millepore," *Tabulipora* (*Cellepora urii*), Fleming. *Annals and Magazine of Natural History, ser. 5* 12: 154–158.

Bryozoan Studies 2022 – Key, Porter & Wyse Jackson (eds)
 ISBN: 978-1-032-59487-3

Colony repair strategies in large trepostome bryozoans from the Upper Ordovician (Katian) of the Cincinnati region, USA

Kate M. Runciman
Department of Earth Sciences, The College of Wooster, Wooster, USA

Mark A. Wilson*
Department of Earth Sciences, The College of Wooster, Wooster, USA
ORCID ID: 0000-0002-4651-0589

Caroline J. Buttler
Department of Natural Sciences, Amgueddfa Cymru – Museum Wales, Cardiff, Wales, UK
ORCID ID: 0000-0002-1577-926X

Shelley A. Judge
Department of Earth Sciences, The College of Wooster, Wooster, USA

ABSTRACT: Calcitic skeletons of trepostome bryozoans from the Upper Ordovician (Katian) of the Cincinnati region (Kentucky and Indiana, USA) record diverse interactions with encrusting and boring organisms. Trepostomes from four units (Bellevue Member, Corryville Formation, Bull Fork Formation, Whitewater Formation) were sectioned and analyzed. Many trace fossils were found, with *Trypanites* borings and bioclaustrations the most common. The borings often contain calcite "ghosts" and were likely excavated mechanically by an organism like a sipunculan or phoronid. Some borings prompted growth reactions in their host bryozoan. Growth reactions sealed the cavity and regained feeding surfaces by: (1) zooids growing laterally over the cavity opening, sealing it with a "roof"; and (2) zooids budding into the cavity then angling upwards. Other bioclaustrated features include cornulitids and a possible sphenothallid.

1 INTRODUCTION

Fossil bryozoans provide a record of the environment they inhabited and the interactions they had with other organisms, thus providing insights into past ecosystems (Ernst 2020; Taylor 2020). In this study, trepostome bryozoans from the Upper Ordovician (Katian) of the Cincinnati region and their internal trace fossils are described and analyzed. The specimens examined contain an abundance of growth responses and trace fossils that indicate the interactions and relationships between trepostomes and other organisms in their environment (see Wyse Jackson & Key 2019). The shape, composition, and nature of these features were here microscopically examined. Evidence of a variety of symbiotic organisms as well as several styles of growth response were found. These growth patterns have implications for understanding trepostome growth and symbiotic processes.

*Corresponding Author: mwilson@wooster.edu

2 STRATIGRAPHY

In this project, trepostome bryozoan samples from four Ordovician stratigraphic units were examined (Table 1). These units are the Bellevue Member of the Grant Lake Formation, the Corryville Formation, the Bull Fork Formation, and the Whitewater Formation, all of which are found within the Cincinnatian Series (Upper Ordovician) of the Cincinnati Arch region (midwestern USA). The Cincinnatian Series is exposed in southwestern Ohio, southeastern Indiana and north-central Kentucky (Figure 1). The Bellevue Member, Corryville Formation, Bull Fork Formation, and Whitewater Formation are fossiliferous and contain abundant trepostome bryozoans (Brett *et al.* 2019; Singh 1979; Utgaard & Perry 1964).

The Cincinnatian Series was deposited during the Katian stage of the Upper Ordovician. At this time, the Cincinnati region was located at a latitude approximately 20-23°S in the southern subtropics and was covered by a shallow epicontinental sea.

The Cincinnati Arch area contained shallow subtidal to peritidal environments, with the subtidal facies found in the north of its modern-day position and the peritidal to the south (Brett *et al.* 2019).

DOI: 10.1201/9781003454915-12

Figure 1. Locality map. Bellevue Member samples (C/W-152) originate from (A), Corryville Formation samples (NMW) from (B), Bull Fork Formation samples (C/W-153) from (C), and Upper Whitewater Formation samples (C/W-148) from (D). Map projection: WGS 84 / UTM zone 16N. Created using QGIS software and data from the National Atlas of the United States (2014a,b) and United States Census Bureau (2018).

3 MATERIALS AND METHODS

Five large colonies (greater than 5 cm in diameter) from the College of Wooster collection, and one from the NMW – Amgueddfa Cymru – National Museum Wales, Cardiff, UK were cut and examined for their growth patterns and evidence of interaction with other organisms. These colonies were collected from the Bellevue Member, Corryville Formation, Bull Fork Formation, and Whitewater Formation at four localities (Table 1, Figure 1). The colonies are all massive trepostomes except for sample C/W-148-2, which is a bifoliate trepostome that overgrew a ramose colony. Five genera and six species were represented in this study (Table 1). It is interesting to note that bioclaustrations such as those studied here are far more common in massive zoaria than in smaller erect forms.

Acetate peels were prepared from trepostome specimens to examine their internal morphology and trace fossils. Peels were prepared in accordance with Wilson & Palmer (1989) and were made using thin acetate (sold as 0.003 inches thick = 0.076 mm).

Acetate peels were examined microscopically using a Nikon Eclipse E400 POL petrographic microscope at various magnifications. Areas of interest were photographed using an Infinity 5 Teledyne Lumenera camera and Infinity Analyze software. Where peels had particular features of

Table 1. Specimens used in this study, along with stratigraphic context and sampling locations.

Colony ID	Formation and Locality	Species
C/W-152-1	Bellevue Member; Boone County, Kentucky (N39.08121°, W84.79230°)	*Amplexopora robusta*
C/W-152-2	Bellevue Member; Boone County, Kentucky (N39.08121°, W84.79230°)	*Amplexopora* (?) *filiasa*
C/W-153-1	Bull Fork Formation; Mason County, Kentucky (N38.58518°, W83.70157°)	*Dekayia stidhami*
C/W-148-1	Upper Whitewater Formation; Wayne County, Indiana (N39.78631°, W84.90318°)	*Monticulipora epidermata*
C/W-148-2	Upper Whitewater Formation; Wayne County, Indiana (N39.78631°, W84.90318°)	*Peronopora pachymura*
NMW 2017.9G.2.2	Corryville Formation; Mason County, Kentucky (N38.609352°, W83.810973°)	*Stigmatella personata*

interest, their associated specimen was mailed to Quality Thin Sections (Phoenix, Arizona), where they were made into thin sections. These thin sections were then microscopically examined in the same manner as the peels.

Repository and institutional abbreviations: C/W – College of Wooster, Ohio, USA; NMW – Amgueddfa Cymru – National Museum Wales, Cardiff, UK.

4 RESULTS

4.1 *Growth patterns*

Within bryozoan colonies, changes in growth were observed in conjunction with damage or disturbance to the calcite skeleton. One such growth pattern involves the colony growing laterally over a disturbance such as a boring, sealing it within the colony skeleton (Figure 2A, C). This process essentially roofs over the disturbance. These "roofing structures" were the most common reaction to damage found in this study and were observed in samples C/W-148-1 (*Monticulipora epidermata*), C/W-152-2 (*Amplexopora* (?) *filiasa*), and C/W-153-1 (*Dekayia stidhami*). These roofing structures are also apparent in a cave-dwelling colony of *Stigmatella personata* (Figure 2C) described by Buttler & Wilson (2018, fig. 7(1)).

A similar growth structure is formed by zooids budding into a cavity, then angling upwards to resume normal growth (Figure 2B). Unlike roofing structures, this type of growth fills in the cavity. Examples of this structure were observed in samples C/W-152-1 (*Amplexopora robusta*) and C/W-152-2 (*Amplexopora* (?) *filiasa*).

4.2 *Borings*

In this examination of Ordovician trepostomes, many macroborings (Wilson 2007) were observed (Figure 3). The most common borings were *Trypanites*, which were seen in every sample. Additionally, several branched cylindrical borings were found in sample C/W-153-1 (*Dekayia stidhami*). A wide range of boring sizes were present in all bryozoan taxa. Though it is difficult to quantify

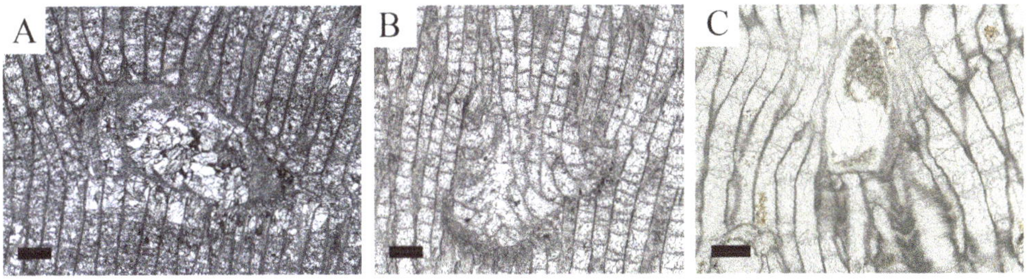

Figure 2. Trepostome growth patterns. A, Zooids grow laterally over the cavity opening, sealing it off with a flat "roof," C/W-152-2 (*Amplexopora* (?) *filiasa*). B, Zooids bud down into the cavity then angle upwards, C/W-152-1 (*Amplexopora robusta*). C, Zooids forming a "roof" with the cavity closed off in a bryozoan growing upside-down from a cave ceiling, NMW 2017.9G.2.2 (*Stigmatella personata*). Scale bars = 0.50 mm.

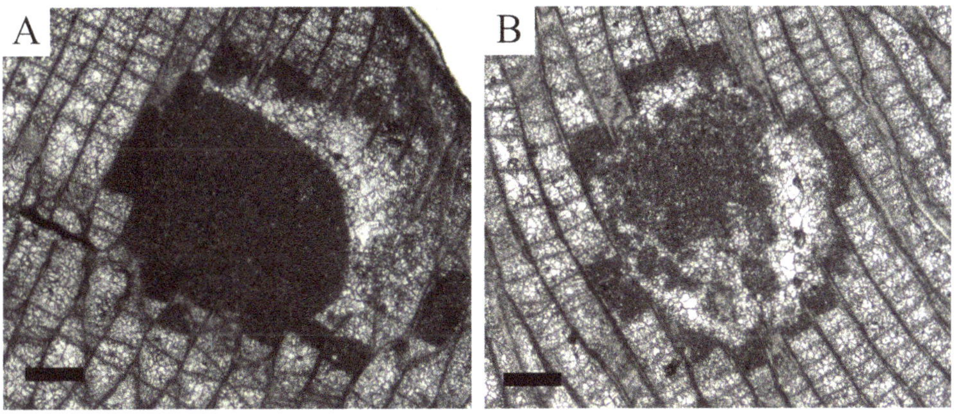

Figure 3. Borings in trepostomes. A, Cylindrical boring (likely (*Trypanites*) cutting through the skeleton, with sediment and sparry calcite infilling; C/W-153-1 (*Dekayia stidhami*). B, Cylindrical boring (likely *Trypanites*) also cutting through the skeleton; peloidal sediment and sparry calcite infilling; C/W-153-1 (*Dekayia stidhami*). Scale bars = 0.50 mm.

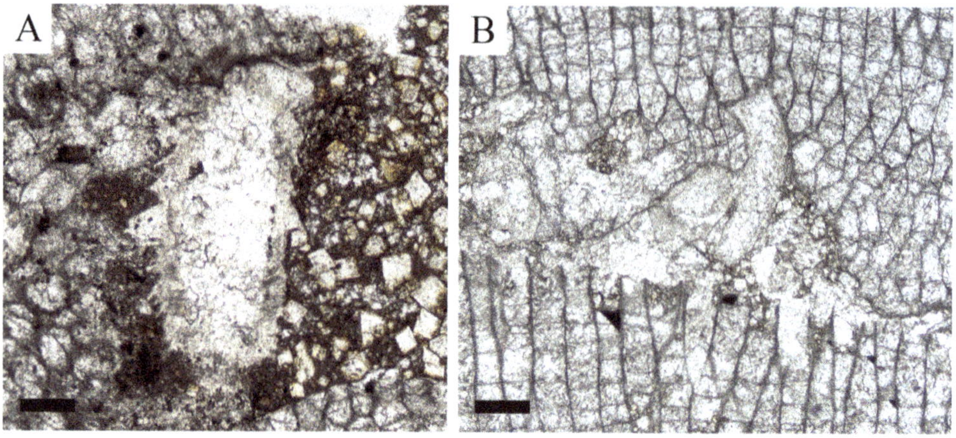

Figure 4. Bioclaustrations contained within trepostome colonies. A, jagged-edged bioclaustration, C/W-148-1 (*Monticulipora epidermata*). B, bioclaustrated tube and holdfast, C/W-152-2 (*Amplexopora* (?) *filiasa*). Scale bars = 0.50 mm.

boring size due to the restrictions of a planar cut through a non-planar boring, the borings appeared to vary from 200 μm to 9000 μm in diameter.

Some borings were associated with changes in bryozoan growth, others were not. Many borings cut roughly through the bryozoan skeleton, with no associated change in zooecial growth. These borings were rimmed by roughly broken zooecial walls, with the boring cavity ending at zooecial diaphragms. Conversely, other borings were associated with changes in growth. These changes include the structures described in section 4.1. Additionally, some borings were encompassed by terminal diaphragms and basal laminae that appear to have developed in response to the boring activity. These structures were accompanied by thickened zooecial walls near the boring.

4.3 *Bioclaustrations*

Jagged-edged bioclaustrations (Palmer & Wilson 1988) were found in sample CW-148-1. These features have relatively thick, sharp irregular walls and their insides are filled with sparry calcite (Figure 4A). They occurred in conjunction with changes in bryozoan growth, including terminal diaphragms, thickened zooecial walls, and roofing structures.

Another feature found in this study was a tube-shaped bioclaustration. Only one example of this structure was observed, found in sample C/W-152-2. It is a holdfast and tube structure, which comprises a flat-topped tube emerging from a small cavity in the colony skeleton (Figure 4B).

4.4 *Infilling materials*

The borings and other disturbances were all infilled. A number of different infilling materials were observed. These materials were sparry calcite cement (all samples), dolomite rhombs (all samples), biosparite (samples C/W-148-1 and C/W-148-2), micrite (all samples), and rarely calcium phosphate (samples C/W-152-2, C/W-153-1). Cavities in the skeleton were observed to contain any one of these materials, or a combination thereof.

Ghosts, or "sparry cement-filled voids within larger matrix-filled borings" (Wyse Jackson & Key 2007: 240) were abundant (Figure 3). Like the borings, the calcite ghosts were typically cylindrical with rounded ends. Some non-cylindrical ghost shapes were found, but these non-cylindrical shapes were unique and were not found repeated within or between colonies.

Sparry calcite was often found in conjunction with other infilling materials. These materials often surrounded calcite ghosts, filling in the space between the ghosts and the boring walls. Sediments also appeared within ghosts.

Phosphate was observed in samples C/W-152-2 and C/W-153-1. It appeared to fill zooid chambers immediately surrounding skeletal disturbances.

4.5 *Self-overgrowths and brown bodies*

Self-overgrowths were common in the trepostomes studied and were observed in every sample. These surfaces occurred where bryozoan growth halted and the zooecial chambers terminated, then the colony overgrew these surfaces (Figure 5A). The earlier growth and overgrowth are frequently separated by a thin layer of sediment.

Brown bodies (Morrison & Anstey 1979) were often seen in the zooecial chambers directly below self-overgrowths. Though brown bodies were typically found beneath self-overgrowths, they were not limited to these settings. The brown bodies were generally located in the centre of the zooecial chamber.

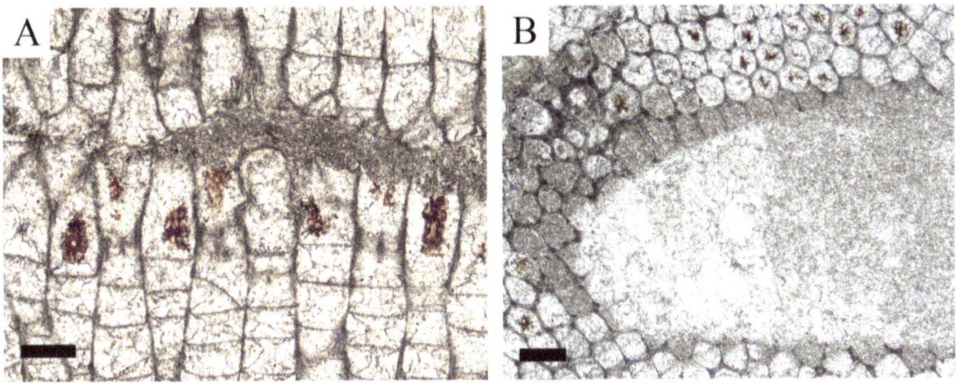

Figure 5. Brown bodies. A, below a self-overgrowth and thin layer of sediment, C/W-153-1 (*Dekayia stidhami*). B, brown bodies in transverse section, C/W-153-1 (*Dekayia stidhami*). Scale bars = 0.50 mm.

5 DISCUSSION

5.1 *Colony disturbance and damage control*

The occurrence of borings that left jagged margins through the trepostome colonies indicate that these borings were excavated mechanically rather than chemically. The lack of bryozoan response to this type of boring suggests that these borings, or at least the parts of the borings through which the colonies were sectioned, were excavated into an area without live zooids.

The presence of colony disturbances that were associated with changes in bryozoan growth suggest that these disturbances affected a living surface of the colony and prompted a growth reaction from the bryozoan. In each example observed, the growth response closed a cavity, allowing the colony to resume normal growth and regain feeding surfaces that had been lost to the disturbance. Except examples where zooids infilled the cavity entirely, the cavities were preserved in the bryozoan skeleton as bioclaustrations.

5.2 *Borings*

The most common ghost shape, cylindrical with rounded ends, indicates that the boring occupants were vermiform organisms. Wisshak *et al.* (2017) argued that a single entrance trace fossil was more likely to have been created by a worm with both the mouth and anus located at its anterior, rather than a worm with mouth and anus at opposing ends. This reasoning holds true for the *Trypanites* borings as well as the *Lapispecus* trace fossils studied by Wisshak *et al.* (2017). In both cases, "the single entrance is in good accordance with the body plan of either a sipunculan (the so-called "peanut worms"; > 100 extant species), or a phoronid (the so-called "horseshoe worms"; ~ 10 extant species)" (Wisshak *et al.* 2017: 173). These two taxa are possible creators of the single-entry *Trypanites* borings found in this study.

Though the majority of borings were unbranched *Trypanites*, several apparently branched borings were also observed. However, these borings likely represent intersecting unbranched borings. Intersecting borings have been observed in trepostome colonies in previous studies but have yet to be identified (Vinn *et al.* 2022).

5.3 *Bioclaustrations*

Some of the trepostome colonies contained jagged-edged bioclaustrations, which often prompted growth responses in the bryozoan. Similar features have been observed in Upper Ordovician corals and interpreted as cornulitids (Dixon 2010). The jagged-edged morphology supports the deduction that these bioclaustrations are of cornulitid with ridged growth increments.

The tube and holdfast structure was likely a small organism that attached to the bryozoan colony. The feature resembles examples from the conulariid genus *Sphenothallus*. The ecology of *Sphenothallus* is poorly known and there is considerable variation in size and proportions within this genus (Neal & Hannibal 2000; Vinn & Kirsimäe 2015). However, the general layout of the observed feature, with gently bending tube connected to a base shaped like an upside-down cup, is consistent with a sphenothallid tube and holdfast (Neal and Hannibal 2000; Peel 2021). This feature is oriented such that the widest part of the holdfast sits directly atop the bryozoan skeleton, which aligns with the orientation in which sphenothallids would encrust their substrates (Van Iten *et al.* 1992). Furthermore, *Sphenothallus* is known to occur in Ordovician strata of the Cincinnati Arch area (Bodenbender *et al.* 1989; Neal & Hannibal 2000). Due to these similarities, it is probable that this structure is a bioclaustration of a sphenothallid tube and holdfast.

5.4 *Sedimentation*

Given the presence of multiple varieties of sediment within individual disturbances, it appears that infilling occurred in several steps (Figure 6). This episodic infilling process is interpreted to have begun with the deposition of sediment while the boring was occupied

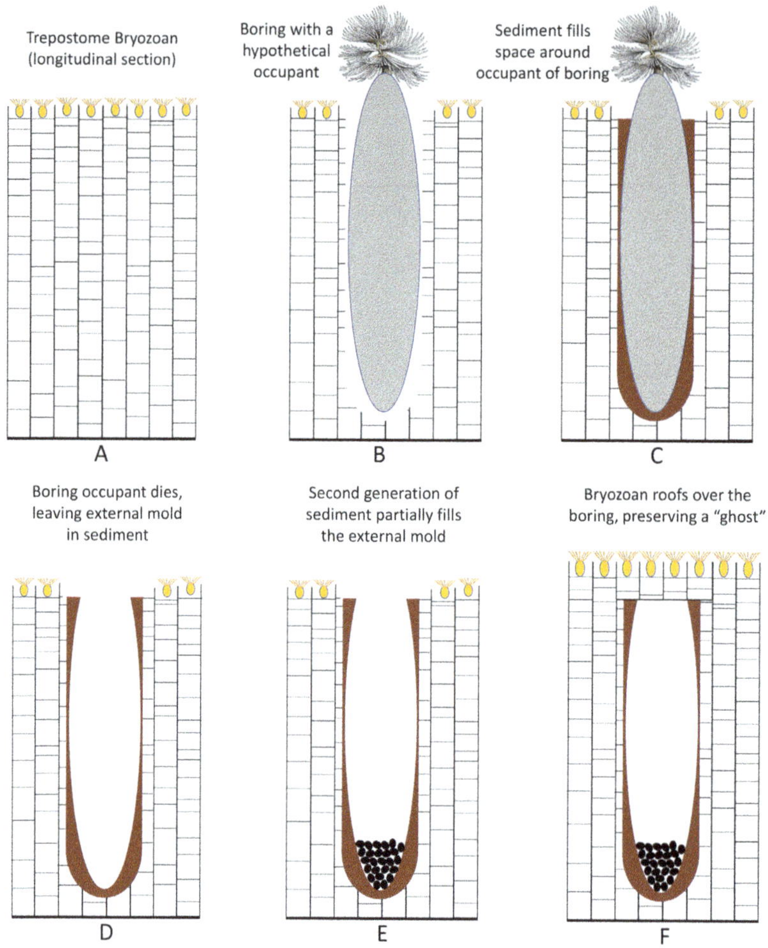

Figure 6. Diagrams illustrating the sedimentation process interpreted from borings with multiple generations of sediment.

(Figure 6C), filling the space around the occupant. Following this deposition, the occupant would die, leaving an external mold in that sediment (Figure 6D). A second generation of sediment could then enter that void and partially fill the mold (Figure 6E). Subsequently, the bryozoan could overgrow the boring (Figure 6F). Eventually, calcite cement could fill the bryozoan and boring, creating a ghost.

Multiple generations of sediment functioned as geopetal structures, as the second generation of sediment collected in the bottom of the boring at the time of deposition. Wyse Jackson & Key (2007) suggest that some trepostome colonies rolled on their substrate during storms. The geopetal structures support this interpretation of trepostome life modes, but the occurrence of these structures was too rare to provide conclusive evidence. Understanding of trepostome life modes could benefit from future investigation of geopetal borings.

The nature of the phosphate observed remains unknown. A variety of Ordovician taxa had phosphatic hard parts or precipitated phosphate (Lowenstam 1972). Therefore, it is possible that this material was organically derived, especially given its association with skeletal disturbances.

5.5 *Brown bodies*

The position of the brown bodies has implications for the cementation processes of the colonies. Had the calcite been deposited in the chambers or progressed laterally through the colonies, the brown bodies would most likely be situated against the chamber wall. However, this was not the case. The brown bodies are located in the centre of the zooecial chamber, which indicates that the calcite cement that now fills the colonies formed first at the zooecial walls and then grew inwards.

6 CONCLUSIONS

1. Examination of the interiors of large trepostome bryozoans from the Upper Ordovician (Katian)

of the Cincinnati region (USA) reveal information about the palaeoecology using bryozoan growth patterns, trace fossils, and sedimentation.
2. Boring and calcite bioclaustrations show a diverse fauna living on and within the colonies when the bryozoan was alive. This includes sipunculan or phoronid worms, cornulitids, and sphenothallids.
3. Self-overgrowths were common within the colonies and often associated with brown bodies. Brown bodies were located at the centre of the zooecial chamber, indicating that the calcite cement formed first at the skeletal walls and built inwards into the chambers.

ACKNOWLEDGEMENTS

This study was part of the senior author's Independent Study project at The College of Wooster. We thank the faculty and staff of the Department of Earth Sciences at Wooster for their encouragement and support. We are particularly grateful for the very helpful reviews by Catherine Reid, Juan Suárez Andrés, and Patrick Wyse Jackson.

REFERENCES

Bodenbender, B.E., Wilson, M.A. & Palmer, T.J. 1989. Paleoecology of *Sphenothallus* on an Upper Ordovician Hardground. *Lethaia* 22: 217–225.

Brett, C.E., Aucoin, C.D., Dattilo, B.F., Freeman, R.L., Hartshorn, K.R., McLaughlin, P.I. & Schwalbach, C. E. 2019. Revised Sequence Stratigraphy of the Upper Katian Stage (Cincinnatian) Strata in the Cincinnati Arch Reference Area: Geological and Paleontological Implications. *Palaeogeography, Palaeoclimatology, Palaeoecology* 540: 1–33.

Buttler, C.J. & Wilson, M.A. 2018. Paleoecology of an Upper Ordovician Submarine Cave-dwelling Bryozoan Fauna and its Exposed Equivalents in Northern Kentucky, USA. *Journal of Paleontology* 92: 568–576.

Dixon, O.A. 2010. Endobiotic Cornulitids in Upper Ordovician Tabulate Corals and Stromatoporoids from Anticosti Island, Quebec. *Journal of Paleontology* 84: 518–528.

Ernst, A., 2020. Fossil Record and Evolution of Bryozoa. *Handbook of Zoology. Bryozoa*. Berlin: De Gruyter, p.11–55.

Lowenstam, H.A. 1972. Phosphatic Hard Tissues of Marine Invertebrates: Their Nature and Mechanical Function, and Some Fossil Implications. *Chemical Geology* 9: 153–166.

Morrison, S.J. & Anstey, R.L. 1979. Ultrastructure and Composition of Brown Bodies in Some Ordovician Trepostome Bryozoans. *Journal of Paleontology* 53: 943–949.

National Atlas of the United States, 2014a, *Cities and Towns of the United States:* https://geodata.lib.berkeley.edu/catalog/stanford-bx729wr3020.

National Atlas of the United States, 2014b, *Global Map: 1:1,000,000 Major Roads, United States*: https://geodata.lib.berkeley.edu/catalog/stanford-tq511pr5084.

Neal, M.L. & Hannibal, J.T. 2000. Paleoecologic and Taxonomic Implications of *Sphenothallus* and *Sphenothallus*-like specimens from Ohio and areas adjacent to Ohio. *Journal of Paleontology* 74: 369–380.

Palmer, T.J. & Wilson, M.A. 1988. Parasitism of Ordovician Bryozoans and the Origin of Pseudoborings. *Palaeontology* 31: 939–949.

Peel, J.S. 2021. Holdfasts of *Sphenothallus* (Cnidaria) from the Early Silurian of Western North Greenland (Laurentia). *GFF* 143: 384–389.

Singh, R.J. 1979. Trepostomatous Bryozoan Fauna from the Bellevue Limestone, Upper Ordovician, in the Tri-state area of Ohio, Indiana and Kentucky. *Bulletins of American Paleontology* 76: 162–280.

Taylor, P.D. 2020. *Bryozoan paleobiology*. Hoboken, New Jersey: John Wiley & Sons.

United States Census Bureau, 2018, *Cartographic boundary files – Shapefile*: https://www.census.gov/geographies/mapping-files/time-series/geo/carto-boundary-file.html.

Utgaard, J. & Perry, T.G. 1964. Trepostomatous Bryozoan Fauna of the Upper Part of the Whitewater Formation (Cincinnatian) of Eastern Indiana and western Ohio. *Indiana Department of Conservation Geological Survey Bulletin* 33: 1–111.

Van Iten, H., Cox, R.S. & Mapes, R.H. 1992. New Data on the Morphology of *Sphenothallus* Hall: implications for its affinities. *Lethaia* 25: 135–144.

Vinn, O., Ernst, A., Wilson, M.A. & Toom, U. 2022. Borings and Bioclaustrations in Bryozoans from the Kunda Regional Stage (Darriwilian; Middle Ordovician) of northern Estonia and NW Russia. *Neues Jahrbuch für Geologie und Paläontologie* 303: 219–225.

Vinn, O. & Kirsimäe, K. 2015. Alleged Cnidarian *Sphenothallus* in the Late Ordovician of Baltica, its Mineral Composition and Microstructure. *Acta Palaeontologica Polonica* 60: 1001–1008.

Wilson, M.A. 2007. Macroborings and the Evolution of Marine Bioerosion. In W. Miller, III (ed), *Trace Fossils: Concepts, Problems, Prospects*: 356–367. Amsterdam: Elsevier.

Wilson, M.A. & Palmer, T.J. 1989. Preparation of Acetate Peels. *The Paleontological Society Special Publications* 4: 142–145.

Wisshak, M., Titschack, J., Kahl, W. & Girod, P. 2017. Classical and New Bioerosion Trace Fossils in Cretaceous Belemnite Guards Characterised via Micro-CT. *Fossil Record* 20: 173–199.

Wyse Jackson, P.N. & Key, M.M., Jr. 2007. Borings in Trepostome Bryozoans from the Ordovician of Estonia: Two Ichnogenera Produced by a Single Maker, a Case of Host Morphology Control. *Lethaia* 40: 237–252.

Wyse Jackson, P.N. & Key, M.M., Jr. 2019. Epizoan and Endoskeletozoan Distribution Across Reassembled Ramose Stenolaemate Bryozoan Zoaria from the Upper Ordovician (Katian) of the Cincinnati Arch region, USA. *Australasian Palaeontological Memoirs* 52: 169–178.

Bryozoan Studies 2022 – Key, Porter & Wyse Jackson (eds)

Distribution patterns of shelf bryozoans around southern Aotearoa New Zealand

Abigail M. Smith*
Department of Marine Science, University of Otago, Dunedin, New Zealand
ORCID ID: 0000-0001-6468-9124

Katerina Achilleos
Department of Biochemistry, University of Otago, Dunedin, New Zealand
ORCID ID: 0000-0002-9732-9683

Dennis P. Gordon
National Institute of Water and Atmospheric Research, Kilbirnie, Wellington, New Zealand.
ORCID ID: 0000-0001-9398-996X

ABSTRACT: The area of southern Aotearoa New Zealand, from Fiordland, Foveaux Strait and Otago southwards to Snares Platform is a well-known bryozoan hotspot, with fewer species known from the shelves around Auckland and Campbell Islands further south. Over 350 bryozoan species have been reported from this region, principally from monographs in the 1980s, augmented by a shelf dredging programme conducted by the University of Otago. We have, to date, 158 University of Otago and 71 NIWA stations in the region from 44°S to 53°S latitude, 165°E to 172°E longitude, in water depths ranging from 6 to 1003 m. A list of 359 bryozoan species was collated from 229 stations, based on published and unpublished distribution data. The 43 commonest species ($\geq$ 15 occurrences) were mapped using ArcGIS® Pro and grouped into four common distribution patterns. Accurate distribution information allows for active conservation of these small but important sessile invertebrates, via habitat protection and management of bottom disturbance.

1 INTRODUCTION

Beginning in the 1960s, intensive collections of benthic organisms were made around New Zealand by the former New Zealand Oceanographic Institute, later subsumed into the National Institute for Water & Atmospheric Research (NIWA), and augmented by studies by University of Otago scientists. They show that southern Aotearoa New Zealand (Fiordland, Foveaux Strait and Otago shelf to Snares Platform) is a recognized bryozoan hotspot. The chief sources of taxonomic data are the monographs by Gordon (1986, 1989), which covered the entire western South Island sea area from intertidal to abyssal depths, recognizing 325 species (13 Ctenostomata, 132 anascan-grade Cheilostomata, 180 ascophoran-grade Cheilostomata). Of the 141 stations cited in the Gordon (1986, 1989) monographs, 14 had more than 50 bryozoan species, and two over 100; most of these bryo-diverse stations were south of 45°S. Additional published works that also included the southern region (e.g. Gordon & Taylor 2001; Taylor *et al.* 1989, 2004; Smith & Gordon 2011), plus unpublished distributional records, have added more taxa, including of Cyclostomata, for a tally of over 350 species.

Nelson *et al.* (1988) reported on the extensive skeletal carbonates blanketing the Snares Platform and extending northward past Stewart Island up to the Otago shelf. While bryozoans formed a major component of all these carbonates (from 30% by weight in Otago to 20% by weight in Foveaux Strait), it was on the Snares Platform that they really dominated, producing >50% of the skeletal carbonate present. Concentrated patchy bryozoan habitat, called 'thickets' were described on the Otago Shelf in water depths from 60 to 90 m (Batson & Probert 2000; Probert *et al.* 1979; Wood *et al.* 2013).

Sediment grab samples and occasional dredgings indicate abundant, diverse and widespread bryozoan colonies thriving on the continental shelves around southern New Zealand, but the details are sketchy. More samples with careful species identification of bryozoans present are needed to understand habitat distribution and

*Corresponding Author: abby.smith@otago.ac.nz

DOI: 10.1201/9781003454915-13

biogeography, and help to answer questions such as: are there identifiably distinct faunas in particular biotopes? And are the bryozoans still thriving on these shelves? Achilleos *et al.* (2019) carried out a biogeographic analysis of the genus *Cellaria* in southern Zealandia, but here we cast the net more broadly, documenting the distribution of the commonest species found around southern Aotearoa New Zealand based on a multi-year bottom-sampling programme collated with known literature.

2 MATERIAL AND METHODS

Between 2008 and 2022, 241 bottom samples were collected by the University of Otago Dredging Programme (UO DP) using a beam trawl deployed from the RV *Polaris II*. The dredging programme includes, to date, 14 different cruises around Aotearoa New Zealand with the majority (158) of stations located south of 44°S (Figure 1, Supplementary Data Table 1). Upon collection, samples were briefly described on board and both

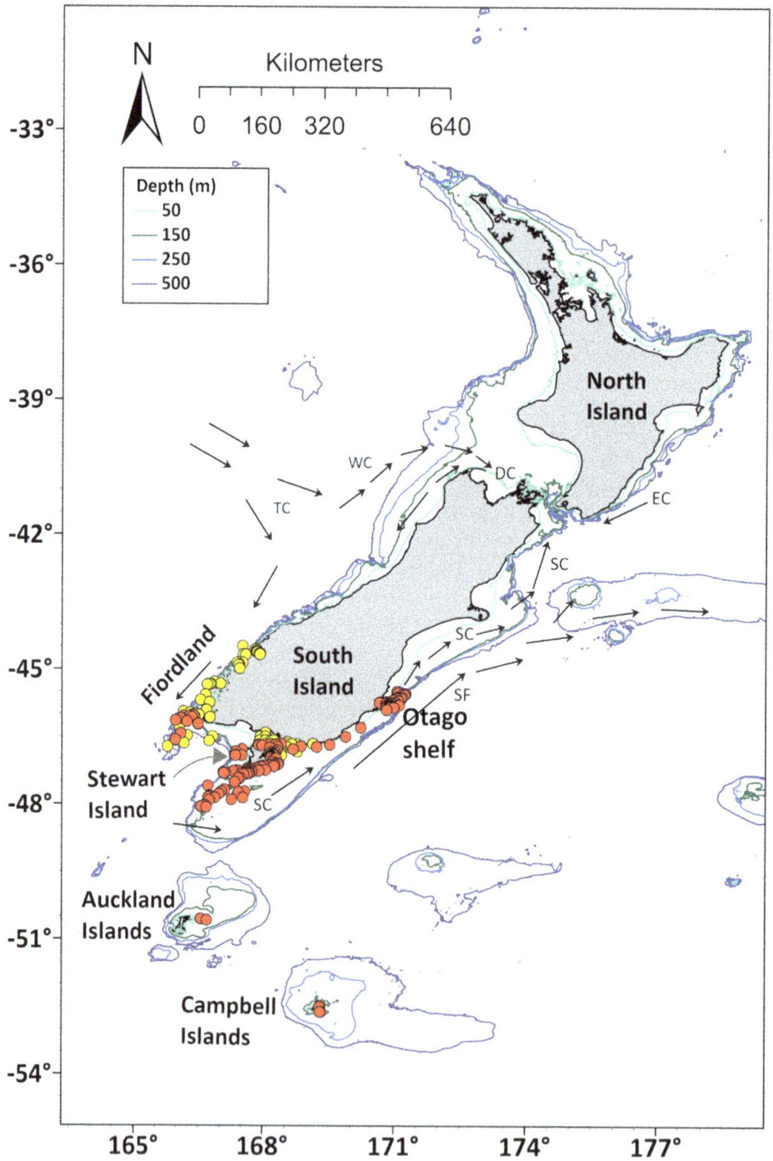

Figure 1. Sampling stations (N = 229) around southern Aotearoa New Zealand. Red circles represent the University of Otago Dredging Programme (2008-2022, N = 158). Yellow circles are the NIWA sample locations (1961 to present, N = 71). EC, Eastern current; SC, Southland Current; SF, Subtropical Front; TC, Tasman Current; WC, Westland Current.

Table 1. The 359 bryozoan species found in dredges from Southern New Zealand, with counts of occurrence at dredge stations.

Order	Species	Stations
Ctenostomatida	*Alcyonidium* cf. *mytili* (Dalyell 1848)	5
Ctenostomatida	*Alcyonidium* spp.	15
Ctenostomatida	*Amathia wilsoni* Kirkpatrick, 1888	1
Ctenostomatida	*Cryptopolyzoon wilsoni* (Dendy 1889)	2
Ctenostomatida	*Elzerina binderi* (Harvey in Busk 1861)	0
Ctenostomatida	*Immergentia* sp.	1
Ctenostomatida	*Penetrantia irregularis* Silén, 1956	11
Ctenostomatida	*Penetrantia parva* Silén, 1946	16
Ctenostomatida	*Triticella periphanta* Marcus, 1921	1
Cheilostomatida	*Adeonellopsis gemina* Liow & Gordon, 2020	3
Cheilostomatida	*Adeonellopsis macewindui* Liow & Gordon, 2020	2
Cheilostomatida	*Adeonellopsis* sp. (*gemina* or *macewindui*)	22
Cheilostomatida	*Aetea* n. sp. cf. *australis* Jullien, 1888	15
Cheilostomatida	*Aetea* sp.	6
Cheilostomatida	*Aetea truncata* (Landsborough 1852)	9
Cheilostomatida	*Aimulosia marsupium* (MacGillivray 1869)	11
Cheilostomatida	*Akatopora circumsaepta* (Uttley 1951)	27
Cheilostomatida	*Alderina gorensis* (Gordon 1986)	6
Cheilostomatida	*Alderina nodulosa* (Uttley 1949)	8
Cheilostomatida	*Allerescha dubia* Gordon 1989	1
Cheilostomatida	*Amastigia fiordica* Gordon 1986	1
Cheilostomatida	*Amastigia funiculata* (MacGillivray 1886)	2
Cheilostomatida	*Amastigia magna* Gordon 1986	1
Cheilostomatida	*Amastigia puysegurensis* Gordon 1986	2
Cheilostomatida	*Amphiblestrum alcimum* Gordon 1984	2
Cheilostomatida	*Amphiblestrum blandum* Gordon 1986	6
Cheilostomatida	*Antarctothoa bathamae* (Ryland & Gordon 1977)	1
Cheilostomatida	*Antarctothoa quadralabia* (Uttley & Bullivant 1972)	2
Cheilostomatida	*Antarctothoa* sp. *delta*? (Ryland & Gordon 1977)	6
Cheilostomatida	*Antarctothoa tongima* (Ryland & Gordon 1977)	17
Cheilostomatida	*Aplousina anxiosa* Gordon, 1986	1
Cheilostomatida	*Arachnopusia perforata* (Maplestone 1909)	1
Cheilostomatida	*Arachnopusia unicornis* (Hutton 1873)	34
Cheilostomatida	*Beania bilaminata* (Hincks, 1881)	9
Cheilostomatida	*Beania cryptophragma* Gordon 1986	9
Cheilostomatida	*Beania elongata* (Hincks 1885)	4
Cheilostomatida	*Beania intermedia* (Hincks 1881)	4
Cheilostomatida	*Beania micheleae* n.sp. Nascimento *et al.* in prep	15
Cheilostomatida	*Beania* n. sp. cf. *cribrimorpha* Gordon, 1984	7
Cheilostomatida	*Beania plurispinosa* Uttley & Bullivant 1972	8
Cheilostomatida	*Beania serrata* Souto, Nascimento, Reverter-Gil & Vieira 2018	7
Cheilostomatida	*Beania spinigera* (Hincks 1885)	10
Cheilostomatida	*Beania stonycha* Gordon, 1986	5
Cheilostomatida	*Beania trampida* Gordon, 1986	1
Cheilostomatida	*Bicellariella ciliata* (Linnaeus 1758)	1
Cheilostomatida	*Bitectipora mucronifera* (Powell 1967)	1

(*continued*)

Table 1. Continued

Order	Species	Stations
Cheilostomatida	*Bitectipora ozalea* (Gordon 1989)	1
Cheilostomatida	*Bitectipora rostrata* (MacGillivray 1887)	9
Cheilostomatida	*Buffonella improvisa* (Uttley & Bullivant 1972)	1
Cheilostomatida	*Buffonella rhomboidalis* Canu & Bassler 1920	2
Cheilostomatida	*Buffonellaria turbula* Gordon, 1989	14
Cheilostomatida	*Bugula* n. sp.	1
Cheilostomatida	*Bugulopsis monotrypa* (Busk 1852)	10
Cheilostomatida	*Caberea boryi* (Audouin 1826)	1
Cheilostomatida	*Caberea* cf. *darwinii guntheri* Hastings, 1943	1
Cheilostomatida	*Caberea helicina* Hastings, 1943	11
Cheilostomatida	*Caberea rostrata* Busk, 1884	4
Cheilostomatida	*Caberea solida* Gordon, 1986	10
Cheilostomatida	*Caberea* sp.	6
Cheilostomatida	*Caberea zelandica* (Gray 1843)	11
Cheilostomatida	*Caleschara denticulata* (MacGillivray 1869)	1
Cheilostomatida	*Calloporina angustipora* (Hincks 1885)	16
Cheilostomatida	*Calwellia gracilis* Maplestone, 1882	6
Cheilostomatida	*Camptoplites asymmetricus* Hastings, 1943	1
Cheilostomatida	*Canda filifera* (Lamarck 1816)	9
Cheilostomatida	*Carbasea indivisa* Busk, 1852	2
Cheilostomatida	*Catenicella* cf. *constans* (Powell 1967)	1
Cheilostomatida	*Catenicella elegans* Busk, 1852	9
Cheilostomatida	*Catenicella* sp.	2
Cheilostomatida	*Caulibugula tuberosa* Hastings, 1939	1
Cheilostomatida	*Cellaria immersa* (Tenison-Woods 1880)	28
Cheilostomatida	*Cellaria pilosa* (Kirchenpauer 1869)	1
Cheilostomatida	*Cellaria* sp.	2
Cheilostomatida	*Cellaria spatulifera?* Achilleos, Gordon & Smith 2020	1
Cheilostomatida	*Cellaria tenuirostris* (Busk 1852)	14
Cheilostomatida	*Celleporaria agglutinans* (Hutton 1873)	27
Cheilostomatida	*Celleporaria emancipata* Gordon, 1989	18
Cheilostomatida	*Celleporaria* sp. nov.	3
Cheilostomatida	*Celleporina grandis* Gordon, 1989	13
Cheilostomatida	*Celleporina hemiperistomata* (Godon 1984)	4
Cheilostomatida	*Celleporina proximalis* (Uttley & Bullivant 1972)	10
Cheilostomatida	*Celleporina sinuata* Gordon, 1989	15
Cheilostomatida	*Celleporina* sp.	1
Cheilostomatida	*Celleporina tubulata* (Uttley & Bullivant 1972)	7
Cheilostomatida	*Chaperia granulosa* Gordon, 1986	18
Cheilostomatida	*Chaperia* n. sp.	5
Cheilostomatida	*Chaperiopsis (Clipeochaperia) chathamensis* Uttley & Bullivant, 1972	2
Cheilostomatida	*Chaperiopsis (Clipeochaperia) funda* Uttley & Bullivant, 1972	11
Cheilostomatida	*Chaperiopsis* cf. *cervicornis* (Busk 1854)	20
Cheilostomatida	*Chaperiopsis incognita* Gordon, 1986	1
Cheilostomatida	*Chaperiopsis lanceola* Hayward & Thorpe, 1988	15
Cheilostomatida	*Chaperiopsis serrata biporosa* (Uttley & Bullivant 1972)	4

(*continued*)

Table 1. Continued

Order	Species	Stations
Cheilostomatida	*Chaperiopsis* sp.	8
Cheilostomatida	*Chaperiopsis spiculata* Uttley, 1949	6
Cheilostomatida	*Chaperiopsis splendida* Gordon, 1986	2
Cheilostomatida	*Chaperiopsis tintinnabula* Hayward & Thorpe, 1988	2
Cheilostomatida	*Chiastosella* cf. *longaevitas* Powell, 1967	1
Cheilostomatida	*Chiastosella* cf. *umbonata* Gordon, 1989	9
Cheilostomatida	*Chiastosella* cf. *watersi* Stach, 1937	17
Cheilostomatida	*Chiastosella dissidens* Gordon, 1989	1
Cheilostomatida	*Chiastosella duplicata* Gordon, 1989	4
Cheilostomatida	*Chiastosella enigma* Brown, 1954	6
Cheilostomatida	*Chiastosella exuberans* Gordon, 1989	2
Cheilostomatida	*Chorizopora* cf. *brongniartii* (Audouin 1826)	5
Cheilostomatida	*Chorizopora* n. sp.	1
Cheilostomatida	*Chronocerastes otakauensis* (Brown 1952)	1
Cheilostomatida	*Corbulella corbula* (Hincks 1880)	16
Cheilostomatida	*Corbulella fossa* (Uttley 1951)	3
Cheilostomatida	*Cornucopina* n. sp.	1
Cheilostomatida	*Cornucopina pectogemma* (Goldstein 1882)	1
Cheilostomatida	*Cornucopina salutans* Gordon, 1986	1
Cheilostomatida	*Cornuticella taurina* (Busk 18520	10
Cheilostomatida	*Cosciniopsis vallata* (Uttley & Bullivant 1972)	15
Cheilostomatida	*Costaticella bicuspis* (Gray 1843)	8
Cheilostomatida	*Costaticella solida* (Levinsen 1909)	6
Cheilostomatida	*Crassimarginatella cucullata* (Waters 1898)	6
Cheilostomatida	*Crassimarginatella* sp.	7
Cheilostomatida	*Crepidacantha crinispina* Levinsen, 1909	27
Cheilostomatida	*Crepidacantha zelanica* Canu & Bassler, 1929	4
Cheilostomatida	*Cribellopora napi* Gordon, 1989	5
Cheilostomatida	*Cribellopora siri* Gordon, 1989	2
Cheilostomatida	*Cribralaria austrinsulensis* Gordon, 1989	1
Cheilostomatida	*Cribricellina cribraria* (Busk 1852)	8
Cheilostomatida	*Crisia tenuis* MacGillivray, 1879	5
Cheilostomatida	*Cryptostomaria crassatina* Canu & Bassler, 1927	1
Cheilostomatida	*Cyclicopora* sp.	1
Cheilostomatida	*Didymosella conchicola* Gordon, 1989	1
Cheilostomatida	*Dimetopia barbata* (Lamouroux 1816)	5
Cheilostomatida	*Dimetopia cornuta* Busk, 1852	7
Cheilostomatida	*Dimetopia* sp. nov.	3
Cheilostomatida	*Dittomesia crispa* Gordon, 1989	1
Cheilostomatida	*Ellisina sericea* (MacGillivray 1890)	8
Cheilostomatida	*Ellisina* sp.	1
Cheilostomatida	*Emma cervicornis* (MacGillivray 1869)	1
Cheilostomatida	*Emma rotunda* Hastings, 1939	2
Cheilostomatida	*Emma triangula* Hastings, 1939	8
Cheilostomatida	*Escharella spinosissima* (Hincks 1881)	11
Cheilostomatida	*Escharoides angela* (Hutton 1873)	8

(continued)

Table 1. Continued

Order	Species	Stations
Cheilostomatida	*Escharoides excavata* (MacGillivray 1860)	8
Cheilostomatida	*Escharoides* sp. nov.	1
Cheilostomatida	*Eurystomella biperforata* Gordon, Mawatari & Kajihara 2002	12
Cheilostomatida	*Euthyroides episcopalis* (Busk 1852)	13
Cheilostomatida	*Euthyroides jellyae* Levinsen, 1909	2
Cheilostomatida	*Exochella armata* (Hincks 1882)	5
Cheilostomatida	*Exochella conjuncta* Brown, 1952	2
Cheilostomatida	*Exochella jullieni* Gordon, 1989	9
Cheilostomatida	*Exochella levinseni* Gordon, 1989	7
Cheilostomatida	*Exochella* sp.	4
Cheilostomatida	*Fasciculipora fruticosa* MacGillvray, 1884	1
Cheilostomatida	*Fasciculipora* sp.	1
Cheilostomatida	*Fenestrulina disjuncta* (Hincks 1885)	1
Cheilostomatida	*Fenestrulina gelasinoides* Gordon, 1984	5
Cheilostomatida	*Fenestrulina incompta* Gordon, 1984	5
Cheilostomatida	*Fenestrulina multicava* Gordon, 1989	1
Cheilostomatida	*Fenestrulina personata* (MacGillivray,1883)	1
Cheilostomatida	*Fenestrulina reticulata* Powell, 1967	5
Cheilostomatida	*Fenestrulina* sp.	11
Cheilostomatida	*Fenestrulina specca* Gordon, 1989	3
Cheilostomatida	*Fenestrulina thyreophora* (Busk 1857)	1
Cheilostomatida	*Fenestulipora cassiformis* (Harmer, 1915)	1
Cheilostomatida	*Figularia carinata* (Waters 1923)	1
Cheilostomatida	*Figularia mernae* Uttley & Bullivant 1972	6
Cheilostomatida	*Foveolaria elliptica* Busk, 1884	3
Cheilostomatida	*Fovoporella spectabilis* (Hincks 1891)	15
Cheilostomatida	*Galeopsis adherens* Gordon, 1989	1
Cheilostomatida	*Galeopsis pentagonus* (d'Orbigny 1842)	1
Cheilostomatida	*Galeopsis polyporus* (Brown 1952)	25
Cheilostomatida	*Galeopsis porcellanicus* (Hutton 1873)	10
Cheilostomatida	*Gregarinidra* cf. *inarmata*	7
Cheilostomatida	*Gregarinidra* cf. *serrata* (MacGillivray 1869)	8
Cheilostomatida	*Gregarinidra* sp.	1
Cheilostomatida	*Hemismittoidea hexaspinosa* (Uttley & Bullivant 1972)	1
Cheilostomatida	*Hiantopora jucunda* Gordon, 1984	5
Cheilostomatida	*Hippadenella clivosa* (Waters 1906)	2
Cheilostomatida	*Hippellozoon novaezelandiae* (Waters 1895)	16
Cheilostomatida	*Hippomenella vellicata* (Hutton 1873)	23
Cheilostomatida	*Hippomonavella flexuosa* (Hutton 1873)	2
Cheilostomatida	*Hippoporina retepora* Gordon, 1989	1
Cheilostomatida	*Hippothoa catophilia* Gordon, 2020	1
Cheilostomatida	*Hippothoa pacifica* Gordon, 1984	3
Cheilostomatida	*Hippothoa watersi* Morris, 1980	7
Cheilostomatida	*Julianca retia* Gordon, 1989	2
Cheilostomatida	*Kenella aliena* Gordon, 1986	1
Cheilostomatida	*Klugerella bifurca* (Powell 1967)	3

(*continued*)

Table 1. Continued

Order	Species	Stations
Cheilostomatida	*Lacerna styphelia* (Gordon 1989)	2
Cheilostomatida	*Leptinatella gordoni* Cook & Bock, 2000	3
Cheilostomatida	*Macrocamera pansa* (Gordon 1989)	1
Cheilostomatida	*Macropora nodulosa* Gordon & Taylor 2008	15
Cheilostomatida	*Maiabuffonella madrecilla* (Gordon 1989)	5
Cheilostomatida	*Malakosaria sinclairii* (Busk 1858)	1
Cheilostomatida	*Margaretta barbata* (Lamarck 1816)	7
Cheilostomatida	*Melicerita chathamensis* Uttley & Bullivant 1972	9
Cheilostomatida	*Menipea vectifera* Harmer, 1923	4
Cheilostomatida	*Menipea vera* Gordon, 1986	4
Cheilostomatida	*Metroperiella* sp.	1
Cheilostomatida	*Micropora gracilis* (Uttley 1949)	5
Cheilostomatida	*Micropora mortenseni* Livingstone, 1929	6
Cheilostomatida	*Micropora* n. sp. cf. *inarmata* Soule, 1959	2
Cheilostomatida	*Micropora* n. sp. cf. *elegans* Maplestone, 1901	8
Cheilostomatida	*Micropora* sp.	18
Cheilostomatida	*Micropora variperforata* Waters, 1887	13
Cheilostomatida	*Microporella* aff. *diademata* (Lamouroux 1825)	7
Cheilostomatida	*Microporella agonistes* Gordon, 1984	31
Cheilostomatida	*Microporella* cf. *discors* Uttley & Bullivant, 1972	5
Cheilostomatida	*Microporella* cf. *intermedia* Livingstone, 1929	7
Cheilostomatida	*Microporella* cf. *marsupiata* (Busk 1860)	2
Cheilostomatida	*Microporella lingulata* Di Martino *et al.*, 2020	10
Cheilostomatida	*Microporella ordo* Brown, 1952	1
Cheilostomatida	*Mobunula bicuspis* (Hincks 1883)	3
Cheilostomatida	*Mollia amoena* Gordon, 1986	1
Cheilostomatida	*Monoporella* cf. *nodulifera* (Hincks 1881)	1
Cheilostomatida	*Mucropetraliella ligulata* Stach, 1936	1
Cheilostomatida	*Mucropetraliella* sp.	1
Cheilostomatida	*Nimba verrucosa* Gordon, 1989	1
Cheilostomatida	*Odontionella cyclops* (Busk 1854)	32
Cheilostomatida	*Odontionella* sp.	2
Cheilostomatida	*Odontionella tessellata* (Hutton 1873)	1
Cheilostomatida	*Odontoporella adpressa* (Busk 1854)	2
Cheilostomatida	*Onchoporoides moseleyi* (Busk 1884)	1
Cheilostomatida	*Opaeophora lepida* (Hincks 1881)	22
Cheilostomatida	*Opaeophora monopia* (Brown 1952)	6
Cheilostomatida	*Orthoscuticella ?fusiformis* (Powell 1967)	3
Cheilostomatida	*Orthoscuticella fissurata* (Levinsen 1909)	11
Cheilostomatida	*Orthoscuticella innominata* Gordon, 1989	7
Cheilostomatida	*Orthoscuticella margaritacea* (Busk 1852)	3
Cheilostomatida	*Orthoscuticella ventricosa* (Busk 1852)	1
Cheilostomatida	*Osthimosia amplexa* Gordon, 1989	3
Cheilostomatida	*Osthimosia avicularis* Gordon, 1989	7
Cheilostomatida	*Osthimosia monilifera* Taylor, Schembri & Cook, 1989	5
Cheilostomatida	*Osthimosia mysterium* Gordon, 1989	1

(*continued*)

Table 1. Continued

Order	Species	Stations
Cheilostomatida	*Osthimosia socialis* Taylor, Schembri & Cook, 1989	9
Cheilostomatida	*Osthimosia turrita* Gordon, 1989	1
Cheilostomatida	*Otionellina affinis* (Cook & Chimonides 1984)	1
Cheilostomatida	*Otionellina* sp.	7
Cheilostomatida	*Otionellina squamosa* (Tenison-Woods 1880)	1
Cheilostomatida	*Otionellina symmetrica* (Cook & Chimonides 1984)	6
Cheilostomatida	*Otionellina zelandica* (Cook & Chimonides 1984)	1
Cheilostomatida	*Paracribricellina cribraria* (Busk 1852)	12
Cheilostomatida	*Parasmittina aotea* (Brown 1952)	12
Cheilostomatida	*Parasmittina delicatula* (Busk 1884)	6
Cheilostomatida	*Parkermavella curvata* (Uttley & Bullivant 1972)	6
Cheilostomatida	*Parkermavella incurvata* (Uttley & Bullivant 1972)	3
Cheilostomatida	*Parkermavella microrhyncha* (Marcus 1921)	1
Cheilostomatida	*Parkermavella punctigera* (MacGillivray 1883)	11
Cheilostomatida	*Parkermavella* sp. nov.	3
Cheilostomatida	*Parkermavella virago* (Gordon 1989)	2
Cheilostomatida	*Patsyella acanthodes* Gordon 1982	5
Cheilostomatida	*Petalostegus bicornis* (Busk 1884)	1
Cheilostomatida	*Phidolopora avicularis* (MacGillivray 1883)	1
Cheilostomatida	*Phonicosia* aff. *circinata* (MacGillivray 1869)	11
Cheilostomatida	*Phonicosia oviseparata* (Brown 1952)	1
Cheilostomatida	*Phonicosia* sp.	1
Cheilostomatida	*Phonicosia styphelia* Gordon, 1989	1
Cheilostomatida	*Prenantia dichotoma* Gordon, 1989	1
Cheilostomatida	*Prenantia firmata* (Waters 1887)	6
Cheilostomatida	*Pterocella scutella* (Hutton 1891)	19
Cheilostomatida	*Reteporella aurantium* Gordon, 2009	1
Cheilostomatida	*Reteporella constricta* (Powell 1967)	2
Cheilostomatida	*Reteporella gracilis* Gordon, 1989	1
Cheilostomatida	*Reteporella ligulata* Gordon, 1989	9
Cheilostomatida	*Reteporella* sp.	6
Cheilostomatida	*Reteporella syrtoxylon* Gordon, 1989	1
Cheilostomatida	*Reteporellina conservatrix* Gordon, 1989	2
Cheilostomatida	*Reteporellina sagitta* Gordon, 1989	1
Cheilostomatida	*Retevirgula acuta* (Hincks 1885)	5
Cheilostomatida	*Retevirgula sejuncta* (MacGillivray 1891)	2
Cheilostomatida	*Rhabdozoum wilsoni* Hincks, 1882	3
Cheilostomatida	*Rhynchozoon paa* Uttley & Bullivant, 1972	2
Cheilostomatida	*Rhynchozoon* sp.	1
Cheilostomatida	*Rhynchozoon zealandicum* Gordon, 2009	10
Cheilostomatida	*Rogicka biserialis* (Hincks 1885)	1
Cheilostomatida	*Scalicella crystallina* (Wyville Thomson 1858)	3
Cheilostomatida	*Schizomavella* cf. *aotearoa* (Gordon 1989)	2
Cheilostomatida	*Schizosmittina cinctipora* (Hincks 1883)	11
Cheilostomatida	*Schizosmittina conjuncta* (Uttley & Bullivant 1972)	1
Cheilostomatida	*Schizosmittina melanobater* Gordon, 1989	10

(*continued*)

Table 1. Continued

Order	Species	Stations
Cheilostomatida	*Schizosmittina* sp.	2
Cheilostomatida	*Sclerodomus gracilis* Gordon, 1988	1
Cheilostomatida	*Scruparia ambigua* (d'Orbigny 1841)	9
Cheilostomatida	*Scrupocaberea ornithorhyncus* (Wyville Thomson 1858)	8
Cheilostomatida	*Selenaria* cf. *varians* Cook & Chimonides 1987	1
Cheilostomatida	*Smittina palisada* Gordon, 1989	5
Cheilostomatida	*Smittina personata* (Hincks 1884)	2
Cheilostomatida	*Smittina purpurea* (Hincks 1881)	3
Cheilostomatida	*Smittina rosacea* Powell, 1967	3
Cheilostomatida	*Smittina torques* Powell, 1967	2
Cheilostomatida	*Smittoidea discoidea* Canu & Bassler, 1923	1
Cheilostomatida	*Smittoidea discoveriae* Powell, 1967	5
Cheilostomatida	*Smittoidea hyalina* Gordon, 1984	2
Cheilostomatida	*Smittoidea magna* Gordon, 1984	1
Cheilostomatida	*Smittoidea maunganuiensis* (Waters 1906)	16
Cheilostomatida	*Smittoidea zealandiae* (Brown 1952)	1
Cheilostomatida	*Steginoporella magnifica* Harmer, 1900	22
Cheilostomatida	*Steginoporella* sp.	4
Cheilostomatida	*Stephanollona scintillans* (Hincks 1885)	14
Cheilostomatida	*Talivittaticella problematica* (d'Hondt 1981)	1
Cheilostomatida	*Taylorus alatus* Perez *et al.* 2021	4
Cheilostomatida	*Taylorus arcuatus* Gordon, 2014	1
Cheilostomatida	*Tricellaria aculeata* (d'Orbigny 1842)	6
Cheilostomatida	*Valdemunitella fraudatrix* Gordon, 1986	2
Cheilostomatida	*Valdemunitella hara* Gordon, 1986	6
Cheilostomatida	*Valdemunitella huttoni* (Brown 1952)	5
Cheilostomatida	*Valdemunitella pyrula* (Hincks 1881)	1
Cheilostomatida	*Valdemunitella* sp.	12
Cheilostomatida	*Valdemunitella* sp.	17
Cheilostomatida	*Valdemunitella valdemunita* (Hincks 1885)	3
Cheilostomatida	*Villicharixa strigosa* (Uttley 1951)	1
Cheilostomatida	*Watersipora subatra* (Ortmann 1890)	2
Cyclostomatida	*Bicrisia biciliata* (MacGillivray 1869)	2
Cyclostomatida	*Bicrisia edwardsiana* (d'Orbigny 1841)	6
Cyclostomatida	*Borgiola* sp.	7
Cyclostomatida	*Cinctipora elegans* Hutton, 1873	53
Cyclostomatida	*Crisia* cf. *acropora* Busk, 1875	2
Cyclostomatida	*Crisia setosa* MacGillivray, 1869	4
Cyclostomatida	*Diaperoecia purpurascens* (Hutton 1873)	21
Cyclostomatida	*Diaperoforma* sp.	1
Cyclostomatida	*Disporella* n. sp.	5
Cyclostomatida	*Disporella* n. sp. 'purple'	10
Cyclostomatida	*Disporella novaehollandiae* (d'Orbigny 1853)	11
Cyclostomatida	*Disporella pristis* (MacGillivray 1884)	25
Cyclostomatida	*Doliocoitis cyanea* Gordon & Taylor, 2001	15
Cyclostomatida	*Entalophoroecia* sp.	4

(*continued*)

Table 1. Continued

Order	Species	Stations
Cyclostomatida	*Eximonea* sp.	2
Cyclostomatida	*Favosipora otagoensis* (Taylor, Schembri & Cook 1989)	1
Cyclostomatida	*Favosipora rosea* Gordon & Taylor, 2001	18
Cyclostomatida	*Filifascigera* n. sp.	2
Cyclostomatida	*Filifascigera pluripora* Canu & Bassler, 1929	1
Cyclostomatida	*Filifascigera* sp.	2
Cyclostomatida	*Hastingsia* sp.	1
Cyclostomatida	*Hemipustulopora* sp.	1
Cyclostomatida	*Hornera* cf. *caespitosa* Busk, 1875	7
Cyclostomatida	*Hornera foliacea* (MacGillivray 1869)	32
Cyclostomatida	*Hornera* n. sp. (delicate)	17
Cyclostomatida	*Hornera robusta* MacGillivray 1883	56
Cyclostomatida	*Idmidronea* sp. ('crassa')	9
Cyclostomatida	*Idmidronea* sp. ('narrow')	1
Cyclostomatida	*Idmidronea* sp. (green)	5
Cyclostomatida	*Liripora* n. sp. 'subcolonies'	1
Cyclostomatida	*Liripora pseudosarniensis* Taylor & Gordon, 2001	4
Cyclostomatida	*Microeciella* cf. *ridleyi* (Borg 1944)	1
Cyclostomatida	*Microeciella* sp.	1
Cyclostomatida	*Nevianipora* sp.	1
Cyclostomatida	*Oncousoecia* sp. (*Microeciella*?)	1
Cyclostomatida	*Plagioecia sarniensis* (Norman 1864)	2
Cyclostomatida	*Plagioecia* sp.	5
Cyclostomatida	*Platonea* or *Idmidronea* sp.	2
Cyclostomatida	*Reptotubigera* sp.	3
Cyclostomatida	*Telopora dentata* (Hutton 1880)	12
Cyclostomatida	*Telopora lobata* (Tenison-Woods 1880)	41
Cyclostomatida	*Tervia* sp.	3
Cyclostomatida	*Tetrocycloecia neozelanica* (Busk 1879)	9
Cyclostomatida	*Tetrocycloecia parapelliculata* (Taylor, Schembri & Cook 1989)	1
Cyclostomatida	*Tetrocycloecia* sp.	6
Cyclostomatida	*Tubulipora anderssoni* Borg, 1926	1
Cyclostomatida	*Tubulipora* sp. "lata"	4
Cyclostomatida	*Tubulipora* sp.	17
Cyclostomatida	*Tubulipora* sp. "alta"	10
Cyclostomatida	*Tubulipora* sp. "epizoica"	1
Cyclostomatida	*Tubulipora* sp. "parallela"	1

living and dead specimens were collected; in addition, bulk reference samples were preserved in 99% molecular-grade ethanol. Additional bryozoan samples were retained dry. Species identification was carried out at the Portobello Marine Laboratory after each expedition and by later visitor researchers who worked with the material. Samples were stored at Portobello Marine Laboratory, Dunedin, Aotearoa New Zealand.

A matrix of species and stations was created based on the work of Achilleos *et al.* (2019, 2020), and updated to include all species found in southern New Zealand (Supplementary Data Table 2). Stations north of 44 °S and those with no bryozoans were discarded, so that 122 Otago stations with bryozoans remained for analysis. From these, 43 species (or species complexes) were identified as common, occurring at more than 15 stations. For these species, presence data from 71 NIWA stations south of 44° S were also included (Gordon 1986, 1989). In total, 229 stations were included in the final dataset (see Supplementary Data Table 2).

Table 2. Distribution categories for 42 bryozoan species around southern Aotearoa New Zealand. Numbers in parentheses indicate the number of stations at which each species was found.

Category	Description	Ctenostomes	Cheilostomes	Cyclostomes
Widespread	Widespread: Everywhere in the study area, from Otago to Fiordland, including (sometimes) the subantarctic islands		*Cellaria immersa* (28) *Chaperiopsis lanceola* (15) *Galeopsis polyporus* (25) *Odontionella cyclops* (32) *Opaeophora lepida* (22) *Steginoporella magnifica* (22)	*Hornera robusta* (56) *Tubulipora* sp. (17)
	Widespread except Fiordland: Otago to Puysegur but not in Fiordland		*Adeonellopsis* spp. (22) *Akatopora circumsaepta* (27) *Arachnopusia unicornis* (34) *Calloporina angustipora* (16) *Celleporaria agglutinans* (27) *Celleporina sinuata* (15) *Crepidacantha crinispina* (27) *Hippomenella vellicata* (23) *Microporella agonistes* (31)	*Cinctipora elegans* (58) *Diaperoecia purpurascens* (21) *Disporella pristis* (25) *Telopora lobata* (41)
Southwestern	Western: Fiordland, Puysegur, Foveaux, Stewart Island, Snares		*Aetea* n. sp. cf. *australis* (15) *Beania micheleae* (15) *Corbulella corbula* (16) *Smittoidea maunganuiensis* (16)	
	Southern: Foveaux, Snares, Puysegur	*Penetrantia parva* (16)	*Celleporaria emancipata* (18) *Chaperia granulosa* (18) *Chaperiopsis* cf. *cervicornis* (20) *Macropora nodulosa* (15)	*Hornera* n. sp. delicate (17)
	Foveaux: Foveaux Strait only		*Antarctothoa tongima* (17) *Fovoporella spectabilis* (15)	
Southeastern	Stewart Island, Foveaux Strait, Snares, Otago Shelf	*Alcyonidium* spp. (15)	*Chiastosella* cf. *watersi* (17) *Consciniopsis vallata* (15) *Hippellozoon novaezelandiae* (16) *Micropora* sp. (18) *Pterocella scutella* (21)	*Doliocoitus cyanea* (15) *Favosipora rosea* (18) *Hornera foliacea* (32)

Distributional data were mapped using ArcGIS® Pro. Broad regional water currents were added on the map by following Beentjes *et al.* (2002). Bathymetric data were sourced from https://www.niwa.co.nz/our-science/oceans/bathymetry/downloadthe-data?sid=16080.

The 43 species distributions were sorted into categories based on their presence and absence patterns and were further correlated to the collection effort bias for certain areas. For example, a single occurrence in an area with high collection effort (e.g., Otago shelf) was given lower weight.

3 RESULTS

In total, 359 species were identified from 229 stations (Supplementary Data Table 2). They included nine ctenostome species, 51 cyclostome species, and 299 cheilostome species (Table 1). The majority of stations (N = 156) had fewer than ten species (Figure 2A), a further 58 stations contained up to 30 species, and 15 stations had exceptional diversity, one (B226) reaching 96 species.

Most species occurred at fewer than five stations (Figure 2B). The commonest 43 species occurred at 15 or more stations and included three ctenostomes, ten cyclostomes, and 30 ascophoran cheilostomes. The three commonest species were cyclostomes: *Cinctipora elegans* (58 records), *Hornera robusta* (56 records), followed by *Telopora lobata* (41 records) (Table 1; see also Supplementary Figures).

Distribution patterns fell into six categories: Widespread (with and without Fiordland), Southwestern (Western, Southern, and Foveaux), and Southeastern, as shown in Table 2. The most common patterns were "Widespread except Fiordland" (9 cheilostomes and 4 cyclostomes) and "Widespread" (6 cheilostomes, 2 cyclostomes) (Table 2). Example

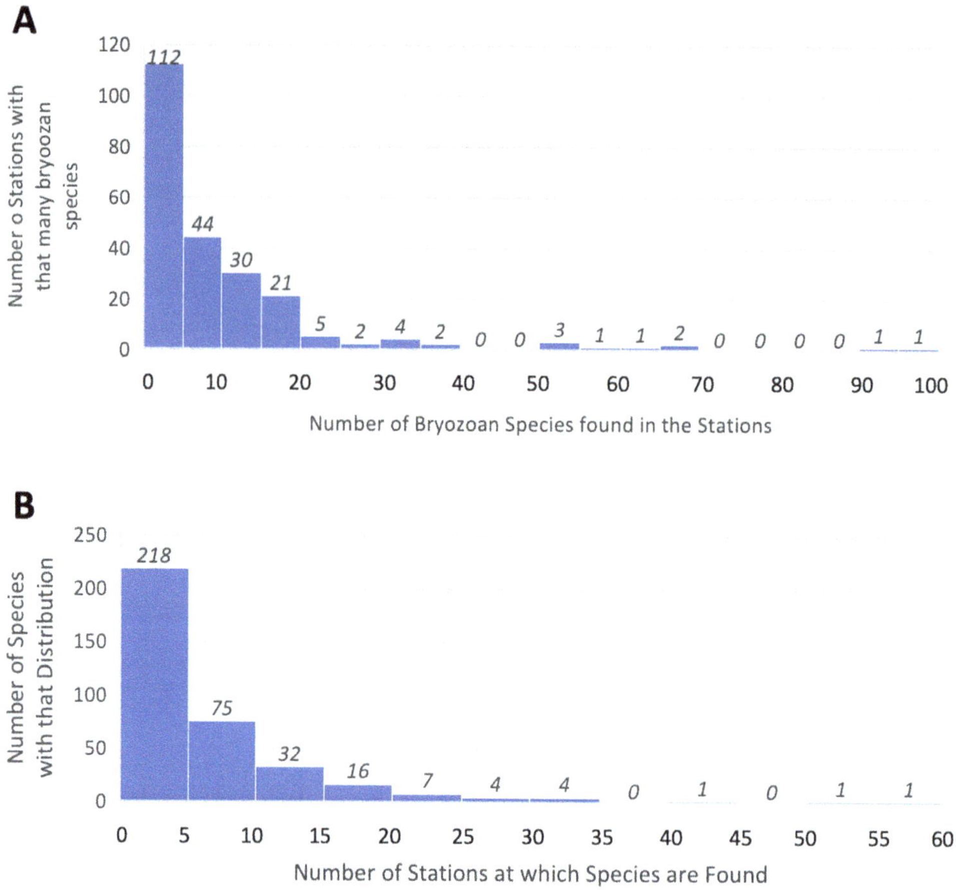

Figure 2. A, Frequency histogram of bryozoan species found at stations, showing that most stations have only a few bryozoan species, but that in a few places, there can be over 90 species. B, Frequency histogram of stations at which bryozoans are found, showing that many species are found only one or two places, while a few are very widespread. Data from Supplementary Data Table 2.

patterns are given in Figure 3, and all 43 maps appear in the Supplementary Data Figs.

4 DISCUSSION

Analysis of presence/absence data is notoriously problematic. While we can be sure about presence, there is much less certainty about absence. A species may not appear in a location because it happened not to be collected, or because it was not noticed or correctly identified. There is no metric here for relative abundance—'presence' can mean that a single specimen was found, or that a dredge was full of them.

Collection effort is also very variable among areas and specimens. For example, some areas were sampled more extensively and frequently (e.g., Otago Shelf) compared to others, inevitably adding a degree of bias to the distribution patterns. Further, the Gordon volumes (1986, 1989) do not include cyclostomes. The OU DP samples, on the other hand, often targeted cyclostomes, especially hornerids, therefore those species would be carefully identified and recorded. Early OU DP dredgings were not systematically sorted immediately after collection, though taxonomists did add some identification data later, using preserved subsamples. Later OU DP dredgings were carefully examined immediately after collection (by Gordon) as part of Oslo-based genetic work ongoing for New Zealand species. Many of the cheilostome species noted here were included in the New Zealand phylogeny of Orr *et al.* (2021).

4.1 *Distribution of southern bryozoans*

Among the commonest species, at least 15 can form large, erect heavily calcified colonies which are important habitat-formers and sediment-producers. Our species distributions are broadly

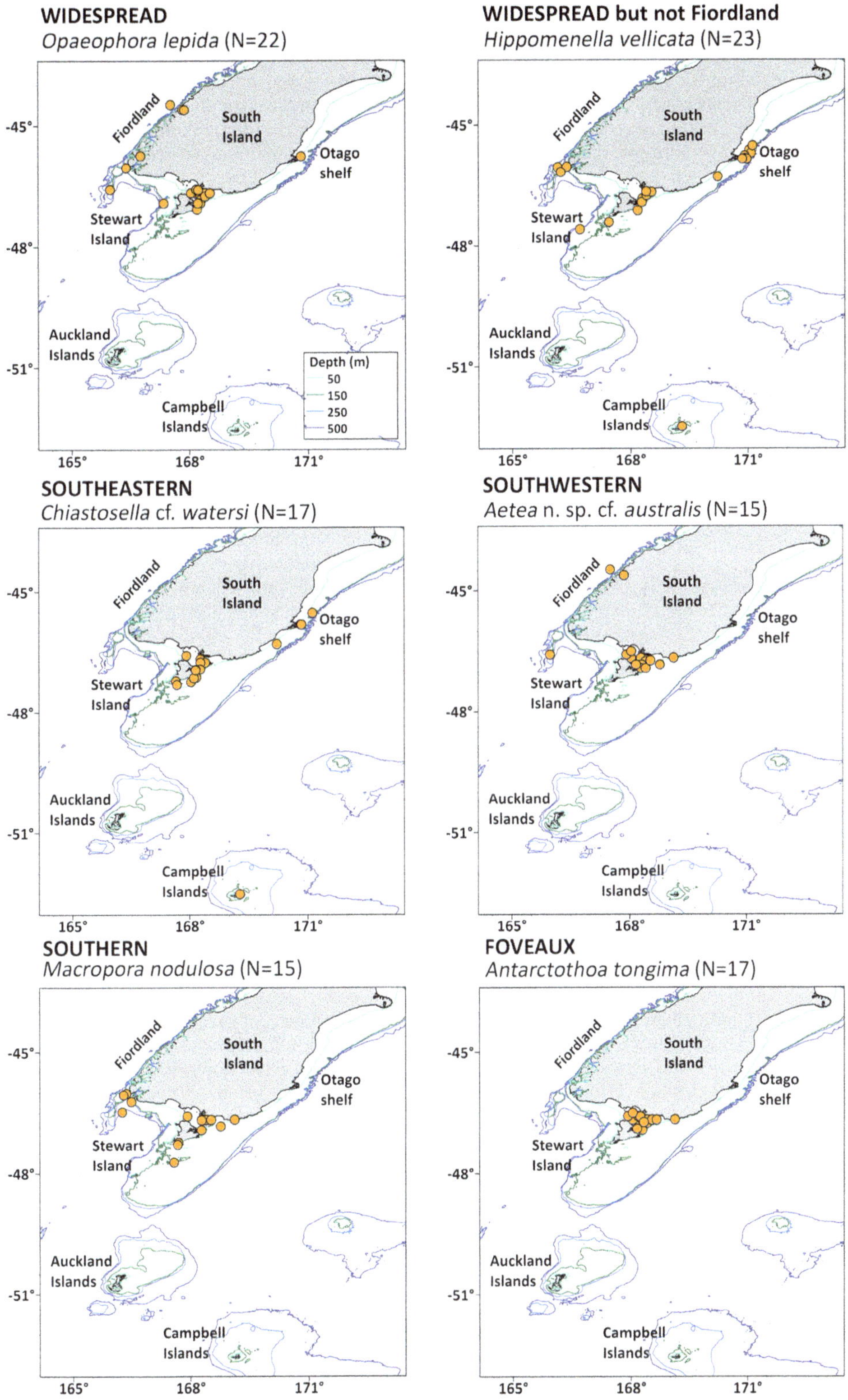

Figure 3. Examples of six of the distribution categories found in bryozoan species in southern Aotearoa New Zealand.

comparable with those produced by Wood *et al.* (2013), but there are a few differences. Based on their predictive model, *A. unicornis* and *C. grandis* should be abundant along the southeastern coast and Otago shelf respectively, but despite the area being one of the most frequently sampled locations, the species have been recorded seldom or not at all. *Cellaria immersa* and *C. tenuirostris* distributions present a few important differences as well, compared to that predicted by the model (see Achilleos *et al.* 2019). *Hornera foliacea* shows a strong southeastern distribution, but in fact the model did not predict any suitable areas of this species around New Zealand. These differences between the distribution patterns presented in this study and the prediction models of Wood *et al.* (2013) are likely due to the 'presence data' used to build the model. Therefore, publishing raw presence/absence data, such as those herein and in Achilleos *et al.* (2019) are important for constructing more accurate prediction models.

4.2 *Conservation of small sessile invertebrates*

The southern Zealandian shelf is clearly a bryozoan hotspot (Table 1), likely due to favourable conditions such as moderate to high energy (Achilleos *et al.* 2019), high nutrients and cool temperatures. Skeletal carbonate sediments blanket the entire region (Nelson *et al.* 1988), to depths of 1-2 km (Gorman, pers. comm., from unpublished boomer survey). These sediments, dominated by bryozoans and molluscs ("bryomol" sediments) suggest that there has been continuous and extensive bryozoan growth and calcification for many thousands of years.

Such productivity may no longer be the case. The experience of those dredging the seafloor is that most hauls comprise grey, dead, ground-up bryozoan fragments, with no sign of the living organisms that produced it. The dredge crews call this kind of sample 'kitty litter'—which is what it looks like. Only a small proportion of dredgings comprise living, intact bryozoans.

Does this mean that the southern shelves are no longer 'carbonate factories'? This concept postulates that there are a few highly productive areas that provide skeletal sediment for the whole shelf. Given current known growth rates of large erect shelf bryozoans (e.g., Smith 2014), a deposit 1500 m thick would take millions of years to produce. Conditions on the Snares Platform have changed greatly during the glacial-interglacial sea-level changes of the Holocene, and it seems unlikely that bryozoan production would be unaffected.

We could argue instead that bryozoans used to be much more productive, and something now is either killing them or slowing them down. The fact that all the most speciose stations were collected in the 1960s, while subsequent nearby dredges in the 2000s find much less, is suggestive. While we are familiar with dramatic changes in the marine coastal environment, the cold-water shelves of southern Aotearoa New Zealand are largely insulated from the coastal pollution, heatwaves, and acidification that can affect nearshore populations. It is unlikely that these factors have turned off the carbonate-production switch among bryozoans living 50-150 m deep.

On the other hand, southern Zealandian shelves are impacted by seafloor fishing practices. Continental shelves around the world are dredged, dragged or trawled every few months, in the search for protein (Amoroso *et al.* 2018). Wood *et al.* (2013) used government fisheries effort data to characterise fishing effort around Aotearoa New Zealand (see their fig. 15). Puysegur Bank, Snares Platform, and especially around Stewart Island/ Foveaux Strait are among the most regularly trawled by commercial fishers, to the complete destruction of the fragile, brittle, erect bryozoans living there (see e.g., Cranfield *et al.* 2003).

Southern-shelf bryozoans would benefit from prohibition of anthropogenic perturbations to seafloor biogenic habitat (Rowden *et al.* 2004). When conservation policy-makers seek advice on circumscribing areas for protection, it is distribution data such as ours that will inform the drawing of boundaries. Under current New Zealand fisheries and conservation legislation, it is nearly impossible to achieve a marine reserve in the dark remote depths of the Snares Platform. A mid-shelf no-take marine reserve has been proposed for the bryozoan meadows off Otago since 2014 (DoC 2020), but it remains undesignated.

Of course, we do not know if simply limiting fishing is sufficient intervention to enable bryozoan populations to recover. In some places, marine reserves have been effective protection for bryozoans (Mello *et al.* 2020), whereas in others, protection has been ineffective or even deleterious (Mello *et al.* 2021a, 2021b).

Conservation, particularly limiting bottom-disturbance, could restore the colourful textural and taxonomic diversity of well-developed bryozoan-dominated biogenic habitats. Failing that, we may have to rely solely on historical collections of spirit-preserved specimens and sediment for information on the once-thriving bryozoan carbonate-producers of southern Aotearoa New Zealand.

4.3 *Looking to the future*

Overall, we recognize that our distributional data present several challenges either as a result of collection bias or because of the limitations of a 'presence/absence' dataset. Nevertheless, the specimens collected during this project have already contributed to several publications (e.g., Achilleos *et al.* 2019; Batson *et al.* 2021; Orr *et al.* 2021, 2022; Tamberg & Smith 2020) and will contribute to and guide future efforts in developing conservation initiatives for these important sessile seafloor invertebrates.

SUPPLEMENTARY DATA

Two data tables and 43 distribution maps underpinning this report are publicly available on Mendeley Data at doi:10.17632/fmkggd9r6z.1

ACKNOWLEGEMENTS

We thank everyone on every cruise, especially *RV Polaris II* skipper Bill Dickson, and crew John Campbell, Phil Heseltine, Evan Kenton, and Mark Elder. We acknowledge the contributions of students and colleagues who braved the Southern Ocean to collect bryozoans (Clare Adams, Peter Batson, Carmen Battez, Rebecca Cumming, Christine Davis, Kirsten Donald, Tyler Feary, Sophie Fern, David Freeman, Chris Garden, Shannon Goodwin, Timo Guiarado, Zoe Henderson, Helen Jenkins, Hannah Mello, Chas Morris, Ania Piwoni-Piórewicz, Jo Porter, Marc Riedi, Thomas Schwaha, Yuta Tamberg, Kerry Walton, Anna Wood) and those who helped with identifications (Seaborne Rust, Andrea Waeschenbach).

The University of Otago Dredging Programme was financially supported throughout the years by the Department of Marine Science, Lee Hsiang Liow, Thomas Schwaha, Andrew Ostrovsky, Hamish Spencer. Data from NIWA were also provided by NIWA Invertebrate Collection manager Sadie Mills. The authors appreciate helpful comments from Marcus Key, Hans de Blauwe and an anonymous reviewer.

REFERENCES

Achilleos, K., Gordon, D.P. & Smith, A.M. 2020. *Cellaria* (Bryozoa, Cheilostomata) from the Deep: New Species from the Southern Zealandian Region. *Zootaxa* 4801: 201–236.

Achilleos, K.A., Smith, A.M. & Gordon, D.P. 2019. The Articulated Bryozoan Genus *Cellaria* in the Southern Zealandian Region: Distribution and Associated Fauna. *Marine Biodiversity* 49: 2801–2812.

Amoroso, R.O., Pitcher, C.R., Rijnsdorp, A.D., McConnaughey, R.A., Parma, A.M., Suuronen, P., Eigaard, O.R., Bastardie, F., Hintzen, N.T., Althaus, F., Baird, S.J., Black, J., Buhl-Mortensen, L., Campbell, A.B., Cetrino, R., Collie, J., Cowan, J.H. Jr, Durholtz, D., Engstrom, N., Fairweather, T.P., Fock, H.O., Ford, R., Gálvez, P.A., Gerritsen, H., Góngora, M.E., González, J.A., Hiddink, J.G., Hughes, K.M., Intelmann, S.S., Jenkins, C., Jonsson, P., Kainge, P., Kangas, M., Kathena, J.N., Kavadas, S., Leslie, R.W., Lewis, S.G., Lundy, M., Makin, D., Martin, J., Mazor, T., Gonzalez-Mirelis, G., Newman, S.J., Papadopoulou, N., Posen, P.E., Rochester, W., Russo, T., Sala, A., Semmens, J.M., Silva, C., Tsolos, A., Vanelslander, B., Wakefield, C.B., Wood, B.A., Hilborn, R., Kaiser, M.J. & Jennings, S. 2018. Bottom Trawl Fishing Footprints on the World's Continental Shelves. *Proceedings of the National Academy of Sciences* 115: E10275–E10282.

Batson, P.B. & Probert, P.K. 2000. Bryozoan Thickets off Otago Peninsula. *New Zealand Fisheries Assessment Report* 2000/46: 1–31.

Batson, P.B., Tamberg, Y., Gordon, D.P., Negrini, M. & Smith, A.M. 2021. *Hornera currieae* n. sp. (Cyclostomatida: Horneridae): A New Bathyal Cyclostome Bryozoan with Reproductively Induced Skeletal Plasticity. *Zootaxa* 5020: 257–287.

Beentjes, M.P., Bull, B., Hurst, R.J., & Bagley, N.W. 2002. Demersal Fish Assemblages Along the Continental Shelf and Upper Slope of the East Coast of the South Island, New Zealand. *New Zealand Journal of Marine and Freshwater Research* 36: 197–223.

Cranfield, H.J., Manighetti, B., Michael, K.P. & Hill, A. 2003. Effects of Oyster Dredging on the Distribution of Bryozoan Biogenic Reefs and Associated Sediments in Foveaux Strait, Southern New Zealand. *Continental Shelf Research* 23: 1337–1357.

DoC (Department of Conservation). 2020. *Proposed Southeast Marine Protected Areas: Consultation Document*. Wellington: New Zealand Department of Conservation.

Gordon, D.P. 1986. Marine Fauna of New Zealand: Bryozoa: Gymnolaemata (Ctenostomata and Cheilostomata Anasca) from the Western South Island Continental Shelf and Slope. *New Zealand Oceanographic Institute Memoir* 95: 1–121.

Gordon, D.P. 1989. Marine Fauna of New Zealand: Bryozoa: Gymnolaemata (Cheilostomida Ascophorina) from the Western South Island Continental Shelf and Slope. *New Zealand Oceanographic Institute Memoir* 97: 1–158.

Gordon, D.P. & Taylor, P.D. 2001. New Zealand Recent Densiporidae and Lichenoporidae (Bryozoa: Cyclostomata). *Species Diversity* 6: 243–290.

Mello, H.L., Smith, A.M., Wood, A.C.L. & Tidey, E.J. 2020. Enhanced Biodiversity and Abundance of Benthic Invertebrate Macrofauna in a New Zealand marine reserve. *Aquatic Conservation* 30: 1854–1867.

Mello, H.L., Smith, A.M. & Wood, A.C.L. 2021a. Voluntary Fishing Restrictions Alone do Not Promote Growth of Bryozoan-dominated Biogenic Habitat on the Otago Shelf, Southeastern New Zealand. *ICES Journal of Marine Science* 78: 1542–1553.

Mello, H.L., Gordon, D.P., Ryland, J.S. & Smith, A.M. 2021b. Protecting the Small: Preliminary Investigation of Bryozoan Community Change in New Zealand's oldest Marine Reserve. *New Zealand Journal of Marine and Freshwater Research*. https://doi.org/10.1080/00288330.2021.1990971

Nelson, C.S., Keane, S.L. & Head, P.S., 1988. Non-tropical Carbonate Deposits on the Modern New Zealand Shelf. *Sedimentary Geology* 60: 71–94.

Orr, R.J.S., Di Martino, E., Gordon, D.P., Ramsfjell, M. H., Mello, H.L., Smith, A.M. & Liow, L.H. 2021. A Broadly Resolved Molecular Phylogeny of New Zealand Cheilostome Bryozoans as a Framework for Hypotheses of Morphological Evolution. *Molecular Phylogenetics and Evolution* 161: 107172.

Orr, R.J.S., Di Martino, E., Ramsfjell, M., Gordon, D.P., Berning, B., Chowdhury, I., Craig, S., Cumming, R.L., Figuerola, B., Florence, W., Harmelin, J.-G., Hirose,

M., Huang, D., Jain, S.S., Jenkins, H.L., Kotenko, O. N., Kuklinski, P., Lee, H.E., Madurell, T., McCann, L., Mello, H.L., Obst, M., Ostrovsky, A.N., Paulay, G., Shunatova, N.N., Smith, A.M., Souto-Derungs, J., Vieira, L., Voje, K.L., Waeschenbach, A., Zágoršek, K., Warnock, R. & Liow, L.H. 2022. Paleozoic Origins of Cheilostome Bryozoans and their Parental Care Inferred by a New Genome-skimmed Phylogeny. *Science Advances* 8: eabm7452.

Probert, P.K., Batham, E.J. & Wilson, J.B. 1979. Epibenthic Macrofauna off Southeastern New Zealand and Mid-shelf Bryozoan Dominance. *New Zealand Journal of Marine and Freshwater Research* 13: 379–392.

Rowden, A.A., Warwick, R.M., & Gordon, D.P. 2004. Bryozoan Biodiversity in the New Zealand Region and Implications for Marine Conservation. *Biodiversity and Conservation* 13: 2695–2721.

Smith, A.M. 2014. Growth and Calcification of Bryozoans in a Changing Ocean. *Biological Bulletin* 226: 203–210.

Smith, A.M. & Gordon, D.P. 2011. Bryozoans of Southern New Zealand: a Field Identification Guide. *New Zealand Aquatic Environment and Biodiversity Report* 76: 1–64.

Tamberg, Y. & Smith, A.M. 2020. In Search of Predictive Models for Stenolaemate Morphometry Across the Skeletal-polypide Divide. *Paleobiology* 46: 218–236.

Taylor, P.D., Gordon, D.P. & Batson, P.B. 2004. Bathymetric Distributions of Modern Populations of Some Common Cenozoic Bryozoa from New Zealand, and Paleodepth Estimation. *New Zealand Journal of Geology & Geophysics* 47: 57–69.

Taylor, P.D., Schembri, P.J. & Cook, P.L. 1989. Symbiotic Associations Between Hermit Crabs and Bryozoans from the Otago Region, Southeastern New Zealand. *Journal of Natural History* 23: 1059–1085.

Wood, A.C.L., Rowden, A.A., Compton, T.J., Gordon, D.P., & Probert, PK. 2013. Habitat-forming Bryozoans in New Zealand: their Known and Predicted Distribution in Relation to Broad-scale Environmental Variables and Fishing Effort. *PLoS One* 8(9): e75160.

Bryozoan Studies 2022 – Key, Porter & Wyse Jackson (eds)
© 2023 The Author(s), ISBN: 978-1-032-59487-3

A review of contact dependent interactions among bryozoans

Sheena E. Stephens-Norton*
Department of Biology, California State Polytechnic University Humboldt, Arcata, California, USA

Sean F. Craig
Department of Biology, California State Polytechnic University Humboldt, Arcata, California, USA
ORCID ID: 0000-0002-0149-3455

ABSTRACT: We review published literature focused on bryozoan interactions and discuss overgrowths, "stalemates," erect bi-laminate growths and fusions. Overgrowths, where one colony grows zooids atop another, have received the most attention in the scientific literature. Stalemates, where two colonies cease growth at the region of contact, have received relatively little research investigating the possibilities for its occurrence. Erect, bi-laminate growth, following inter- and/or intraspecific contact between colonies, has also received little research, but could reflect mutualistic behavior of encrusting species to increase feeding and reproduction. Fusion of self-self-contacts appears common, yet non-self-fusions to form natural chimeras may also occur and could represent cooperative behavior among genetically related individuals to enhance survival, growth and/or reproduction. In-depth analyses of the costs and benefits of these behaviors in bryozoans are needed to further understand the evolution of these alternative behaviors as well as their contribution to the abundance and diversity of bryozoans in marine habitats.

1 INTRODUCTION

Succession in many marine communities includes larval settlement and growth of bryozoans. Depending on an individual species' life history traits, bryozoan larvae have been shown to have settlement preferences for different substrates and/or biofilms (Lidgard & Jackson 1989; McKinney & McKinney 1993), light levels (McKinney & McKinney 1993), flow rates (Okamura & Partridge 1999), types of non-bryozoan neighbors (Barnes 2002; Bishop 1989; Jackson 1979), as well as conspecific relatives (Buss 1981; Keough 1984; McKinney & McKinney 1993). Where these larvae settle and grow, either on a vertical or horizontal gradient, will influence surrounding populations and their community dynamics. Following bryozoan larval settlement and metamorphosis, colony growth can lead to a shift in life history traits, from tactics that emphasize survival and growth to those which emphasize reproduction. For example, if a bryozoan colony invests more energy into growing quickly, there may be less energy reserved for early reproduction, which could lead to lower initial reproductive output. These so-called "trade-offs" imply that energy primarily utilized on one life-history trait may come at the expense of another life history trait (but note that "trade-offs" do not necessarily exist in modular organisms). "Trade-offs" may also be seen in energy used for calcification, creation of offensive/defensive polymorphs (e.g., avicularia), zooid size, feeding rate and many other features (Liow *et al.* 2019, 2017; McKinney 1993; Okamura & Partridge 1999; Pratt 2004) as a colony grows.

Bryozoan growth often results in contact with other organisms, including other bryozoans, during which these "trade-offs" and species-specific life history traits may influence how colonies interact. When an individual bryozoan colony contacts another bryozoan of the same (intraspecific) or different species (interspecific), it may react by exhibiting one (or more) previously described behaviors, resulting in (1) overgrowth and/or reciprocal overgrowth, (2) fusion, (3) stalemate/standoff or (4) erect, bi-laminate growth. Overgrowths occur between two (or more) colonies when a zooid of one colony grows directly on top of the lophophoral aperture of a zooid from a second colony, preventing the second zooid from feeding (Figure 1). Reciprocal overgrowths occur when one colony overgrows its neighbor along a portion of their contact zone, while the neighbor in turn overgrows it along a different region of contact (Taylor 2016). Fusions occur when tissues from two or more (self or non-self) distinct colony growing edges merge into a single growth front. Fusions are sometimes indicated by a "Y" shaped kenozooid (or zooid: Craig 1994) that results from the budding of tissues from two separate colonies which coalesce into a single zooid (Figure 2).

*Corresponding Author: ses797@humboldt.edu

DOI: 10.1201/9781003454915-14

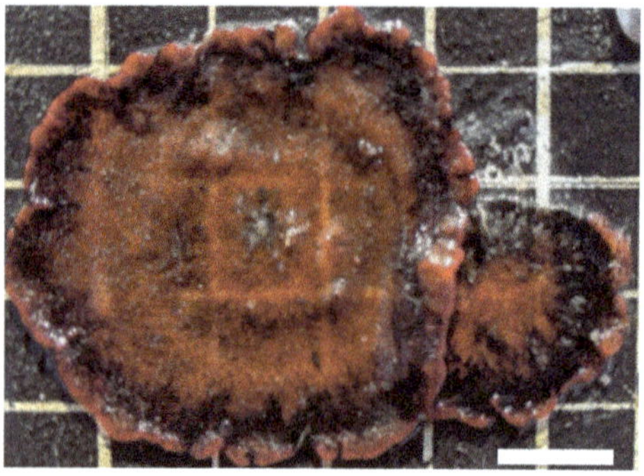

Figure 1. *Watersipora* sp. demonstrating overgrowth as one colony grows on top of another, smaller colony. Scale bar=1 cm.

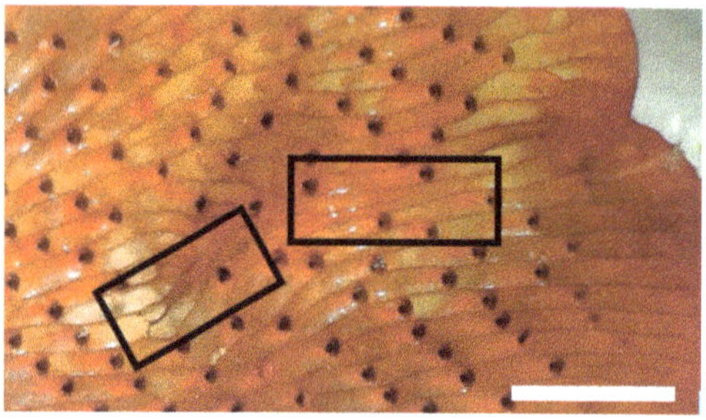

Figure 2. *Watersipora* sp. demonstrating self-fusion after losing a portion of the colony. Black boxes indicate "Y" shaped zooids that begin from separate zooecia, but merge into a single zooecium. Scale bar=2 mm.

Stalemates, also termed standoffs, occur when the growing edges of two separate colonies meet, but cease growth along the area(s) of contact (Figure 3). Erect, bilaminar growths are sheet-like growths that occur when two separate, encrusting colonies meet and redirect growth upwards, off the benthos into the water column, using the zooids of the contacting colony as a growing surface to ultimately form a 3-dimensional structure (Figure 4). Although these types of interactions

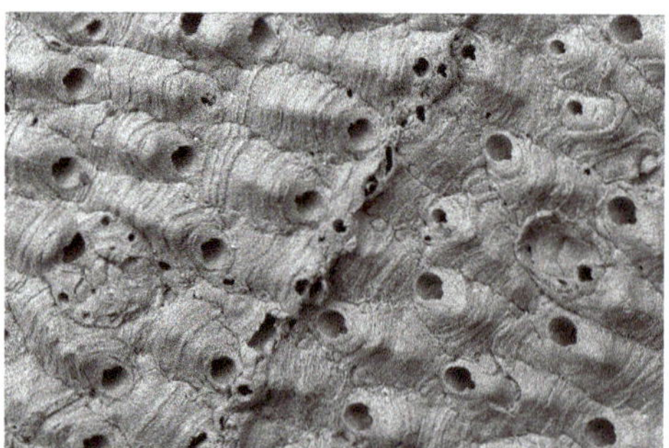

Figure 3. Fossilized *Antarctothoa tongima* demonstrating a stalemate/standoff interaction of equally sized zooids. Photograph by Paul Taylor. Scale bar=100 μm.

Figure 4. *Watersipora* sp. demonstrating erect, bilaminant sheet growth as two separate colonies use each other as "substrate" and grow upwards. White-lined scaling squares=1 cm^2.

strongly influence community structure, for example by creating 3-dimensional habitat that is often colonized by many other organisms (Vieira *et al.* 2018), relatively little is known about most of these behaviors.

This review will cover the importance of individual colony interactions between bryozoans exhibiting these four key interactions [1) fusions, 2) overgrowths and/or reciprocal overgrowths, 3) stalemates/standoffs and 4) erect, bilaminar growths], and ask what is already known and what remains to be determined.

2 FUSIONS

Fusions occur most often when the growing edges of the same colony encounter one another, merging together, sometimes after breakage or growth around an obstacle. These contacts are termed self-self fusions. Other, perhaps less common, fusions can occur between genetically related colonies, such as between siblings or half-siblings (Craig & Wasson 2000). Thus far, the literature suggests that fusions rarely occur in contacts between unrelated individuals, but there have not been enough studies to fully support this conclusion. One hypothesis to explain fusion is that an allorecognition system exists which restricts fusion to genetically related colonies, which must share at least one (or more) alleles at one or several genetic loci in order to fuse, as has been demonstrated in the colonial sea squirt *Botryllus schlosseri* (Pallas, 1766), (Grosberg & Quinn 1986) and the colonial hydroid *Hydractinia symbiolongicarpus* (Buss & Yund, 1989), (Grosberg *et al.* 1996). In some bryozoans, unique pore plates have been found adjoining the skeletal walls of fused colonies (Craig & Wasson 2000), but relatively few cases of fusion have been examined and even fewer have used the Scanning Electron Microscope (SEM) to look for these connections. Therefore, it is often difficult to determine whether or not two or more colonies simply appear to be fused, but lack the necessary connections to facilitate sharing of nutrients and formation of a resulting chimera.

Fusion may nevertheless represent a significant advantage to colony survival if it enables sharing of nutrients between colonies, resulting in a "larger" chimera that is less likely to suffer mortality (due to starvation, predation, or other size-dependent factors). Upon fusion, a colony could increase its effective size by connecting with a neighboring colony, gaining access to shared nutrients. Nutrient sharing within a chimera could then lead to additional reproductive advantages including earlier sexual reproduction and/or an increase in the rate and/or number of larvae produced. Other advantages to chimera formation may also exist, such as the possibility that fused colonies may together become superior competitors against their smaller neighbors, however we have found no studies that focus on this topic.

In contrast, fusion could be disadvantageous if, following connection to form a physiologically integrated chimera, an internal cellular takeover of the "weaker" colony occurs by the "stronger" colony, whose cells might be able to take over the connected neighbor's reproductive system to their own advantage, as has been shown to occur in the colonial ascidian, *Botryllus schlosseri* (Grosberg & Quinn, 1986). In addition, unequal sharing of energetic resources within a chimera may yield an advantage to only one genetic individual within it (Grosberg & Quinn 1986).

Because very little research has been done to examine bryozoan fusion events, many of these may simply be "apparent fusions" in which the two (or more) colonies involved are not physically or physiologically connected chimeras. It is also important to determine whether fusions of bryozoan colonies are limited by genetic relatedness (or other factors), and whether germ line parasitism is even possible in these organisms (Craig 1994). A better understanding of fusions in terms of when and where they occur, however, might lead us to completely rethink our current understanding of bryozoan population dynamics, because what we currently classify as single, genetically distinct colonies might (at least in some cases) actually represent multiple, genetically unified chimeras (Craig 1994) which could, for example, lead to the important formation of 3-dimensional habitats for other marine species to live on (or in).

3 OVERGROWTHS

Overgrowths are arguably the best studied interactions between marine bryozoans in both extant and fossil communities. In general, ecologists and evolutionary biologists see this type of competitive interaction as a highly important driver of evolutionary adaptation as well as population and community structure; therefore, overgrowths have received the vast majority of attention in the scientific literature (Barnes & Dick 2000; Buss 1979; Liow *et al.* 2016, 2017, 2019; López Gappa 1989; McKinney 1992; Rubin 1985; Russ 1982; Turner & Todd 1994). Overgrowths are especially important for paleo-bryozoologists as the calcified skeletons of bryozoans fossilize remarkably well, and iterative growth of modular zooids in extant bryozoan colonies is relatively easy to follow, making observations of these interactions during a given "snapshot" in time possible in both living (Barnes 2006) and fossil (Taylor 2016) species. However, since overgrowth is also the "easiest" interaction for paleo-bryozoologists to classify, much research has focused on it, placing a heavy emphasis on this particular interaction in fossil and extant bryozoan communities which could over-emphasize its importance in marine community dynamics.

Different life history traits, such as larval substrate preference, could also affect our understanding of bryozoan interactions in both fossil and extant species. Substrate preferences of bryozoans may have important consequences for their competitive ability. In fossil and extant species, most research is done on bryozoans that settle on hard substrates, such as rocks or shells, because they fossilize well and/or are easier to track on solid substrates that do not degrade. Species who preferentially settle on softer substrates (e.g., algae or seagrasses), however, do not fossilize well, in part due to relatively low calcification which makes these colonies more flexible as their substrate bends in flowing water (Liow *et al.* 2019; Ward & Thorpe, 1989). Because soft substrates degrade more rapidly than harder ones, bryozoan species that settle on algae may be under strong selection to grow quickly, reproduce early and put less energy into competition, unlike hard substrate colonizers which must compete for space over much longer time periods. Since most published research on bryozoans has focused on solid substrates, knowledge of bryozoan interactions may therefore miss important behaviors and/or life history traits exhibited on soft substrates which could influence the type of contact-dependent interactions seen as well as their prevalence in both extant and fossil communities.

Bryozoan species that settle in areas with high disturbance rates may have life history traits similar to those which prefer soft substrates, because they too must grow and reproduce quickly before being scraped away (e.g., by sea ice or seasonal rock tumbling). Disturbance could therefore favor species that are not fully "equipped" for physical competition (e.g., they may have smaller zooid sizes and/or lack spikes or spines to compete with their neighbors) (Liow *et al.* 2019) but grow and reproduce rapidly. Because most bryozoan research is done on hard-substrates that include species with highly competitive overgrowth traits, other species may be erroneously grouped into the "poor competitor" life history "bin" and ranked as "losers" when, in fact, they may be superior on soft substrates. Basing our understanding of bryozoan life history traits mostly on overgrowth competition on hard substrates could therefore create a research bias in our understanding of bryozoan interactions in general as well as the importance of particular contact-dependent behaviors in population growth and evolution.

For example, poor competitors are often the first to colonize a new substrate, where they grow rapidly and reproduce before being overgrown by superior competitors during the latter stages of community succession (Barnes 2002; López Gappa 1989). However, poor competitors are sometimes more abundant within an area, despite their lower competitive ability (Barnes 2006; Liow *et al.* 2016).

Even in habitats where interspecific competition is hierarchical, especially for encrusting species at higher latitudes, competitively dominant species are rarely capable of fully monopolizing the available substrata (Barnes & Kukliński, 2003; Centurión & López Gappa 2011). In the Antarctic, for example, the most abundant species are often poorer competitors, relying on rapid colonization and growth rate for survival, whereas good competitors are relatively rare (Barnes & Rothery 1996). In contrast, Centurión & López Gappa (2011) indicated that the large majority of relatively rare species were on larger, less disturbed substrates while more common species were found primarily on smaller, easily disturbed substrata. Hence other factors, such as tolerance of sedimentation (Clark *et al.* 2017), may strongly influence bryozoan morphological and life-history traits as well as their community structure. In a study by Barnes & Rothery (1996) in the Antarctic, colony thickness and/or vertical relief was found to be the most advantageous trait for superior competitive ability. They suggested that substrates at different depths could be made up of genetically distinct populations and that the level of relatedness between bryozoans may play a key role in behavioral interactions, as close kin may be less likely to compete for space via overgrowth interactions.

Studies of bryozoan overgrowth interactions have shown three traits to be important among the successful "winners" of competition over millennia (Liow *et al.* 2019): (1) zooid size (bigger is generally better), (2) multilayered growth (thicker is better) and (3) the presence of spikes (which may also improve defensive outcomes). These authors showed that average zooid size in fossil species has fluctuated in response to common external factors (i.e., temperature) over the course of 2-million years, but species-specific competitive abilities have generally remained stable through time (Liow *et al.* 2017). Larger zooids may also be favored by natural selection due to their ability to gather more food compared to smaller colonies (Buss 1979; Lidgard & Jackson 1989). Smaller zooids may be limited to smaller prey sizes due to smaller feeding organs (lophophores) that result in less prey captured (intake of smaller, lower energy prey), while larger zooids may be more limited by metabolic exchange between zooids rather than their ability to take in more prey items (Liow *et al.* 2017; Pratt 2004).

However, Okamura & Partridge, (1999) found that within the encrusting sheetlike species *Membranipora membranacea* (Linnaeus, 1767), colonies in higher flow environments trended towards smaller, shorter zooids, fewer feeding tentacles and shorter distances between excurrent chimneys, yet maintained the same growth rate as colonies with larger zooids in lower flow. Okamura & Partridge (1999) suggest that this plasticity in zooid size creates equal feeding success in smaller and larger zooids in faster and slower flow rates, respectively. No other studies were found in our review in which colonies were reared in different flow rates to examine whether different zooid sizes can be formed plastically. Additionally, as far as we are aware, no studies have placed such colonies (with different zooid sizes) into competitive interactions with one another within and across different flow regimes to study the importance of zooid size in overgrowth competition. Nevertheless, this would help further determine the role of zooid size in competitive interactions of extant species. Such studies could be modeled after the excellent work of Harvell (1992) on the costs and benefits of inducible spine production in the bryozoan *Membranipora membranacea*.

Overgrowth interactions could have several effects on individuals as well as their populations: the "loser" in overgrowth competition may be completely overgrown and smothered and killed, leading to population decline of a poor competitor. However, the process of contact by a superior competitor may also trigger the "losing" colony to release larvae sooner (Harvell & Grosberg 1988), possibly increasing its fecundity. This could enable the genes of "losing" colonies to be maintained within a population(s) through time, presuming their larvae can settle, grow, and reproduce prior to being outcompeted themselves. The "winning" colony may therefore gain surface area on the substrate as it grows over the top of a "losing" colony and is therefore able to acquire space resources that the "loser" once had, theoretically also reproducing over longer time periods to further its own genetic contribution into future generations. Both species could potentially continue for many generations, and we might therefore begin to see evolution favoring differentiation of their life history traits that leads to their coexistence.

4 STALEMATES/STANDOFFS

Very little is known about stalemate/standoff interactions between bryozoans. These behaviors are sometimes mentioned in studies focused on bryozoan overgrowths, but are seldom discussed in detail. This interaction appears to occur when contacting colonies are roughly equal in total size and have roughly the same sized zooids, and colonies matching this description rarely initiate overgrowth interactions (Centurión & López Gappa 2011; Liow *et al.* 2016, 2017). Centurión & López Gappa (2011) noted nearly all resulting outcomes of intraspecific (same species) interactions were stalemates. In the fossil record (2-million years to the Recent), there appears to be spatial clustering of conspecifics and the resultant intraspecific standoffs are much more common than interspecific stand-offs, suggesting that these interactions may be predictable (Liow *et al.* 2016, 2017). In the fossil record, stalemates are assumed when stand-offs occur because it is thought that stalemates do not occur if one of the colonies is dead; meaning that the dead colony would be overgrown by the living colony (Liow et al. 2016). Although stand-offs have been found mostly in intraspecific interactions, both intra- and interspecific stand-offs do occur, suggesting this behavior is more complex than currently understood (Barnes & Kukliński 2003).

Even though stand-offs are likely to include costs for the interacting colonies involved (Barnes & Neutel 2016), we have yet to discover what those costs truly are (other than "giving up" potential growing space). Possible reasons that some interactions result in "stalemates" include: 1) There may be a possible "sizing up" period before competitive interactions in which colonies "gauge" the competitive ability of their opposition, 2) Colonies that do not engage in competitive overgrowth may do so because their effort would be equally matched by the competitor, resulting in a draw, which would be a poor use of limited energy, 3) A colony may recognize the neighbor as close kin without (yet) initiating fusion, 4) Colonies may redirect energy resources into growth in other directions where open substrates exist (especially on short lived soft substrates where time for growth is limited), 5) Lophophoral interactions or allelopathic chemical defenses may simply deter or prevent overgrowth (Barnes 2002; Liow *et al.* 2019; Ward & Thorpe 1989) and/or 6) there is some form of chemical signaling between bryozoans in general (Alino *et al.* 1992; Barnes 2002; Barnes & Kukliński 2003; Craig & Wasson 2000; Liow *et al.* 2019; Romano 1990) that regulates these stand-offs. In this review, we found no studies that looked specifically at stand-offs among bryozoans, however.

In cases of stand-offs, a possible influence on population structure is that both colonies may continue to live without zooid mortality even though chemical warfare could be occurring, as it does in corals (Alino *et al.* 1992; Romano 1990). Contact during stand-offs may result in the triggering of reproductive output from both colonies, instead of just the "losing" colony as mentioned (above) in competitive overgrowths. This could mean that both colonies would release more larvae into the population and continue to increase their population size through time, despite the lack of growing space gained along the line of contact.

Within the literature we reviewed, stalemate/standoff contacts were often associated with the term "ties," but this term is highly problematic when trying to understand this behavior and its influence on population structure. As defined by Barnes & Kukliński (2003), "ties" are "scored when mutual, but equal, overgrowth took place or skeletal build-up, but without any overgrowth (sometimes termed a standoff), occurred." Elsewhere in their paper, they state that these "ties" could also include fusion events. The problem with the use of this terminology is that competitive overgrowths, poorly understood fusions, and even less understood standoffs have been lumped together into the same bin of data. This is a problem because it does not allow for clear focus on the evolution of these behaviors from an individual, evolutionary standpoint. Why would a bryozoan use different behaviors to achieve the same results? For example, if competition results in just a "tie," why would both colonies spend their energy trying to overgrow their neighbor, rather than choosing a stand-off or redirecting growth elsewhere? If these outcomes are just "ties," what would drive these different types of behaviors? From a community ecology point of view, the definition of "ties" may be sufficient to describe the outcome (in terms of community structure and/or species richness), but only at that specific moment or "snapshot" in time.

Barnes & Kukliński (2003) do, however, show the overall importance of intraspecific interactions in areas such as the Arctic and Antarctic. Clark *et al.* (2017) and Centurión & López Gappa (2011) found nearly all polar interactions recorded were intraspecific. Centurión & López Gappa (2011) specifically found that almost all intraspecific interactions involved a single, sheetlike species, *Harmeria scutulata* (Busk, 1855). Nearly all interactions between *H. scutulata* colonies resulted in "tied outcomes (standoffs)." It is unclear, however, whether these interactions were properly categorized as standoff interactions, and not lumped into the umbrella definition of "ties," because Centurión & López Gappa (2011) follow the definition of "ties" given by Barnes & Kukliński, (2003). The success of *H. scutulata* was attributed to the high amount of disturbance seen in this species' habitat and its relatively high growth rate, when compared to that of its' competitors. Additionally, "ties" were generally more common between species that had similar competitive abilities versus those that had large differences in win/loss ratios (Centurión & López Gappa 2011). Other life history traits which may also influence the "success" of this species were not mentioned, however.

5 BILAMINATE GROWTHS

Erect, bilaminate, sheet-like growths appear to be the least discussed type of interaction seen among bryozoans in the literature and are mentioned in passing, if at all. Craig & Wasson (2000) described the bilaminar upright growth of the encrusting bryozoan *Hippodiplosia insculpta* (Hincks, 1883), seen in both self-self and non-self interactions. No communication plates/pores were found between the two opposing, upright sheet-like growths of these colonies, so there was no indication of nervous communication or coordinated translocation between them. However, growth and shape of these colonies formed upright, bilaminate growths that appeared to be synchronized with equally matched growing edges on either side.

In the fossil record, bilaminate growth was found to be associated with a thickening in calcification to support 3-dimensional colony formation (Barnes & Whittington 1999). Erect, calcified colonies are often brittle and prone to breakage during periods of high-water flow or resulting physical damage from various objects (e.g., ice and debris). Fragmentation could be beneficial for dispersal and colonization of these bryozoans so long as the resulting fragments can re-attach to a new substrate. This method of reproduction (via fragmentation) has been shown to be highly effective for certain species in the Antarctic where disturbances are common, due to ice scraping (Barnes & Whittington 1999).

We have found few studies concerning the reasons for the appearance of upright bilaminate growth in bryozoans, hence we offer the following thoughts for consideration: 1) Two (or more) colonies may simply be using that portion of their own colony (or the neighboring colony) as a substrate for outward, upright growth into the water column, perhaps to increase feeding success (for example, by creating eddies that enhance food capture) (Jackson 1979), 2) Upright bilaminate growth may represent a mutualistic behavior in which each colony acts as a substrate for the other when space on the primary substrate is limited, allowing both colonies to further increase in size (Craig & Wasson 2000), 3) Two (or more) competitive, equally matched colonies (i.e., same zooid and colony size, growth rate, etc.) may be attempting to overgrow one another, but by building upon each other; the height of one colony may simply hinder the overgrowth ability of the neighbor during "battle" (Jackson 1979; Liow *et al.* 2019).

The effects of upright bilaminate growth on population dynamics are unknown, but it seems that both colonies may still be able to produce as many, if not more, larvae to increase their genetic representation in the next generation due to increased feeding surface area up in the water column. Upright growth could also allow them to increase prey capture and bring more nutrients into the colony. It is thought that upright bilaminate growths occur mostly in intraspecific contacts, but they have also been observed in contacts between species as noted by Liu *et al.* (2017), who studied the interactions between *Cryptosula pallasiana* (Moll, 1803) and *Watersipora* (Neviani 1896), sp. on a boat hull in China. Liu *et al.* (2017) documented a case of bilaminar growth between these two species (Figure 5). On close inspection, it can be seen that *C. pallasiana* is also overgrowing the same *Watersipora* sp. colony at a different area of contact (Figure 5B). This may suggest that this particular bilaminate growth interaction falls under category 3 above, where *Watersipora* sp. is itself defending against overgrowth by *C. pallasiana* on one growing edge, but is failing to defend itself from overgrowth on another growing edge. These images better fit the definition of reciprocal overgrowths (Taylor 2016) and underscore the complexity of bryozoan interactions and the challenges for future studies which seek to understand them better.

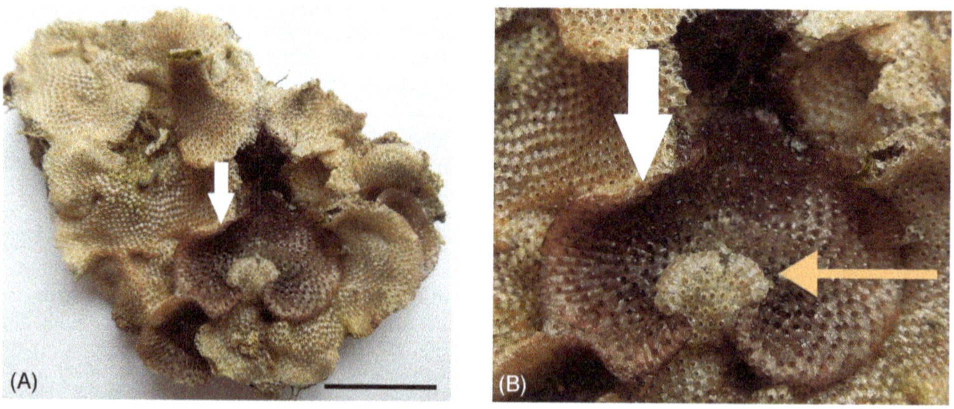

Figure 5. Example of interspecific bilaminant growth as indicated by the white arrow, between *Cryptosula pallasiana* (pale tan coloration) and *Watersipora* sp. (darker, red/orange coloration). Bilaminant interaction (white arrow) and overgrowth interaction (orange arrow) of *C. pallasiana* overgrowing *Watersipora* sp. can clearly be seen in B. Photograph by Hulian Liu. Modified from Liu *et al.* (2017). Scale bar=1 cm.

6 CONCLUSIONS

In conclusion, studies of the physical interactions between bryozoans upon contact have mainly focused on the mechanics of overgrowth due to competition, which has received the vast majority of focus in the scientific literature. Very few studies have examined fusions, stalemates/standoffs and erect, bilaminate sheet-like growths following intra- or interspecific contact between bryozoan colonies. The complex answers to the questions of "when" and "why" these behaviors occur are therefore difficult to obtain, particularly when research continues to focus almost entirely on overgrowth competition. This research focus may create a bias in our understanding of bryozoan interactions because it does little to explore the multiple factors, including life history trade-offs and genetic relatedness issues, which might drive the evolution of the different behaviors exhibited upon contact in bryozoans. More research needs to be done to examine colony contacts in extant species over extended periods of time, from settlement to senescence and colony mortality, documenting frequently (more than once or twice a month) the costs and benefits of these behaviors in order to better capture the subtle interactions between colonies and the resultant changes in their survival, growth, and reproduction. The mindset of bryozoan research needs to be adjusted to think more about colony interactions as complex social behaviors rather than simply scored as static "snapshots" of community structure. Few studies have truly examined bryozoans and their multiple potential behaviors following contact as complex interactions with multiple potential outcomes that alter their growth, reproduction and survival through time. Future studies could be modeled after the excellent work of Harvell (1986, 1992) to examine the costs and benefits of inducible defenses (spines) in the bryozoan *Membranipora membranacea* which grows on kelp blades. Understanding the relative frequency of the behaviors discussed herein on different substrates, as well as the resultant outcomes in terms of individual colony successes (or failures) will allow for a better understanding of the evolution of bryozoan context-dependent behaviors. This will also further enhance our scientific ability to predict the dynamics of marine populations and communities through time as well as in the fossil record. There have been multiple calls for time series data from natural communities for this type of information, which is thus far highly lacking (Barnes & Kukliński 2003; Liow *et al.* 2017; Wootton & Emmerson 2005). As Barnes (2002) stated, "clearly, spatial competition involves more than a simple mechanical "showdown" and as emphasized by Liow *et al.* (2017), there is a "cornucopia" of research avenues still to be explored.

ACKNOWLEDGEMENTS

The authors would like to give thanks to the International Bryozoology Association (IBA) for their generous travel funding (to Sheena Stephens-Norton) as well as the opportunities given at the 19th International Bryozoology Conference for collegial discussion which were sorely needed after several years of a pandemic which relegated us to our individual homes. We would also like to thank the Council on Ocean Affairs, Science & Technology (COAST) for funding the travel and board expenses of Sheena Stephens-Norton. Thanks also goes to the Department of Biological Sciences at the California State Polytechnic University-Humboldt for guidance and research support. We are most grateful to Paul Taylor and Kamil Zagorsek for providing Figures 3 and 5 respectively. We thank both Mary Spencer Jones and Mali Ramsfjell for their editorial comments on an earlier draft of this paper. Finally, special thanks go to Patrick Wyse Jackson and the IBA conference team for their warmth and hospitality, and to John Norton for his patience, love and understanding as his fiancée completed her first published paper.

REFERENCES

Alino, P.M., Sammarco, P.W. & Coll, J.C. 1992. Competitive Strategies in Soft Corals (Coelenterata, Octocorallia). IV. Environmentally Induced Reversals in Competitive Superiority. *Marine Ecology Progress Series* 81: 129–145.

Barnes, D.K. 2002. Clade Perseverance from Mesozoic to Present: a Multidisciplinary Approach to Interpretation of Pattern and Process. *The Biological Bulletin* 203(2): 161–172.

Barnes, D.K. 2006. Temporal–spatial Stability of Competition in Marine Boulder Fields. *Marine Ecology Progress Series* 314: 15–23.

Barnes, D.K. & Dick, M.H. 2000. Overgrowth Competition Between Clades: Implications for Interpretation of the Fossil Record and Overgrowth Indices. *The Biological Bulletin* 199(1): 85–94.

Barnes, D.K. & Kukliński, P. 2003. High Polar Spatial Competition: Extreme Hierarchies at Extreme Latitude. *Marine Ecology Progress Series* 259: 17–28.

Barnes, D.K.A., & Neutel, A.M. 2016. Severity of Seabed Spatial Competition Decreases Towards the Poles. *Current Biology* 26(8): R317–R318.

Barnes, D.K. & Rothery, P. 1996. Competition in Encrusting Antarctic Bryozoan Assemblages: Outcomes, Influences and Implications. *Journal of Experimental Marine Biology and Ecology* 196(1–2): 267–284.

Barnes, D.K. & Whittington, M. 1999. Biomechanics and Mass Mortality of Erect Bryozoans on a Coral Reef. *Journal of the Marine Biological Association of the United Kingdom* 79(4): 745–747.

Bishop, J.D.D. 1989. Colony Form and the Exploitation of Spatial Refuges by Encrusting Bryozoa. *Biological Reviews* 64(2): 197–218.

Buss, L.W. 1979. Bryozoan Overgrowth Interactions—the Interdependence of Competition for Space and Food. *Nature* 281(5731): 475–477.

Buss, L.W. 1981. Group living, Competition, and the Evolution of Cooperation in a Sessile Invertebrate. *Science* 213(4511): 1012–1014.

Centurión, R. & López Gappa, J.J. 2011. Bryozoan Assemblages on Hard Substrata: Species Abundance Distribution and Competition for Space. *Hydrobiologia* 658: 329–341.

Clark, G.F., Stark, J.S. & Johnston, E.L. 2017. Tolerance Rather than Competition Leads to Spatial Dominance of an Antarctic Bryozoan. *Journal of Experimental Marine Biology and Ecology* 486: 222–229.

Craig, S.F. 1994. Intraspecific Fusion in the Encrusting Bryozoan *Fenestrulina* sp. In P.J. Hayward, J.S. Ryland, and P.D. Taylor (eds). *Biology and Paleobiology of Bryozoans. Proceedings of the 9th International Bryozoology Conference*: 51–54. Olsen and Olsen, Fredensborg.

Craig, S.F. & Wasson, K. 2000. Self/non-self Recognition and Fusion in the Bryozoan *Hippodiplosia insculpta*. In *Proceedings of the 11th International Bryozoology Association Conference:* 189–196. Balboa: Smithsonian Tropical Research Institute.

Grosberg, R.K., Levitan, D.R. & Cameron, B.B. 1996. Evolutionary Genetics of Allorecognition in the Colonial Hydroid *Hydractinia symbiolongicarpus*. *Evolution* 50(6): 2221–2240.

Grosberg, R.K. & Quinn, J.F. 1986. The Genetic Control and Consequences of Kin Recognition by the Larvae of a Colonial Marine Invertebrate. *Nature* 322 (6078): 456–459.

Harvell, C.D. 1986. The Ecology and Evolution of Inducible Defenses in a Marine Bryozoan: Cues, Costs, and Consequences. *American Naturalist* 128(6): 810–823.

Harvell, C.D. 1992. Inducible Defenses and Allocation Shifts in a Marine Bryozoan. *Ecology* 73 (5): 1567–1576.

Harvell, C.D. & Grosberg, R.K. 1988. The Timing of Sexual Maturity in Clonal Animals. *Ecology* 69(6): 1855–1864.

Jackson, J.B.C. 1979. Overgrowth Competition Between Encrusting Cheilostome Ectoprocts in a Jamaican cryptic Reef Environment. *The Journal of Animal Ecology* 48(3): 805–823.

Keough, M.J. 1984. Kin-recognition and the Spatial Distribution of Larvae of the Bryozoan Bugula Neritina (L.). *Evolution* 38(1): 142–147.

Lidgard, S. & Jackson, J.B.C. 1989. Growth in Encrusting Cheilostome Bryozoans: I. Evolutionary trends. *Paleobiology* 15(3): 255–282.

Liow, L.H., Di Martino, E., Voje, K.L., Rust, S. & Taylor, P.D. 2016. Interspecific Interactions Through 2 Million Years: Are Competitive Outcomes Predictable? *Proceedings of the Royal Society B: Biological Sciences* 283(1837): 20160981.

Liow, L.H., Di Martino, E., Krzeminska, M., Ramsfjell, M., Rust, S., Taylor, P.D. & Voje, K.L. 2017. Relative Size Predicts Competitive Outcome Through 2 Million Years. *Ecology Letters* 20(8): 981–988.

Liow, L.H., Reitan, T. Voje, K.L., Taylor, P.D. & Di Martino, E. 2019. Size, weapons, and armor as predictors of competitive outcomes in fossil and contemporary marine communities. *Ecological Monographs* 89(2): e01354.

Liu, H., Zágoršek, K., Wang, S., Ma, S. & Taylor, P.D. 2017. Interactions Between *Cryptosula* and *Watersipora* (Bryozoa: Cheilostomata) on a Ship's Hull in Qingdao Harbour (South Yellow Sea) after Five and a Half Years of Immersion. *Chinese Journal of Oceanology and Limnology* 35(5): 1179–1188.

López Gappa, J.J. 1989. Overgrowth Competition in an Assemblage of Encrusting Bryozoans Settled on Artificial Substrata. *Marine Ecology Progress Series. Oldendorf* 51(1): 121–130.

McKinney, F.K. 1992. Competitive Interactions Between Related Clades: Evolutionary Implications of Overgrowth Interactions Between Encrusting Cyclostome and Cheilostome Bryozoans. *Marine Biology* 114: 645–652.

McKinney, F.K. 1993. A Faster-paced World?: Contrasts in Biovolume and Life-process Rates in Cyclostome (Class Stenolaemata) and Cheilostome (Class Gymnolaemata) Bryozoans. *Paleobiology* 19 (3): 335–351.

McKinney, F.K. & McKinney, M.J. 1993. Larval Behaviour and Choice of Settlement Site: Correlation with Environmental Distribution Pattern in an Erect Bryozoan. *Facies* 29: 119–131.

Okamura, B. & Partridge, J.C. 1999. Suspension Feeding Adaptations to Extreme Flow Environments in a Marine Bryozoan. *The Biological Bulletin* 196(2): 205–215.

Pratt, M.C. 2004. Effect of Zooid Spacing on Bryozoan Feeding Success: is Competition or Facilitation More Important? *The Biological Bulletin* 207(1): 17–27.

Romano, S.L. 1990. Long-term Effects of Interspecific Aggression on Growth of the Reef-building Corals *Cyphastrea ocellina* (Dana) and *Pocillopora damicomis* (Linnaeus). *Journal of Experimental Marine Biology and Ecology* 140(1–2): 135–146.

Rubin, J.A. 1985. Mortality and Avoidance of Competitive Overgrowth in Encrusting Bryozoa. *Marine Ecology Progress Series. Oldendorf* 23(3): 291–299.

Russ, G.R. 1982. Overgrowth in a Marine Epifaumal Community: Competitive Hierarchies and Competitive Networks. *Oecologia* 53: 12–19.

Taylor, P.D. 2016. Competition Between Encrusters on Marine Hard Substrates and Its Fossil Record. *Paleontology* 59(4): 481–497.

Turner, S.J. & Todd, C.D. 1994. Competition for Space in Encrusting Bryozoan Assemblages: The Influence of Encounter Angle, Site and Year. *Journal of the Marine Biological Association of the United Kingdom* 74(3): 603–622.

Vieira, E.A., Dias, G.M. & Flores, A.A. 2018. Adding Early-stage Engineering Species Affects Advanced-stage Organization of Shallow-water Fouling Assemblages. *Hydrobiologia* 818: 211–222.

Ward, M.A. & Thorpe, J.P. 1989. Assessment of Space Utilisation in a Subtidal Temperate Bryozoan Community. *Marine Biology* 103: 215–224.

Wootton, J.T. & Emmerson, M. (2005). Measurement of Interaction Strength in Nature. *Annual Review of Ecology, Evolution, and Systematics* 36: 419–444.

Bryozoan Studies 2022 – Key, Porter & Wyse Jackson (eds)

Enigmatic structures in palaeozoic fenestrate bryozoans: the case of *Fenestella sculptilis*

Juan Luis Suárez Andrés*
SONINGEO, Avda. La Cerrada, 10. Maliaño, Cantabria, Spain
ORCID ID: 0000-0002-8843-1256

Patrick N. Wyse Jackson
Department of Geology, Trinity College, Dublin, Ireland
ORCID ID: 0000-0001-5605-0670

Consuelo Sendino
Earth Sciences, Natural History Museum, Cromwell Road, London, UK
ORCID ID: 0000-0003-4823-2145

ABSTRACT: The fenestrate bryozoan *Fenestella* (*s.l.*) *sculptilis* Ulrich, 1886 from the Middle Devonian of the USA developed enigmatic channel-like structures on the obverse colony surface. These structures resemble the auxiliary tubes described in the literature as characteristic of the type D brood chambers, known only from the coeval *Polypora* (*s.l.*) *shumardi* Prout, 1858. Study of silicified specimens of *F. sculptilis* via optical microscopy and SEM shows type A brood chambers and other morphological differences with the type D brood chambers of *Polypora* (*s.l.*) *shumardi*. A detailed description of the channels present in *Fenestella* (*s.l.*) *sculptilis* is provided. Despite the scarcity of available data, a tentative discussion is undertaken regarding the nature of this singular feature as structures associated with breeding or, alternatively, the result of symbiotic interaction.

1 INTRODUCTION

The interpretation of inflated structures developed by Palaeozoic fenestrate bryozoans as brood chambers relies on their morphological similarity to structures present in Recent tubuliporates. Such structures have been observed, described and illustrated in different Palaeozoic fenestrate bryozoans (McKinney & Wyse Jackson 2015). Bancroft (1986) made a comprehensive review of the literature and defined four types of brood chambers, which he called type A-D ovicells. This classification was not modified until McKinney & Wyse Jackson (2015) added a type E brood chamber.

Fenestella (*s.l.*) *sculptilis* Ulrich, 1886, a rare Middle Devonian species from the Falls of the Ohio, USA, shows conspicuous grooves that resemble the auxiliary tubes of type D brood chambers described by Stratton (1981). These tubes are currently known only from *Polypora* (*s.l.*) *shumardi* Prout, 1858 (Figure 1A). Additionally, together with the channels, *Fenestella* (*s.l.*) *sculptilis* contains conventional type A chambers. The channel-like structures observed in *Fenestella* (*s.l.*) *sculptilis* are difficult to interpret; the purpose of this paper is to describe them in detail and then discuss their possible function as brood chambers and the suitability of alternative interpretations, including symbionts.

2 MATERIAL AND METHODS

The present study is based on direct observation of five silicified fragmentary specimens of *Fenestella sculptilis* Ulrich, 1886 from the Middle Devonian Hamilton Group, Falls of the Ohio, USA, NHMUK PI D5257 and the review of previous literature. The specimens belong to the collection of fossil bryozoans provided by E. O. Ulrich to the Natural History Museum (NHM London) and have been studied under optical microscopy and imaged using a low vacuum scanning electron microscope (SEM), FEI Quanta 650 FEG at the NHM, London.

The taxonomic status of *Fenestella sculptilis* should be revised, but it is beyond the scope of this work. Regarding Ulrich's (1886) description of *F. sculptilis*, quoted below, the author most probably created the

*Corresponding Author: juanl_suarez@yahoo.es

DOI: 10.1201/9781003454915-15

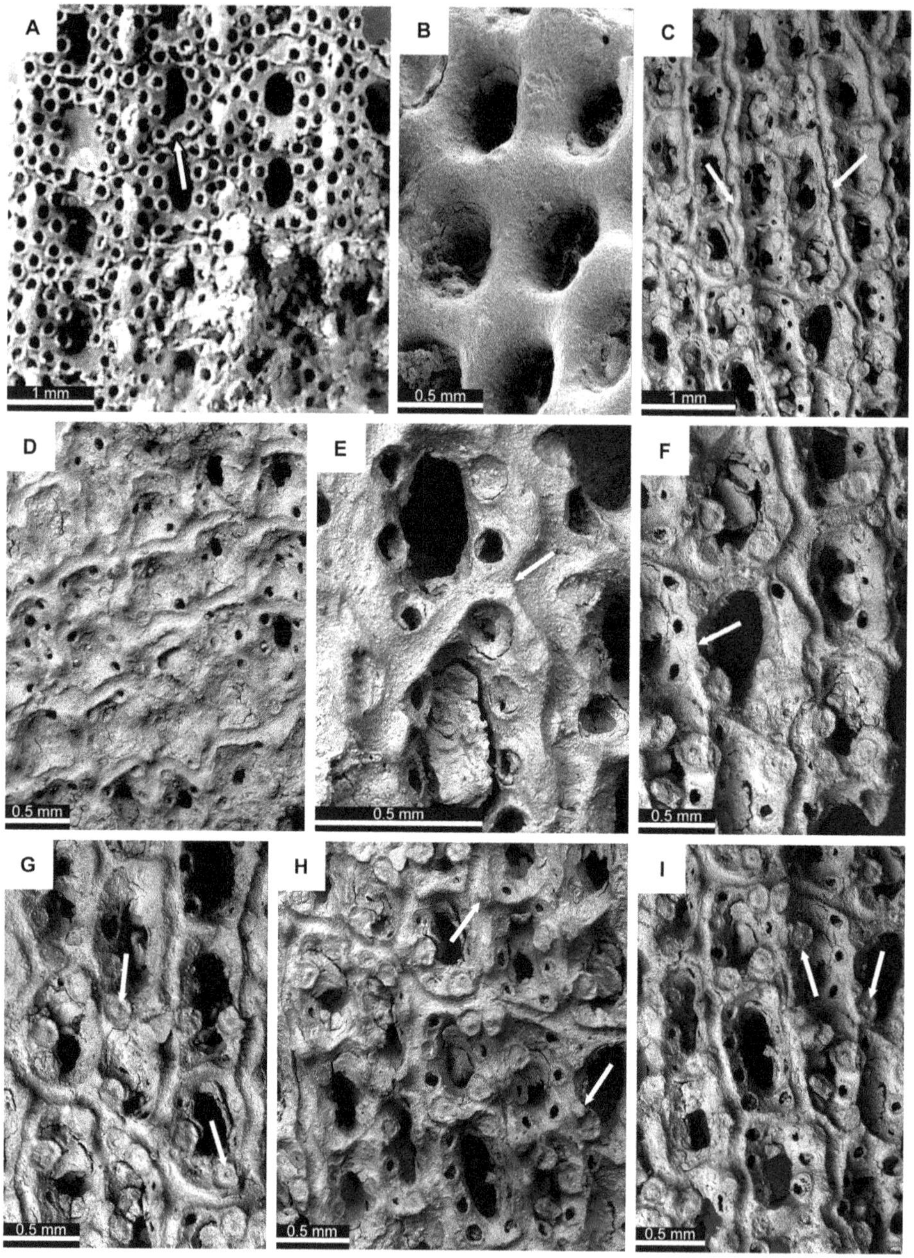

Figure 1. (opposite). A, de-roofed type D brood chambers (arrow) and associated tubes in *Polypora shumardi* Prout, 1858, Hamilton Group, Eifelian, Devonian, Falls of the Ohio, Indiana, USA, FMNH (UC) 57424. B-I, Scanning electron micrographs of *Fenestella sculptilis* Ulrich, 1886 NHMUK PI D5257. B, reverse surface of a colony fragment showing absence of grooves. C, obverse surface of a colony fragment with channels running longitudinally on branches (arrows) and linked by transverse channels along dissepiments. D, detail of a colony fragment of *F. sculptilis* showing a network of channels apparently oriented predominantly transverse to branches. E, detail of channels at different depths in branches and dissepiments. The arrow points a connection of a shallow transverse channel with a sharply deeper longitudinal one. F, channels on branches and dissepiments with no abrupt change in depth. The arrow points to a segment of a branch in which the keel of *F. sculptilis* can be observed due to the absence of a channel. G, detail of a fragment with branches and dissepiments dominated by channels, and abundant A-type brood chambers (arrows). H, colony fragment with A-type brood chambers and relatively few developed channels. Such areas allow for the observation of normal characters of this species such as the blunt, large nodes (arrows) that have been obliterated by the channels in other parts of the meshwork. I, detail of a specimen with abundant A-type brood chambers. Some of them are tangential to longitudinal or transverse channels, but others (arrows) are not close to these structures.

species on the sole basis of silicified specimens. Ulrich (1886) must have selected exclusively specimens showing the channels that he interpreted to be a specific character. The holotype of this species is silicified; it belonged to Mr G.K. Greene's private collection (Ulrich 1886, p. 2) and is presently housed at the American Museum of Natural History, New York (AMNH-FI 25362). Three colony fragments are hosted unregistered at the Smithsonian Institution, USNM, and they are also silicified.

3 DESCRIPTION

Ulrich (1886, p. 11) in his description of the new species *Fenestella sculptilis* stated: "This species is remarkable in having a median groove instead of a keel on the poriferous side of the branches. This character gives the zoarium such a distinctive aspect that comparisons with other species are rendered quite unnecessary. A somewhat similar channelling is usually present in an associated species of *Polypora*. The latter is very common, and may be the *P. shumardi* of Prout."

The channel-shaped structures described by Ulrich (1886) make up a network of connected concave elongate depressions, straight to wavy in outline, on the obverse surface of the colony's meshwork. They occur both on the branches, running between the two rows of zooecial apertures, and on dissepiments. No channels have been observed on the reverse surface (Figure 1B). It is not surprising that Ulrich (1886) interpreted them as a distinctive character of the species, as the real keel of the branches can be rarely observed due to the conspicuous presence of these channels. The pattern defined by these structures may be either predominantly longitudinal (Figure 1C) or roughly transversal to branches (Figure 1D).

These elongate depressions are usually about 0.1 mm wide, and their apparent depth is not consistent on branches and dissepiments, sometimes looking more superficial and narrower (Figure 1D, E). Their base is a concave, smooth wall. As observed at junctions, the bases of channels may be at different depths being those on dissepiments higher than the ones on branches (Figure 1E) or most commonly, flush (Figure 1F). Occasionally apparently incipient channels can be observed on dissepiments between adjacent tubes (Figure 1G). The base of these is topographically higher than those of the longitudinal channels that bind them and may represent younger stages of development of the structure. Autozooecial apertures are placed on the steep flanks of branches, commonly aside the edges of the channels. The lateral walls of channels can occasionally cut the distal tube of autozooecial chambers, but it cannot be clearly stated if some zooids were effectively connected with the presumed tubes or this is a preservation artefact (Figure 1E).

4 DISCUSSION

Polypora shumardi Prout, 1858 is a common species in the Middle Devonian beds of Indiana and Kentucky (Ulrich 1886). Ulrich (1890, p. 586) described the dissepiments as "depressed, rounded, often with a sharply margined irregular excavation or channel". Stratton & Horowitz (1977, p. 33) included in their diagnosis the presence of "strong shallow groovelike channels laterally traversing the obverse dissepiments and branches". Stratton (1981) interpreted the channels as structures associated to presumed brood chambers placed in the dissepiments, based on ten silicified specimens from the neotypic suite. Similar tubular structures present in the cystoporate family Rhinoporidae have been interpreted as brood chambers (Buttler 1991a, 1991b).

Though the channel width is about 0.1 mm in both species, there are some differences between the structures on *P. shumardii* and *F. sculptilis*. According to Stratton (1981), the former are short, about 0.4 mm long, and usually end at the summit of branches; this morphology being different from the much longer channels of *F. sculptilis*. Suggesting that *F. sculptilis* had auxiliary tubes functionally equivalent to those of *P. shumardi* would require a plausible explanation for the widespread channels of the former, particularly those parts running longitudinally for at least 2 mm along branches. Structures associated with reproduction would probably be predominantly transverse to branches (D. Gordon, *pers. comm.*), connecting the central area of branches with the fenestrules, which is the case for the tubes developed in some specimens of *P. shumardi*. The structures in *P. shumardi* are clearly connected with the bowl-shaped voids placed in dissepiments and named type D brood ovicells by Bancroft (1986), but no type D brood chambers have been observed in dissepiments of *F. sculptilis*. On the other hand, some specimens of the latter show abundant collapsed type A brood chambers (Figure 1G, H), observed as shallow, concave dish-like structures randomly located in both zooecial rows. There is no evident connection between the type A brood chambers and the channels in *F. sculptilis*. No channel has been observed to cross a brood chamber; instead, the chambers placed closest to a channel commonly keep a well-defined edge. Moreover, these type A brood chambers do not have channels (Figure 1I), which may support the hypothesis that these structures in *F. sculptilis* are not directly related to reproduction.

Stratton (1981) noticed that only approximately half of the specimens of *P. shumardi* showed channels, which he interpreted as only a part of the studied population achieved maturity for brooding before dying. Nevertheless, the presence of channels in only some of the specimens is consistent with other potential causes different from the brooding function.

Another plausible cause for the network of channels developed on some specimens of *F. sculptilis* is the occurrence of a symbiotic organism. These structures cannot be considered as borings; voids produced by boring organisms have no skeletal walls, unless the affected area is repaired by the infected host after the perforation occurred. The preservation of the silicified specimens prevents any microstructural or ultrastructural observations, but the morphology of superficial features is finely preserved. The channels have well-defined skeletal walls that are apparently continuous with the extrazooidal skeleton of the bryozoan colony. Boring activity occurring on *F. sculptilis* with the extension observed in the network of channels would have caused damage to the many intersected zooecial chambers, possibly producing at least the death of the affected area. Encrusting organisms such as hederellids, which are a common component of Devonian benthic faunas (Wilson & Taylor 2001) are also excluded, as they have their own skeleton and a distinctive morphology. The channels, or tubes if we assume that they were originally enveloped by a thin wall, were more possibly due to interaction of the bryozoan host with a symbiotic organism.

Symbiotic intergrowths have already been reported in fenestrate bryozoans. For example, the ichnogenus *Caupokeras* McKinney, 2009 is known to occur on different genera of Emsian-Eifelian bryozoans, most of them fenestrates (McKinney 2009; Suárez Andrés 2014), and it has been interpreted as the result of bioclaustration (Palmer & Wilson 1988) of a symbiont modular organism, possibly a hydroid, by the bryozoan hosts (McKinney 2009). Nevertheless, the channels of *F. sculptilis* cannot be readily assigned to *Caupokeras*. The latter typically consists of a characteristic system of tubes, which in fenestrate bryozoans were developed on extrazooidal tissues, most commonly on the obverse colony surface. These tubes have a terminal aperture and, except for those developed in trepostomes (Suárez Andrés *et al.* 2022), the walls are thick, and the whole section of the tubes is usually well preserved despite their position being exposed to mechanical damage. This particular difference might be explained by the fact that *Caupokeras calyptos* McKinney, 2009 developed on fenestrate bryozoans with a protective superstructure and common thickenings of the laminar skeleton. Thickened laminar skeleton among the Fenestellidae is usually modest and mostly confined to the reverse surface of branches. Thus, it is possible that the observed channels were originally tubes with a very thin mineralized wall. Notwithstanding this, the channels on the obverse of *F. sculptilis* do not show the same arrangement as the tubes of *Caupokeras*. The former seem to be constrained by the morphology of the host meshwork and no evident termination has been observed. In contrast, the latter commonly developed on the superstructure showing a distinct habit strikingly different from a fenestrate meshwork, and the terminal tubes were suspended over the zooecial apertures of their hosts or exceptionally, over fenestrules on the reverse surface of the host (Suárez Andrés 2014).

Should the structures in *F. sculptilis* represent the bioclaustration of a symbiont, there is not enough evidence for a reasonable speculation on the hypothetical position of the symbiont apertures and the interaction with the host feeding function. It is only clear that the occurrence of these structures, were they channels or tubes, must have disrupted the normal activity and arrangement of the bryozoan feeding currents according to the model by Cowen & Rider (1972) and Taylor (2012, fig. 6b). Thus, it may be speculated that the symbiont was a parasite. Nevertheless, the output of symbiotic interactions cannot be asserted with certainty in fossil specimens, and the organism that triggered the structures of *F. sculptilis* may have been a commensal (E. Håkansson, *pers. comm.*). The morphology and position of these channels cannot be compared to any other structure reported in Palaeozoic bryozoans.

Some modern cheilostome bryozoans host specific symbiotic hydroids of the family Zancleidae (Boero *et al.* 2000; Puce *et al.* 2007). In these cases, the hydroids are smaller and their hosts produce thin-walled tubes with terminal apertures. Intricate intergrowths have also been reported in worms symbiotic with sponges (Aguado *et al.* 2022), but these are also tubular structures inside the host body with terminal apertures on its surface. The hypothesis suggesting that the channels of *F. sculptilis* were originally tubes with a very thin wall is not unlikely and cannot be rejected, as some kind of bryozoan host reaction to infestation can be expected. It might be also speculated that the potential symbiont settled on the obverse surface of the host and was surrounded only by soft tissues of the bryozoan, but there is no evidence supporting this idea.

5 CONCLUSIONS

The Devonian fenestrate *F. sculptilis* developed enigmatic structures unique among Palaeozoic bryozoans consisting of a set of channel-like grooves occurring on the obverse surface of colony branches and dissepiments. Due to the observed differences, it cannot be stated that the structures developed by *F. sculptilis* are the same or equivalent to those of the coexistent *P. shumardi*, which have been interpreted as auxiliary tubes connected to brood chambers. The interpretation of the causes and functions of the structures observed in *F. sculptilis* remains dubious due to the lack of unequivocal evidence supporting the reproductive function hypothesis. These channel-like structures may have been a result of symbiotic intergrowth, but their morphology and silicified preservation do not allow for a reasonably solid interpretation of the nature of the symbiont, and any possible effects of its presence on the host functions or the type of host-symbiont interaction.

ACKNOWLEDGEMENTS

This research received support from the SYNTHESYS Project http://www.synthesys.info/ which is financed by European Community Research Infrastructure Action under the FP7 "Capacities" Programme, through a grant awarded to J.L.S.A. in 2012 (GB-TAF 2129) to visit the Natural History Museum, London. Dennis P. Gordon (NIWA, NZ) is thanked for helpful discussion on the possible interpretation of the studied structures as associated to brood chambers. This manuscript benefitted from suggestions by Eckart Håkansson (University of Western Australia, Perth) during the IBA Meeting, regarding the possible nature of the host-symbiont interaction. Bushra M. Hussaini (American Museum of Natural History, NY) is acknowledged for confirmation of the holotype of *F. sculptilis* hosted at the AMNH. Kathy A. Hollis and Mark S. Florence (Smithsonian Institution, USNM) are thanked for providing information regarding unregistered specimens of *F. sculptilis* in their collections. Marcus M. Key, Jr. (Dickinson College, Carlisle, PA) and the reviewers Catherine M. Reid (University of Canterbury, Christchurch) and Hans Arne Nakrem (NHM, University of Oslo) are thanked for their constructive comments.

REFERENCES

Aguado, M.T., Ponz-Segrelles, G., Glasby, C.J., Ribeiro, R.P., Nakamura, M., Oguchi, K., Omori, A., Kohtsuka, H., Fisher, C., Ise, Y., Jimi, N., & Miura, T. 2022. *Ramisyllis kingghidorahi* n.sp., A New Branching Annelid from Japan. *Organisms Diversity and Evolution* https://doi.org/10.1007/s13127-021-00538-4

Bancroft, A.J. 1986. Ovicells in the Palaeozoic Bryozoan order Fenestrata. *Palaeontology* 29: 155–164.

Boero, F., Bouillon, J. & Gravilli, C. 2000. A Survey of *Zanclea, Halocoryne* and *Zanclella* (Cnidaria, Hydrozoa, Anthomedusae, Zancleidae) with Description of New Species. *Italian Journal of Zoology* 7: 93–124.

Buttler, C.J. 1991a. A New Upper Ordovician Bryozoan Fauna from the Slade and Redhill Beds, South Wales. *Palaeontology* 34(1): 77–108.

Buttler, C.J. 1991b. Possible Brooding Structures in Rhinoporid Cystoporate Bryozoans. In F.P. Bigey (ed), *Bryozoaires Actuels et Fossiles: Bryozoa Living and Fossil*: 61–70. Nantes: Bulletin de la Société des Sciences Naturelles de l'Ouest de la France, Mém. HS 1.

Cowen, R. & Rider, J. 1972. Functional Analysis of Fenestellid Bryozoan Colonies. *Lethaia* 5: 145–164.

McKinney, F.K. 2009. Bryozoan-hydroid Symbiosis and a New Ichnogenus, *Caupokeras. Ichnos* 16: 193–201.

McKinney, F.K. & Wyse Jackson, P.N. 2015. Part G, Revised, Vol. 2, Chapter (A: Order Fenestrata: Morphology and Growth. *Treatise Online* 66: 1–91, 54 figs. Lawrence: Paleontological Institute, The University of Kansas.

Palmer, T.J. & Wilson, M.A. 1988. Parasitism of Ordovician Bryozoans and the Origin of Pseudoborings. *Palaeontology* 31(4): 939–949.

Prout, H. 1858. Second Series of Descriptions of Bryozoa from the Paleozoic Rocks of the Western States and Territories. *Academy of Sciences of St. Louis Transactions* 1: 266–274.

Puce, S., Bavestrello, G., Di Camillo, C.G. & Boero, F. 2007. Symbiotic Relationships Between Hydroids and Bryozoans. *Symbiosis* 44: 137–143.

Stratton, J.F. 1981. Apparent Ovicells and Associated Structures in the Fenestrate Bryozoan *Polypora shumardii* Prout. *Journal of Paleontology* 55(4): 880–884.

Stratton, J.F. & Horowitz, A.S. 1977. *Polypora* M'Coy from the Devonian of Southern Indiana. *State of Indiana Department of Natural Resources, Geological Survey Bulletin* 56: 1–48.

Suárez Andrés, J.L. 2014. Bioclaustration in Devonian Fenestrate Bryozoans. The Ichnogenus *Caupokeras* McKinney, 2009. *Spanish Journal of Palaeontology* 29 (1): 5–14.

Suárez Andrés, J.L, Sendino, C. & Wilson, M.A. 2022. *Caupokeras badalloi*, A New Ichnospecies of Impedichnia from the Lower Devonian of Spain. Palaeoecological Significance. *Historical Biology* 34 (1): 62–66.

Taylor, P.D. 2012. A New Bryozoan Genus from the Jurassic of Switzerland, with a Review of the Cribrate Colony-form in Bryozoans. *Swiss Journal of Palaeontology* 131: 201–210.

Ulrich, E.O. 1886. Descriptions of New Silurian and Devonian fossils. *Contributions to American Palaeontology* 1(1): 3–35.

Ulrich, E.O. 1890. Palaeozoic Bryozoa. *Illinois Geological Survey* 8: 285–688, pl. 29–78.

Wilson, M.A. & Taylor, P.D. 2001. "Pseudobryozoans" and the Problem of Encruster Diversity in the Paleozoic. *PaleoBios* 21 (Supplement to No. 2): 134–135.

Bryozoan Studies 2022 – Key, Porter & Wyse Jackson (eds)

Bryozoans from the Turonian Carlile Shale of the Western Interior Seaway in Kansas, USA

Paul D. Taylor*
Natural History Museum, London, UK
ORCID ID: 0000-0002-3127-080X

Roger J. Cuffey†
Department of Geosciences, Pennsylvania State University, University Park, Pennsylvania, USA

ABSTRACT: Cretaceous rocks of the epicontinental Western Interior Seaway (WIS) of North America have yielded few bryozoans, despite the profusion of other marine groups. Nevertheless, stratigraphical and palaeoecological papers published by the late Don Hattin mention and figure abundant bryozoans from the Turonian Fairport Member of the Carlile Shale in west-central Kansas. Taxonomic study of material collected by Hattin from seven localities reveals the presence of just three bryozoan species, one cyclostome (*Oncousoecia defluxima* sp. nov.) and two cheilostomes (*Iyarispora hattini* sp. nov. and *Marginaria sparsa* sp. nov.). All encrust the exterior surfaces of inoceramid bivalves. Colonies of the oligoserial cyclostome *O. defluxima* are notable in exhibiting profuse evidence for *in-vivo* damage and reparative growth. Whereas previously described cheilostomes from the WIS are overwhelmingly non-ovicellate malacostegines, including the Carlile Shale electrid *I. hattini*, *M. sparsa* is unusual in being only the second known WIS ovicellate cheilostome. The dominance of malacostegines in the WIS compared to Late Cretaceous deposits elsewhere is postulated to result from their superior dispersal and ability to recolonize after deoxygenation events in this epicontinental water mass.

ZooBank registration: zoobank.org:pub:B211B8CC-F58A-49A3-B3D3-E4DEF49A5896

1 INTRODUCTION

In spite of being widespread and intensively studied, the Cretaceous sedimentary rocks of the Western Interior Seaway (WIS) have yielded few bryozoans (Cuffey 1994). Those that do occur are from widely scattered local areas and separated horizons, as recorded in several studies (Allan & Sanderson 1945; Cuffey *et al.* 1981; Frey & Larwood 1971; Nye & Lemone 1978; Scott 1970; Taylor & Cuffey 1992; Taylor & Rogers 2021; Toots & Cutler 1962; Wilson & Taylor 2013). All of these papers include species identifications, taxonomic descriptions and figures. In addition, other references include only names, figures or revisions (e.g., Bergquist 1944; Cuffey 1990; Cuffey & Scott 1992; Duke 1985, p. 171, fig. 4-26B, C; Fürsich & Kirkland 1986; Gill & Cobban 1966; Hartman & Kirkland 2002, p. 286, fig. 11; Kues 1983, 1986, 1989; Roberts *et al.* 2008; Voigt 1971, Thompson 1982, p. 619–620, pl. 386). Although several colleagues have alerted us to yet more possible occurrences, the overall impression is that bryozoans are both uncommon and low in diversity in the Cretaceous of the WIS, at least in comparison with the Gulf and Atlantic Coastal Plains (Cheetham 1954; Cheetham *et al.* 2006; Laughbaum 1960; Martha *et al.* 2019a, 2019b; McKinney & Taylor 2016; Taylor & McKinney 2006; Thomas & Larwood 1956).

The Carlile Shale outcrops in the Great Plains from northern New Mexico up through the Dakotas (Hattin 1962, p. 22). In west-central Kansas, it outcrops across a SW-NE strip (Figure 1A). Turonian bryozoans collected from localities scattered along this outcrop belt (Hattin 1962, p. 9, table 1, pl. 1) are the focus of this paper. Previously identified in older references (e.g., Hattin 1962) erroneously as belonging to un-named species of the cyclostome genus *Proboscina* and the cheilostome genus *Membranipora*, we show here that one cyclostome and two cheilostome genera are represented among Hattin's material and introduce a new species for each genus present. Our results aim to enhance knowledge of the sparse Cretaceous bryozoans recorded from the Western Interior Seaway and provide formal taxonomic names that can be used in future palaeoecological and other studies.

*Corresponding Author: p.taylor@nhm.ac.uk
†Deceased 1 January 2022

DOI: 10.1201/9781003454915-16

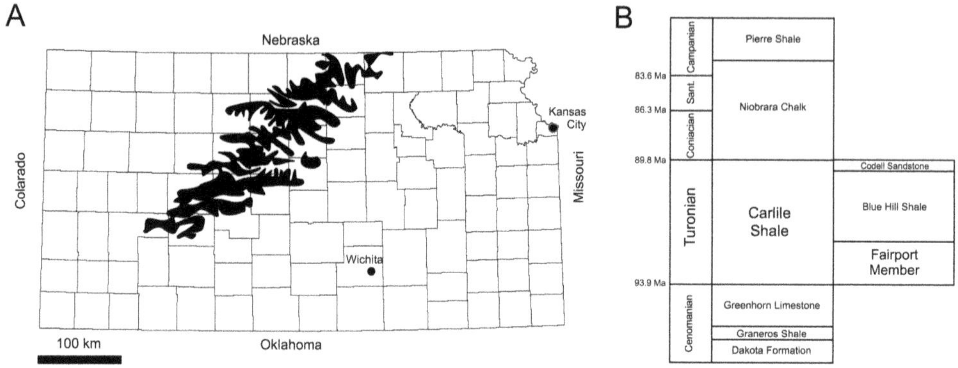

Figure 1. Geographical and geological context for the bryozoans described. A, county map of Kansas showing the outcrop of the Carlile Shale (after Hattin 1962). B, stratigraphy of the Kansas Upper Cretaceous (based on McIntosh *et al.* 2016); all bryozoans in this study come from the Fairport Member of the Carlile Shale. Dates come from the 2021 International Chronostratigraphic Chart of the International Commission on Stratigraphy (http://www.stratigraphy.org/ICSchart/ChronostratChart2021-10.pdf).

2 GEOLOGICAL SETTING

The west-central part of North America was submerged during the Late Cretaceous by a wide epeiric sea extending from the Gulf of Mexico to the Arctic Ocean (Bell *et al.* 2014, p. 188; Williams & Stelck 1975, p. 11). At the latitude of the central plains, the seaway stretched from Utah and Wyoming across to Missouri, Iowa and Minnesota (Hattin 1982, p. 60; Hattin & Hirt 1991, p. 554). The eastern shores were subdued, tranquil and contributed mostly fine-grained terrigenous sediments to the nearshore, while the western shores were subject to tectonism, volcanism and coarser sedimentation. In the centre of the seaway, on the broad flat shelf of present-day Kansas, carbonate sediments accumulated, especially fine-grained, chalk-like carbonates.

Fossils are common in these sediments, the best known of which are marine reptiles (particularly mosasaurs) and molluscs (especially bivalves and ammonites) (e.g., Everhart 2017). Surprisingly, however, some marine groups which one might have expected to encounter, including bryozoans, brachiopods and echinoderms are rare (Kauffman 1975, p. 185; Reeside 1957, p. 512). The reason is unknown but may relate to the very soft substrates, highly turbid waters, or salinity fluctuations and extremes.

The Carlile Shale in Kansas consists of two members: a lower chalky shale (Fairport Member) and an upper grey shale (Blue Hill Member); further west, these are overlain by additional sandy members (Codell Sandstone Member and Juana Lopez Calcarenite Member) (Hattin 1962, pl. 1, 1986, p. 348, McIntosh *et al.* 2016; Figure 1B; Shimada & Bell 2006, p. 591). Sedimentation in the Kansas Cretaceous has been interpreted cyclically, with the Fairport Member being at the start of a regional regression, but still deposited far offshore (Hattin 1986, p. 348).

All of the bryozoans described here come from the Fairport Member, which averages 38 m in thickness and consists of grey, olive-grey, or whitish-grey, interbedded thin chalks, shaly chalk, and chalky shale, some beds fossiliferous, others seemingly barren (Hattin 1962, table 1, pl. 1, figs 2A, 3B, 4A-C; Hattin & Hirt 1991, p. 554; Moore *et al.* 1951, pp. 23–24). The Fairport Formation is dated biostratigraphically as early middle Turonian (McIntosh *et al.* 2016).

3 MATERIAL & METHODS

The specimens examined for the present study were collected over many years (1951-1977) by D. E. Hattin and his students (including RJC). Only a few Carlile localities yielded bryozoans (Hattin 1962, p. 147–149, table 1; Hattin & Cobban 1965), as listed below:

KC-4: NE¼ NW¼ sec 6, T 22 S, R 24 W and SW¼ SE¼ sec 31, T 21 S, R 24 W, Hodgeman Co.; road bank and bluff along E-W county road, ~13.7 km NW of Jetmore; middle Fairport.

KC-11: SW¼ SW¼ sec 33, T 16 S, R 19 W, Rush Co.; bluffs and cut bank on S side of Big Timber Creek, 6.4 km N of Hargrave; middle Fairport.

KC-25: SW ¼ sec 29 and SE ¼ SE ¼ sec 30, T 11 S, R 16 W, Ellis Co.; stream banks, gullies, and bluffs in S slope of Saline River valley, ~ 8.1 km W and 1.6 km N of Fairport; upper Fairport (and overlying Blue Hill).

KC-26: SE¼ sec 29, T 11 S, R 16 W, Ellis Co.; stream banks and bluff of small southern tributary to Saline

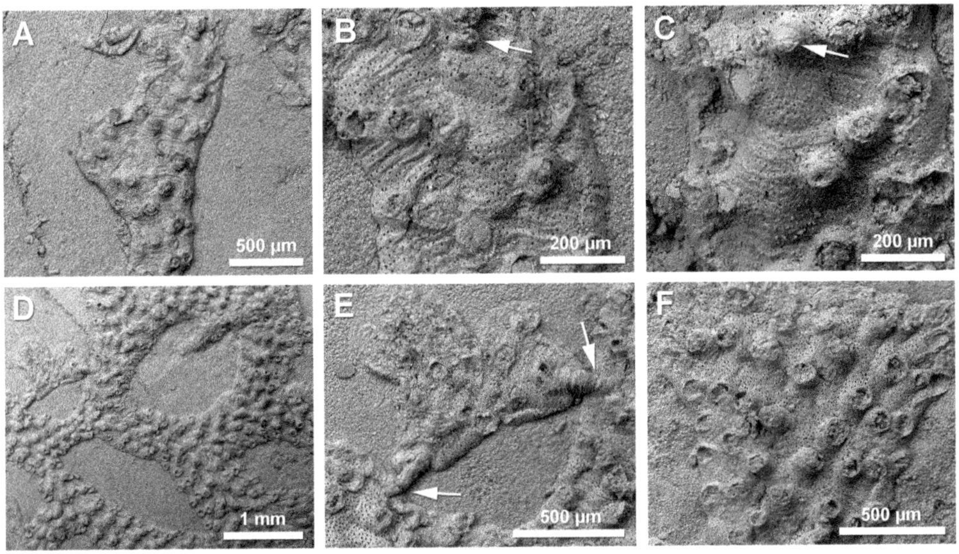

Figure 2. Scanning electron micrographs of *Oncousoecia defluxima* sp. nov. from the Carlile Shale of Kansas. A-B, holotype, KUMIP 585750; A, oligoserial encrusting branch; B, gonozooid (ooeciopore arrowed). C, KUMIP 585752, gonozooid (ooeciopore arrowed). D-F, KUMIP 585754; D, colony with relatively broad encrusting branches; E, two coalescing lateral branches (origins arrowed); E, flabellate encrusting branch prior to bifurcation.

River, 7.2 km W and 1.6 km N of Fairport; lower and middle Fairport (and underlying Greenhorn) (bryozoans in middle Fairport).

KC-30: NE¼ sec 1, T 13 S, R 16 W, Ellis Co.; cut bank on S side of small creek, 9.7 km N and 3.2 km E of Walker; middle Fairport.

KC-41: SW¼ sec 2, T 2 S, R 5 W, Republic Co.; excavation and gulley in S bluff of Republican River valley, ~ 13.7 km N and 4 km E of Courtland; upper Fairport (and overlying Blue Hill); immediately adjacent roadcut in same quarter-section, labelled "near 41", middle Fairport.

KC-53: NE¼ NE¼ sec 30, T 11 S, R 16 W, Ellis Co.; bluff of small tributary to Saline River, ~ 8.9 km WNW of Fairport; upper Fairport.

The bryozoans encrust broken fragments of inoceramid shells, some extending over serpulid tubes attached to the same inoceramids. Surface preservation is variable, ranging from poor to moderately good. Specimens were initially examined using a binocular microscope. A selection of the better-preserved bryozoans representing all of the provisional morphospecies evident was then made for SEM study. A LEO 1455VP SEM was employed operating at low vacuum to image uncoated specimens using backscattered electrons. All measurements were subsequently made from electronic SEM images using ImageJ.

Specimens have been deposited in the University of Kansas Museum of Natural History invertebrate paleontology collections (KUMIP).

4 SYSTEMATIC PALEONTOLOGY

Order Cyclostomata Busk, 1852
Suborder Tubuliporina Milne Edwards, 1838
Family Oncousoeciidae Canu, 1918
Genus *Oncousoecia* Canu, 1918

Type species. Tubulipora lobulata Canu, 1918.

Oncousoecia defluxima sp. nov.
Figures 2–3

Etymology. defluximus L. for fragmented, referring to the broken-up state of the encrusting colonies.

Type material. Holotype: KUMIP 585750, KC-53. Paratypes: (1) KUMIP 585751, KC-53, on same shell as holotype; (2) KUMIP 585752, KC-11; (3) KUMIP 585753, KC-53; (4) KUMIP 585754, KC-53; (5) KUMIP 585755, KC-41.

Description. Colony encrusting, ramose, branch multiplication through bifurcation and lateral branching (Figure 2D, E); branches oligoserial (Figure 2A), straight to curved, about 0.3 -1.2 mm wide, tending to increase in width towards bifurcations, sometimes flabellate (Figure 2F), 4-8 autozooids across the branch except for lateral branch origins which are uniserial;

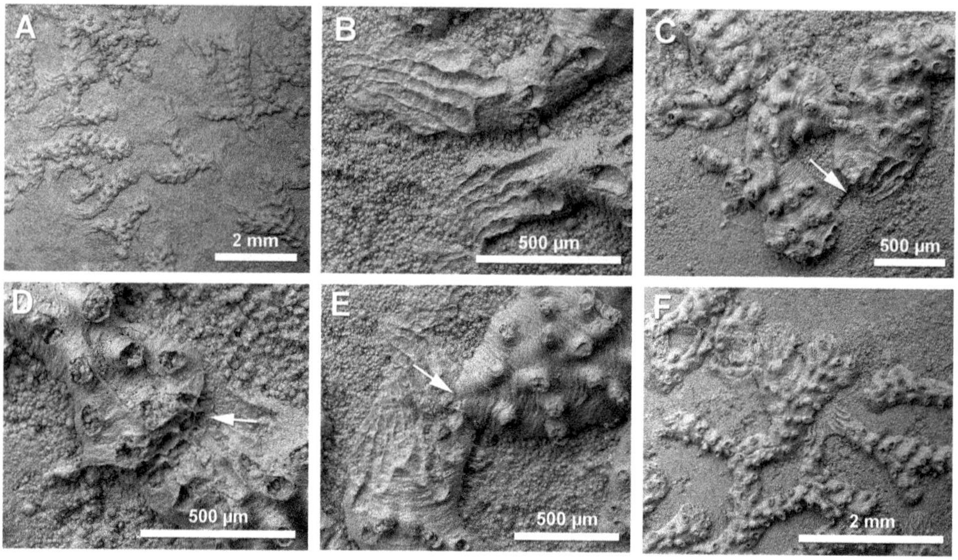

Figure 3. Scanning electron micrographs of *Oncousoecia defluxima* sp. nov. from the Carlile Shale of Kansas; KUMIP 585755. A, fragmented encrusting colony. B, destroyed proximal ends of two encrusting branches. C, curved lateral branch growing onto patch of substrate (arrowed) once occupied by the proximal part of the parent branch. D, fractured encrusting branch developing a new growing edge (arrowed) budding zooids over the area of damage. E, lateral branch (arrowed) developing close to a damaged part of a parent branch. F, branches overgrowing damaged older branches.

surface profile low and flat, in some instances slightly convex and raised medially. Early astogeny and ancestrula not observed. Branches typically disjunct with plentiful evidence for *in vivo* breakage and regenerative growth (Figure 3D-F).

Autozooids small, length 244-421 μm (mean 313 μm; SD 57 μm; 3 colonies, 20 zooids), width 79-161 μm (mean 119 μm; SD 26 μm; 3 colonies, 20 zooids), curving slightly away from branch midlines and with apertures arranged in variably defined transverse series (Figure 2A, F); zooidal boundaries distinct, marked by a thin groove; frontal walls moderately convex, especially distally, and rugose, crossed by irregular growth bands; pseudopores small, 7-9 um in length, often teardrop-shaped; peristomes low; apertures circular, length 56-88 μm (mean 71 μm; SD 9 μm; 3 colonies, 20 zooids), width 54-94 μm (mean 73 μm; SD 11 μm; 3 colonies, 20 zooids), tending to be slightly larger in medial than lateral zooids; terminal diaphragms occasionally visible below peristomial rim, presence of pseudopores uncertain.

Gonozooids located close to branch midline (Figure 2B, C), longitudinally elongate, narrow, 447-623 μm long by 187-243 μm wide (3 determinations), edges sometimes indented by autozooidal peristomes; frontal wall densely pseudoporous; ooeciostome short; ooeciopore terminal, smaller than an autozooidal aperture, 37-65 μm long by 43-75 m wide (3 determinations).

Remarks. This species was referred to in older publications as *Proboscina* sp. (Hattin 1962, 1986; Hattin & Cobban 1965; Hattin & Hirt 1991). Preliminary study suggested the presence of two oligoserial cyclostome species in the Carlile Shale, one typically *Oncousoecia*-like with low profile branches and unaligned apertures (cf. Taylor & Zatoń 2008), and the second with branches of higher profile and aligned apertures on either side of a median summit recalling the Jurassic-Cretaceous genus *Idmonea* (cf. Walter 1970). Some specimens show branches of these two morphologies growing side-by-side. However, the distinction between these morphotypes vanishes when a large suite of specimens is studied: autozooid size and the gonozooids are identical in the two morphotypes, and all intermediate stages appear to exist between them. Therefore, we conclude that they represent variants of a single species.

Compared with other North American Cretaceous species of *Oncousoecia*, the new species has less robust branches and autozooids about half the size of those in *O.*? *nonomologabili* Taylor & McKinney, 2006 from the Maastrichtian of New Jersey. Another species, *O. khirar* Martha *et al.*, 2019a, from the Albian of Texas, also has larger autozooids, and the gonozooid is somewhat broader and more subdued.

By far the most common species present among the specimens studied, *O. defluxima* is present at all seven of the sampled localities. As with the other two bryozoan species found in the Carlile Shale, colony origins have not been

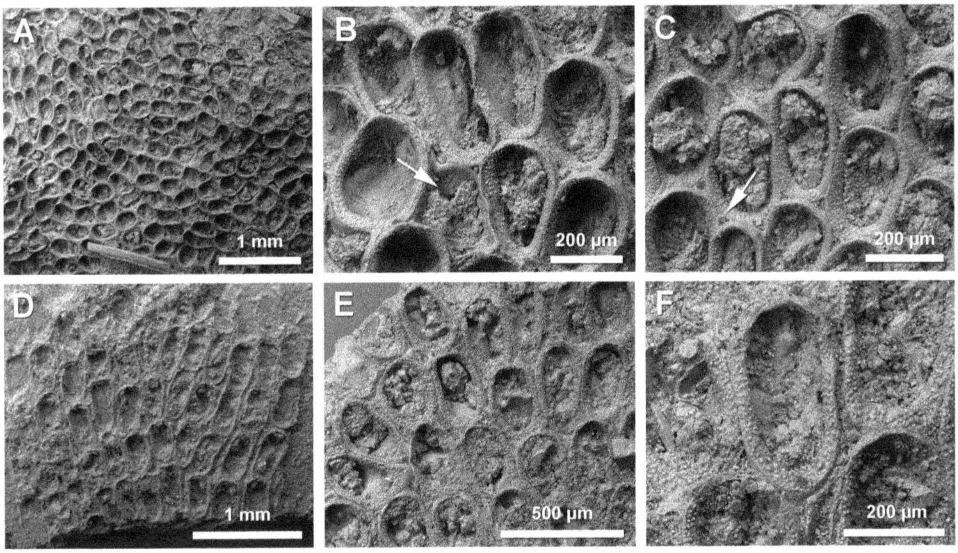

Figure 4. Scanning electron micrographs of *Iyarispora hattini* sp. nov. from the Carlile Shale of Kansas. A-C, holotype, KUMIP 585756; A, multilamellar colony; B, autozooids and a large kenozooid (arrowed); C, oblique view of autozooids, some with small kenozooids at their corners (arrowed). D, KUMIP 585757, colony on the edge of a shell. E-F, KUMIP 585758; E, group of autozooids; F, detail of an autozooid.

observed and characters of early astogeny, including the ancestrula, remain unknown. When traced back in a proximal direction, branches invariably terminate in broken ends (Figure 3B). The disjunct nature of the specimens, with abundant evidence of reparative growth following damage, are notable features of *O. defluxima*, surpassing levels recorded in other species of *Oncousoecia* and indeed in most cyclostomes (cf. Taylor *et al.* 2018). Adventitious lateral branches (Figure 2E) are abundant, and it is possible that at least some of these represent a form of regenerative growth (Figure 3E). The majority of the lateral branches originate from autozooidal apertures on the edge of an existing branch but some emerge between apertures. Those of the first type resemble lateral branches described from the Jurassic cyclostome *Idmonea snehi* Wilson *et al.*, 2014, while formation of the second type probably involved resorption of skeleton to form a window (see Batson *et al.* 2020).

Distribution. Carlile Shale, Fairport Member localities KC-4, KC-11, KC-25, KC-26, KC-30, KC-41, KC-53.

Order Cheilostomata Busk, 1852
Suborder Malacostegina Levinsen, 1902
Family Electridae d'Orbigny, 1851
Genus *Iyarispora* Martha, Taylor & Rader, 2019b

Type species. Iyarispora ikaanakiteeh Martha, Taylor & Rader, 2019b.

Iyarispora hattini sp. nov.
Figure 4

Etymology. Named for Don Hattin (1929-2016) who collected the specimens described here and in honour of his research on the Cretaceous of Kansas.

Type material. Holotype: KUMIP 585756, KC-53. Paratypes: (1) KUMIP 585757, KC-11; (2) KUMIP 585758, KC-53.

Description. Colony encrusting, multiserial, sheet-like, unilamellar (Figure 4D) or occasionally multilamellar (Figure 4A). Early astogeny and ancestrula not observed.

Autozooids rounded rectangular in shape (Figure 4B, C, E, F), moderately small, length 315-483 µm (mean 378 µm; SD 54 µm; 3 colonies, 20 zooids), width 186-285 µm (mean 229 µm; SD 24 µm; 3 colonies, 20 zooids), separated by grooves. Opesia elongate elliptical, occupying almost all the frontal surface, tending to be squared-off distally. Gymnocyst lacking or very narrow and mostly visible proximally; spines absent. Mural rim raised. Cryptocyst steeply sloping, narrow, pustulose, thinning distally. Pore chambers, intramural buds, closure plates and ovicells not observed, the last of these presumed to be lacking.

Kenozooids present at the corners of some autozooids, subtriangular and small, about 40 µm in width (Figure 4C); occasional larger kenozooids with concave sides (Figure 4B).

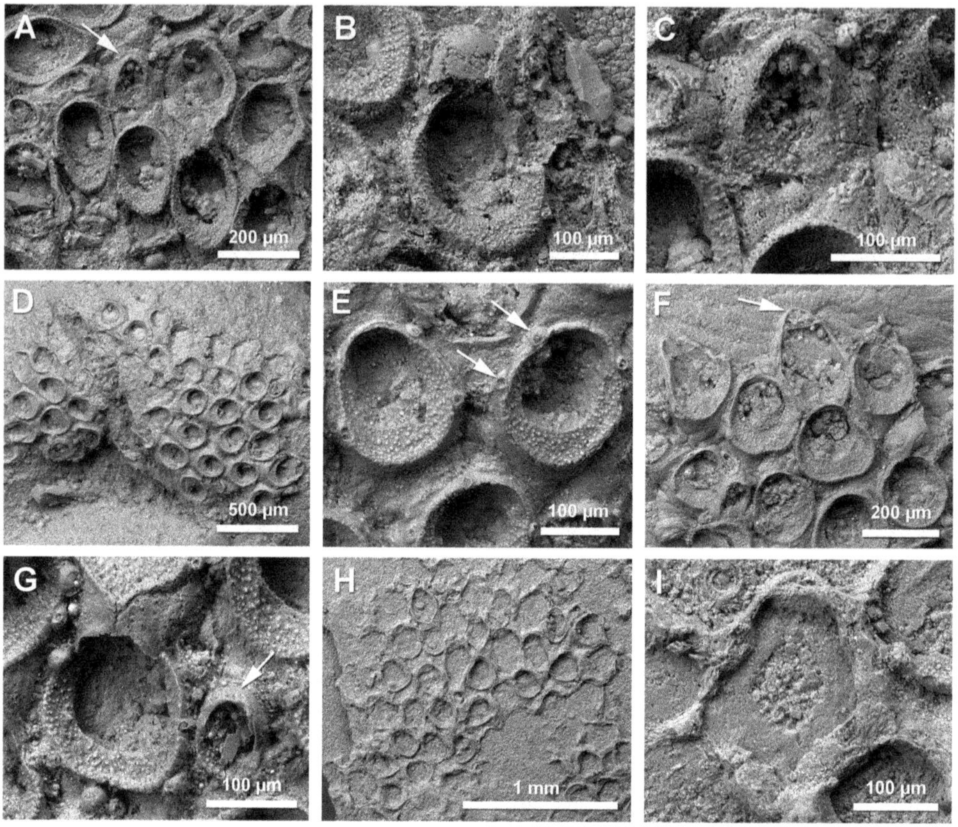

Figure 5. Scanning electron micrographs of *Marginaria sparsa* sp. nov. from the Carlile Shale of Kansas. A-C, holotype, KUMIP 585759; A, autozooids and a few avicularia (arrowed); B, ovicellate autozooid; C, avicularium. D-E, KUMIP 585762; D, small colony; E, autozooids with two pairs of spine bases (arrowed). F, KUMIP 585764, growing edge showing large distal pore chambers (arrowed). G, KUMIP 585760, oblique view of an ovicellate zooid and avicularium (arrowed). H-I, KUMIP 585763; H, typical preservation; I, worn zooid with uncalcified window in the basal wall exposing the rough surface of the substrate beneath the zooid.

Remarks. Generic placement of this species is somewhat problematical. Compared with the Campanian-Recent genus *Conopeum* Gray, 1848, the Carlile species lacks a distinct groove between gymnocyst and cryptocyst (see Gordon *et al.* 2020; Taylor & Rogers 2021). In the Albian-Ypresian genus *Kenocharixa* Dick *et al.*, 2018, kenozooids are conspicuously developed at the corners of the autozooids whereas they are sporadic in the Carlile species. It is with some reservation that we place the new species in *Iyarispora* as the distinctive porous closure plates present in this genus have not been seen. Compared with the Campanian species *Conopeum flumineum* Taylor & Rogers, 2021 from Montana further north in the WIS, the Carlile Shale species has somewhat smaller zooids (mean length 378 µm vs. 444 µm) but more importantly lacks a cryptocystal groove. Three Campanian-Maastrichtian species of *Iyarsipora* were originally assigned to *Conopeum* by Taylor & McKinney (2006): *I. spissamentum*, *I. nelsoni* and *I. paranelsoni*. All have closure plates and larger autozooids and, while the proximal cryptocyst in the first two is broader than that of *I. hattini*.

Distribution. Carlile Shale, Fairport Member localities KC-11, KC-25, KC-41, KC-53.

Suborder Flustrina Smitt, 1867
Family Calloporidae Norman, 1903
Genus *Marginaria* Roemer, 1840

Type species. *Cellepora elliptica* von Hagenow, 1839.

Marginaria sparsa sp. nov.
Figure 5

Etymology. In reference to the sparse avicularia compared to other species belonging to the genus.
Type material. Holotype: KUMIP 585759, KC-41. Paratypes: (1) KUMIP 585760, on same shell as holotype; (2) KUMIP 585761, on same shell as holotype; (3) KUMIP 585762, KC-41; (4) KUMIP 585763, KC-11; (5) KUMIP 585764, KC-41.

Description. Colony encrusting, multiserial, sheet or spot-like (Figure 5D), unilamellar, generally patchily preserved (Figure 5H). Early astogeny and ancestrula not observed.

Autozooids oval in shape (Figure 5A), small, length 222-387 µm (mean 292 µm; SD 39 µm; 5 colonies, 20 zooids), width 153-255 µm (mean 203 µm; SD 23 µm; 5 colonies, 20 zooids), separated by grooves. Opesia elongate elliptical, occupying less than half of the frontal surface, length 186-234 µm (mean 210 µm; SD 14 µm; 5 colonies, 20 zooids), width 140-220 µm (mean 178 µm; SD 17 µm; 5 colonies, 20 zooids). Gymnocyst well developed proximally. Mural rim crenulated, raised into a sharp ridge, slightly indented by one or two pairs of spine bases, a distolateral pair and mediolateral to proximolateral pair (Figure 5E), lacking in some zooids. Cryptocyst sloping, deep and almost shelf-like proximally, narrowing distally, pustulose. Distolateral pore chamber large. Basal wall containing an uncalcified window in at least some zooids (Figure 5I). Intramural buds rare, closure plates not observed. Ovicells small (Figure 5G), wider than long, mostly immersed beneath the gymnocyst of the distal zooid, exposed ooecia with a smooth, non-porous ? ectooecium (Figure 5B), about 61-86 µm long by 102-120 µm wide.

Avicularia interzooidal (Figure 5A, C, F, G), scattered and significantly outnumbered by the autozooids, small, about 209-238 µm long by 154-169 µm wide; opesia with a pustulose, shelf-like cryptocyst; rostrum of similar length to opesia but slightly narrower, arch-shaped, with a well-rounded end (Figure 5C); combined opesia and rostrum length 121-130 µm, width 83-100 µm; one observed example containing a possible intramural bud.

Remarks. The avicularia of this new species recall those of *Wilbertopora hoadleyae* Cheetham *et al.*, 2006 from the Upper Albian of Texas in their size relative to the autozooids and the well-rounded rostrum. They differ, however, in the opesia being wider than the rostrum and having a pustulose cryptocyst. Some zooids of *M. sparsa* reveal uncalcified windows in the basal wall (Figure 5I), a trait rarely recorded among Cretaceous cheilostomes but which seems to be commoner in the post-Cretaceous. The ovicell is usually enveloped by the gymnocystal calcification of the distal zooid (Figure 5G), the ooecial surface becoming visible in abraded examples where this has been worn away (Figure 5B).

In most species of *Marginaria*, avicularia are more abundant than in *M. sparsa*, often equalling or surpassing the number of autozooids. Avicularia are more numerous in the Campanian-Maastrichtian species described by Taylor & McKinney (2006) from the eastern USA, none of which has autozooids with such a well-developed proximal cryptocyst as *M. sparsa*. An un-named species from the Cenomanian of Devon, UK (Taylor & Martha 2017) is comparable with *M. sparsa* in having few avicularia but the autozooids possess 10-12 spine bases compared with 0-4 in the Carlile Shale species.

Distribution. Carlile Shale, Fairport Member localities KC-11, KC-41, KC-53.

5 DISCUSSION

All of the bryozoans are associated with fragments of a bivalve identified by Hattin (1962) as *Inoceramus cuvieri* Sowerby. Intact specimens of *I. cuvieri* are said by Hattin & Hirt (1991) to reach up to a metre in size in the Carlile Shale. Some of the bryozoans encrusted serpulid worm tubes growing on the inoceramid fragments but most grew directly on the shell surfaces of the bivalve. Whereas Hattin & Hirt (1991) recorded a few examples of gooseneck barnacles in the Carlile Formation apparently attached to the interior surfaces of inoceramid shells, all of the bryozoans observed here encrusted exterior surfaces, which are distinguished by exhibiting growth banding compared to the smooth interior surfaces. While the apparent preference for inoceramid exteriors may reflect in-vivo encrustation and the inaccessibility of tissue-covered interior surfaces, it is possible that what appear to be interior surfaces actually represent the boundaries between the preserved, outer calcite parts of the shells and the non-preserved, inner aragonitic parts which were lost during post-burial diagenesis and never formed exposed surfaces that could be encrusted by bryozoans on the seafloor (cf. Håkansson *et al.* in press). Should the latter be the case, the apparent preference for colonization of shell exteriors may be an artifact, although Hattin & Hirt (1991) believed that encrustation of the inoceramid shells from the Carlile Shale mostly occurred while the host bivalves were still alive.

Unusually high levels of disturbance indicated by branch breakage and reparative growth are evident in *Oncousoecia defluxima* sp. nov. None of the colonies observed are complete and early astogenetic stages have not been seen. Instead, most consist of disjunct patches of zooids (Figure 3A), broken proximally but with distal growing edges still usually intact (Figure 3F). Evidence for fragmentation of colonies during life is abundant. Lateral branches originating on the side of damaged branches can be seen to grow onto areas of the substrate formerly covered by older branches belonging to the same colony (Figure 3C). In one colony, a new growing edge is observed budding zooids distally over the remnants of destroyed zooids in a branch that had been severed (Figure 3D). All of the examples of reparative growth observed involve either lateral branches (Figure 3E) or originate from the broken distal ends of branches; none originate from branch proximal ends. This suggests a strong polarity in growth. The cause of branch breakage is unclear but is more likely to be biological grazing than abiotic factors in view of the tranquil palaeoenvironment (e.g. Hattin & Hirt 1991) and the lack of evidence for abrasive coarse sediment particles. Unfortunately, the grazing trace ichnogenera *Gnathichnus* (produced by echinoids) and *Radulichnus* (produced by polyplacophoran and gastropod molluscs) have not been observed here on shell surfaces encrusted by the bryozoans.

Most of the relatively few cheilostomes described previously from the Late Cretaceous of the WIS are of malacostegine skeletal grade, lacking ovicells for larval brooding and which instead can be inferred to have possessed planktotrophic larvae of cyphonautes type. These include *Eokotosokum bicystosum* (Allan & Sanderson, 1945), *Heteroconopeum ramosum* (Toots & Cutler, 1962), *Herpetopora shawi* (Frey & Larwood, 1971), *Villicharixa lintonensis* (Cuffey *et al.*, 1981), *Conopeum flumineum* Taylor & Rogers, 2021, and *Iyarispora hattini* n. sp. To these six species must be added an undescribed species of *Iyarispora* associated with the ammonite *Placenticeras* from the Campanian Bearpaw Shale of Alberta, Canada (PDT unpublished). In contrast, the only ovicellate cheilostomes from the WIS are *Dysnoetopora demissa* (White 1879), redescribed by Toots & Cutler (1962), and *Marginaria sparsa* sp. nov. described in the current paper. Thus, more than three-quarters of cheilostome species known from the WIS are malacostegines, an unusually high proportion. By comparison, of the 94 cheilostome species described by Taylor & McKinney (2006) from the Gulf and Atlantic coastal plains, 11 (12%) are malacostegines, a proportion that is also typical of Late Cretaceous faunas globally, including both northern and southern hemispheres (Taylor 2019). The failure of ovicellate cheilostomes to prosper in the WIS is unclear but one possible explanation relates to fluctuating oxygen levels in this epicontinental basin. During the Late Cretaceous, the WIS experienced periodic deoxygenation (Lowery *et al.* 2018). This would potentially have caused extinctions of local benthic communities containing bryozoans, demanding recolonization after oxygen levels rose back to normal levels. Malacostegines with planktotrophic larvae capable of long distance dispersal may have had the advantage over neocheilostomes with their short-lived larvae in their capacity to recolonize benthic habitats at these times.

ACKNOWLEDGEMENTS

The late Donald E. Hattin (Indiana University, Bloomington) is thanked for sending his collection of bryozoan-encrusted shells from the Carlile Formation to us for study. Curator Bruce S. Lieberman and Collection Manager Natalia López Carranza of the University of Kansas Biodiversity Institute (KUBI) Division of Invertebrate Paleontology are acknowledged for accessing the material. Critical reviews by Eckart Håkansson, Kamil Zagorsek and Marcus Key led to improvements in this paper.

REFERENCES

Allan, J.A. & Sanderson, J.O.G. 1945. Geology of Red Deer and Rosebud Sheets, Alberta. *Report, Research Council of Alberta* 13: 1–115.

Batson, P.B., Tamberg, Y., Taylor, P.D., Gordon, D.P. & Smith, A.M. 2020. Skeletal Resorption in Bryozoans: Occurrence, Function and Recognition. *Biological Reviews* 95: 1341–1371.

Bell, P.R., Fanti, F., Mitchell, M.T. & Currie, P.J. 2014. Marine Reptiles (Plesiosauria and Mosasauridae) from the Puskwaskau Formation (Santonian-Campanian), West-central Alberta. *Journal of Paleontology* 88: 187–194.

Bergquist, H.R. 1944. Cretaceous of the Mesabi Iron Range, Minnesota. *Journal of Paleontology* 18: 1–30.

Busk, G. 1852. An Account of the Polyzoa, and Sertularian Zoophytes, Collected in the Voyage of the Rattlesnake, on the Coasts of Australia and the Louisiade Archipelago. In J. MacGillivray (ed.) *Narrative of the Voyage of HMS Rattlesnake:* 343–402. London: T and W. Boone.

Canu, F. 1918. Les Ovicelles des Bryozoaires Cyclostomes. Études sur Quelques Familles Nouvelles et Anciennes. *Bulletin de la Société Géologique de France, Série 4* 16: 324–335.

Cheetham, A.H. 1954. A New Early Cretaceous Cheilostome Bryozoan from Texas. *Journal of Paleontology* 28: 177–184.

Cheetham, A.H., Sanner, J., Taylor, P.D. & Ostrovsky, A.N. 2006. Morphological Differentiation of Avicularia and the Proliferation of Species in the Mid-Cretaceous *Wilbertopora* Cheetham, 1954 (Bryozoa: Cheilostomata). *Journal of Paleontology* 80: 49–71.

Cuffey, R.J. 1990. Cretaceous Bryozoans on Baculite Mesa, Colorado. *Geological Society of America, Abstracts with Programs* 22 (6): 7.

Cuffey, R.J. 1994. Cretaceous Bryozoan Faunas of North America – Preliminary Generalizations. In P.J. Hayward, J.S. Ryland & P.D. Taylor (eds), *Biology and Palaeobiology of Bryozoans*: 55–56. Fredensborg: Olsen & Olsen.

Cuffey, R.J. & Scott, R.W. 1992. Bryozoans and Sclerosponges in Early Cretaceous Patch Reefs of the Upper Mural Limestone (lower Albian) along the Arizona–Sonora Border (United States–Mexico). *Geological Society of America, Abstracts with Programs* 24 (6): 7.

Cuffey, R.J., Feldmann, R.M. & Pohlable, K.E. 1981. New Bryozoa from the Fox Hills Sandstone (Upper Cretaceous, Maestrichtian) of North Dakota. *Journal of Paleontology* 55: 401–409.

Dick, M.H., Sakamoto, C. & Komatsu, T. 2018. Cheilostome Bryozoa from the Upper Cretaceous Himenoura Group, Kyushu, Japan. *Paleontological Research* 22: 239–264.

Duke, W.L. 1985. *Sedimentology of the Upper Cretaceous (Turonian) Cardium Formation in Outcrop in Southern Alberta*. Unpublished Ph.D. thesis, McMaster University, Hamilton; 724 pp.

Everhart, M.J. 2017. *Oceans of Kansas*. Second Edition. Indiana University Press, Bloomington; 704 pp.

Frey, R.W. & Larwood, G.P. 1971. *Pyripora shawi*, New Bryozoan from the Upper Cretaceous of Kansas (Niobrara Chalk) and Arkansas (Brownstown Marl). *Journal of Paleontology* 45: 969–976.

Fürsich, F.T. & Kirkland, J.I. 1986. Biostratinomy and Paleoecology of a Cretaceous Brackish Lagoon. *Palaios* 1: 543–560.

Gill, J.R. & Cobban, W.A. 1966. The Red Bird Section of the Upper Cretaceous Pierre Shale in Wyoming. *U.S. Geological Survey Professional Paper* 393A: 1–73.

Gordon, D.P., Sutherland, J.E., Perez, B.A., Waeschenbach, A., Taylor, P.D. & Di Martino, E. 2020. The Bryozoan Genus *Conopeum* (Electridae) in New Zealand, with Description of a New Species and Discussion of the Morphological and Genetic Characters of *Conopeum seurati* (Canu, 1928). *Journal of Natural History 54: 947–970.*

Gray, J.E. 1848, List of the Specimens of British Animals in the Collections of the British *Museum. Part I. Centrionae or Radiated Animals. London: Trustees of the British Museum*, 173 pp.

Hagenow, F. v. 1839. Monographie der Rugen'schen Kreideversteinerungen. I Abt. Phytolithen u. Polyparien. *Neues Jahrbuch für Geognosie, Geologie und Petrefaktenkunde, Stuttgart* 1839: 253–296.

Håkansson, E., Gordon, D.P. & Taylor, P.D. in press. Bryozoa from the Maastrichtian Korojon Formation, Western Australia. *Fossils and Strata*.

Hartman, J.H. & Kirkland, J.I. 2002. Brackish and Marine Mollusks of the Hell Creek Formation of North Dakota: Evidence for a Persisting Cretaceous Seaway. *Geological Society of America Special Paper* 361: 271–296.

Hattin, D.E. 1962. Stratigraphy of the Carlile Shale (Upper Cretaceous) in Kansas. *Kansas Geological Survey Bulletin* 156: 1–155.

Hattin, D.E. 1982. Stratigraphy and Depositional Environment of Smoky Hill Chalk Member, Niobrara Chalk (Upper Cretaceous) of the Type Area, Western Kansas. *Kansas Geological Survey Bulletin* 225: 1–108.

Hattin, D.E. 1986. Carbonate Substrates of the Late Cretaceous Sea, Central Great Plains and Southern Rocky Mountains. *Palaios* 1: 347–367.

Hattin, D.E. & Cobban, W.A. 1965. *Upper Cretaceous Stratigraphy, Paleontology, and Paleoecology of Western Kansas: A Field Conference Guidebook for the Annual Meetings*, the Geological Society of America and Associated Societies, Kansas City, Missouri, 1965. Lawrence and New York: Kansas Geological Survey and the Geological Society of America, 69 pp.

Hattin, D.E. & Hirt, D.S. 1991. Paleoecology of Scalpellomorph Cirripeds in the Fairport Member, Carlile Shale (Middle Turonian) of Central Kansas. *Palaios* 6: 553–563.

Kauffman, E.G. 1975. Dispersal and Biostratigraphic Potential of Cretaceous Benthonic Bivalvia in the Western Interior. *Geological Association of Canada Special Paper* 13: 163–194.

Kues, B.S. 1983. Bryozoan and Crustacean from Fruitland Formation (Upper Cretaceous) of New Mexico. *New Mexico Geology* 5: 52–55, 68.

Kues, B.S. 1986. Paleontology and Correlation of a Lower Cretaceous (Albian) Outlier in Roosevelt County, New Mexico. *New Mexico Geology* 8: 88–94.

Kues, B.S. 1989. Taxonomy and Variability of three *Texigryphaea* (Bivalvia) Species from their Lower Cretaceous (Albian) Type Localities in New Mexico and Oklahoma. *Journal of Paleontology* 63: 454–483.

Laughbaum, L.R. 1960. A Paleoecologic Study of the Upper Denton Formation, Tarrant, Denton, and Cooke counties, Texas. *Journal of Paleontology* 34: 1183–1197.

Levinsen, G.M.R. 1902. Studies on Bryozoa. *Videnskabelige Meddelelser fra den Naturhistoriske Forening i Kjobenhavn* 54: 1–31.

Lowery, C.M., Leckie, R.M., Bryant, R., Elderbak, K., Parker, A., Polyak, D.E., Schmidt, M., Snoeyenbos-West, O. & Sterzinar, E. 2018. The Late Cretaceous Western Interior Seaway as a Model for Oxygenation Change in Epicontinental Restricted Basins. *Earth-Science Reviews* 177: 545–564.

Martha, S.O., Taylor, P.D. & Rader, W.L. 2019a. Early Cretaceous Cyclostome Bryozoans from the Early to Middle Albian of the Glen Rose and Walnut formations of Texas, USA. *Journal of Paleontology* 93: 244–259.

Martha, S.O., Taylor, P.D. & Rader, W.L. 2019b. Early Cretaceous gymnolaemate bryozoans from the early to middle Albian of the Glen Rose and Walnut formations of Texas, USA. *Journal of Paleontology* 93: 260–277.

McIntosh, A.P., Shimada, K. & Everhart, M.J. 2016. Late Cretaceous Marine Vertebrate Fauna from the Fairport Chalk Member of the Carlile Shale in Southern Ellis County, Kansas, U.S.A. *Transactions of the Kansas Academy of Science* 119: 222–230.

McKinney, F.K. & Taylor, P.D. 2016. The Premier North American Maastrichtian Bryozoan Fauna: Coon Creek, *Tennessee. Bulletin of the Alabama Museum of Natural History* 33: 1–6.

Milne Edwards, H. 1838. Mémoire Sur les Crisies, les Hornères et Plusieurs Autres Polypes Vivants Ou fossiles dont l'organisation est Analogue à Celle des Tubulopores. *Annales des Sciences Naturelles, Série 2* 9: 193–238.

Moore, R.C., Frye, J.C., Jewett, J.M., Lee, W. & O'Connor, H.G. 1951. The Kansas Rock Column. *Kansas Geological Survey Bulletin* 89: 1–132.

Norman, A. 1903. Notes on the Natural History of East Finmark, Polyzoa. *Annals and Magazine of Natural History, Series 7* 11: 567–598.

Nye, O.B. & Lemone, D.V. 1978. Multilaminar Growth in *Reptomulticava texana*, a New Species of Cyclostome Bryozoa. *Journal of Paleontology* 52: 830–845.

Orbigny, A.D. d' 1851–54. *Paléontologie Française. Description des Mollusques et Rayonées Fossils. Terrains Crétacés. Tome 5 Bryozoaires*. Paris: Victor Masson, 1192 pp.

Reeside, J.B. 1957. Paleoecology of the Cretaceous Seas of the Western Interior of the United States. *Geological Society of America Memoir* 67: 505–542.

Roberts, E.M., Tapanila, L. & Mijal, B. 2008. Taphonomy and Sedimentology of Storm-generated Continental Shell beds: A Case Example from the Cretaceous Western Interior Basin. *Journal of Geology* 116: 462–479.

Roemer, F.A. 1840–41. *Die Versteinerungen des norddeutschen Kreidegebirges*. Hannover: Hahn'sche Hofbuchhandlung, 145 pp.

Scott, R.W. 1970. Paleoecology and Paleontology of the Lower Cretaceous Kiowa Formation, *Kansas. University of Kansas Paleontological Contributions Article* 52: 1–94.

Shimada, K. & Bell, G.L. 2006. *Coniasaurus* Owen, 1850 (Reptilia: Squamata), from the Upper Cretaceous Niobrara Chalk of Western Kansas. *Journal of Paleontology* 80: 589–593.

Smitt, F.A. 1867. Kritisk Förteckning Öfver Skandinaviens Hafs-Bryozoer. II. *Öfversigt af Kongliga Vetenskaps-Akademiens Förhandlingar* 23: 395–534.

Taylor, P.D. 2019. A Brief Review of the Scanty Fossil Record of Cretaceous Bryozoans from Gondwana. *Australasian Palaeontological Memoirs* 52, 147–154.

Taylor, P.D. & Cuffey, R.J. 1992. Cheilostome Bryozoans from the Upper Cretaceous of the Drumheller Area, Alberta, Canada. *Bulletin of the British Museum (Natural History) (Geology Series)* 48: 13–24.

Taylor, P.D., Di Martino, E. & Martha, S.O. 2018. Colony Growth Strategies, Dormancy and Repair in Some Late Cretaceous Encrusting Bryozoans: Insights into the Ecology of the Chalk Seabed. *Palaeobiodiversity and Palaeoenvironments* 99: 425–446.

Taylor, P.D. & Martha, S.O. 2017. Cenomanian Cheilostome Bryozoans from Devon, England. *Annales de Paléontologie* 103: 19–31.

Taylor, P.D. & McKinney, F.K. 2006. Cretaceous Bryozoa from the Campanian and Maastrichtian of the Atlantic and Gulf Coastal Plains, United States. *Scripta Geologica* 132: 1–346.

Taylor, P.D. & Rogers, R.R. 2021. A New Cheilostome Bryozoan from a Dinosaur Site in the Upper Cretaceous (Campanian) Judith River Formation of Montana. *Journal of Paleontology* 95: 965–973.

Taylor, P.D. & Zatoń, M. 2008. Taxonomy of the Bryozoan genera *Oncousoecia*, *Microeciella* and *Eurystrotos* (Cyclostomata: Oncousoeciidae). *Journal of Natural History* 42: 2557–2574.

Thomas, H.D. & Larwood, G.P. 1956. Some 'Uniserial' Membraniporine Polyzoan Genera and a New American Albian Species. *Geological Magazine* 93: 369–376.

Thompson, I. 1982. *The Audubon Society Field Guide to North American Fossils*. New York: Alfred A. Knopf, 846 pp.

Toots, H. & Cutler, J.F. 1962. Bryozoa from the 'Mesaverde' Formation (Upper Cretaceous) of Southeastern Wyoming. *Journal of Paleontology* 36: 81–86.

Voigt, E. 1971. The Cheilostomate Nature of the Alleged Cyclostomatous Bryozoan Genus *Dysnoetopora. Lethaia* 4: 79–100.

Walter, B. 1970. Les Bryozoaires Jurassiques en France. *Documents des Laboratoires de Géologie de la Faculté des Sciences de Lyon* 35 [for 1969], 1–328.

White, C.A. 1879. Paleontological Papers no. 11: Remarks upon Certain Carboniferous Fossils From Colorado, Arizona, Idaho, Utah, and Wyoming, and Certain Cretaceous Corals From Colorado, Together with Descriptions of New Forms. *Bulletin of the United States Geological and Geographical Survey* 5: 209–221.

Williams, G.D. & Stelck, C.R. 1975. Speculations on the Cretaceous Palaeogeography of North America. *Geological Association of Canada Special Paper* 13: 1–20.

Wilson, M.A. & Taylor, P.D. 2013. Palaeoecology, Preservation and Taxonomy of Encrusting Ctenostome Bryozoans Inhabiting Ammonite Body Chambers in the Late Cretaceous Pierre Shale of Wyoming and South Dakota, USA. In A. Ernst, P. Schäfer & J. Scholz (eds), *Bryozoan Studies 2010*: 419–433. Heidelberg: Springer.

Wilson, M.A., Bosch, S. & Taylor, P.D. 2014. Middle Jurassic (Callovian) Cyclostome Bryozoans From the Tethyan Tropics (Matmor Formation, Southern Israel). *Bulletin of Geosciences* 90: 51–63.

Bryozoan Studies 2022 – Key, Porter & Wyse Jackson (eds)
© 2023 The Author(s), ISBN: 978-1-032-59487-3

Shallow coral reef bryozoans of Caracol Bay, Haiti

Judith E. Winston*
Smithsonian Marine Station, Fort Pierce, FL, USA
ORCID ID: 0000-0002-2843-7859

Jeremy B.C. Jackson
American Museum of Natural History, Division of Paleontology, New York, NY, USA
ORCID ID: 0000-0002-6080-2254

ABSTRACT: Fifty-eight species of coral reef associated bryozoans, three cyclostomes and 55 cheilostomes were collected in May of 1981, during a research cruise organized by Walter Adey of the Smithsonian's National Museum of Natural History. The area studied, Caracol Bay, on the north coast of Haiti, is now part of the Three Bays National Park. Sampling for bryozoans was limited to 4 days and 7 sites between depths of 3 to 23 metres where bryozoans were collected by SCUBA diving. Species richness was exceptionally high for such a small area, yielding 43 percent of the total reef associated bryozoan diversity obtained from a decade of intensive collecting all around Jamaica. Five new species are described: *Reptadeonella haitiensis*, *Celleporaria soleata*, *Petraliella calceata*, *Schizoporella trimaxillata*, and *Hippaliosina caracolensis*. One additional species, *Gemellitheca multilamellosa*, described by Canu & Bassler (1928a) based on a Panamanian fossil specimen, is redescribed here with SEM illustration.

1 INTRODUCTION

In comparison with many other tropical marine regions the Caribbean Sea has a long history of study. For some organisms, e.g., corals, mollusks, and reef fish, scientific knowledge is considerable. It would seem reasonable that at least the marine life around the large islands of the Greater Antilles would be well known by now, but especially for many of the non-commercially valuable, yet ecologically important groups, including bryozoans, this is not the case.

1.1 *Cuba*

Ferdinand Canu and Raymond S. Bassler's 1928 report on the *Fossil and Recent Bryozoa of the Gulf of Mexico* was based primarily on collections deposited in the United States National Museum (now the National Museum of Natural History) resulting from the 19th century dredgings of the United States Fish Commission steamer *Albatross* between 1883 and 1888. Seventeen stations yielding bryozoans in their publication were identified as being located off the northern coast of Cuba (N. of Habana, Cuba, NW of Habana, off Habana, or North of Cuba). They ranged in depth from 53–708 m. Substrata included corals and algal nodules, but all were from depths below the shallow coral reefs that are of most interest to conservation and management efforts.

1.2 *Puerto Rico*

Raymond C. Osburn's study of the bryozoans of Puerto Rico was published in 1940. It was based on collections made at 50 stations during the summer of 1915. A few of the stations were from shallow reef or patch reef habitats to a depth of 55 m, but most of the collections were from very shallow depths: harbor pilings, mangroves and beach drift that could be collected by hand.

1.3 *Jamaica*

A study of the taxonomy of coral reef associated bryozoans of Jamaica, based on collections made by SCUBA and submersible from the early 1970s to 1980 (Winston & Jackson 2021) covered 14 sites around the island's coasts. It yielded 132 species of cyclostomes and cheilostomes, more than half of them (56%) new.

The following results include a checklist of the species found at Caracol Bay and a comparison with results from the older studies on bryozoans of the adjacent Greater Antillean islands. Forty six (81%) were previously described from Jamaica. (Winston & Jackson 2021). We also describe the five new species found, *Celleporaria soleata*, *Hippaliosina caracolensis*, *Petraliella calceata*, *Reptadeonella haitiensis*, and *Schizoporella trimaxillata*, and redescribe one Canu & Bassler species, *Gemellitheca multilamellosa* (Canu & Bassler 1923) which, despite forming large colonies, does not seem to have been studied and illustrated since the original description was published.

*Corresponding Author: judithewinston@gmail.com

DOI: 10.1201/9781003454915-17

2 MATERIALS AND METHODS

Caracol Bay is located on the northern coast of Haiti (Figure 1A). It is the center bay of three adjoining north coast bays, Limonade, Caracol and Fort Liberté created as a Marine Protected Area by the Haitian government, and now called Three Bays National Park. The entire park covers 75,618 ha and contains significant biodiversity, in riparian, mangrove, and shallow marine (including coral reef) ecosystems, despite long-term human settlement and activity (Kramer et al. 2016).

The bryozoans were collected from a 20 km long line of barrier coral reef, broken by a channel leading into Caracol Bay. The north coast of Haiti has a warm humid tropical climate. Temperatures range from 23°–35°C and the wind direction is almost always northeast. Wet periods occur in May–June

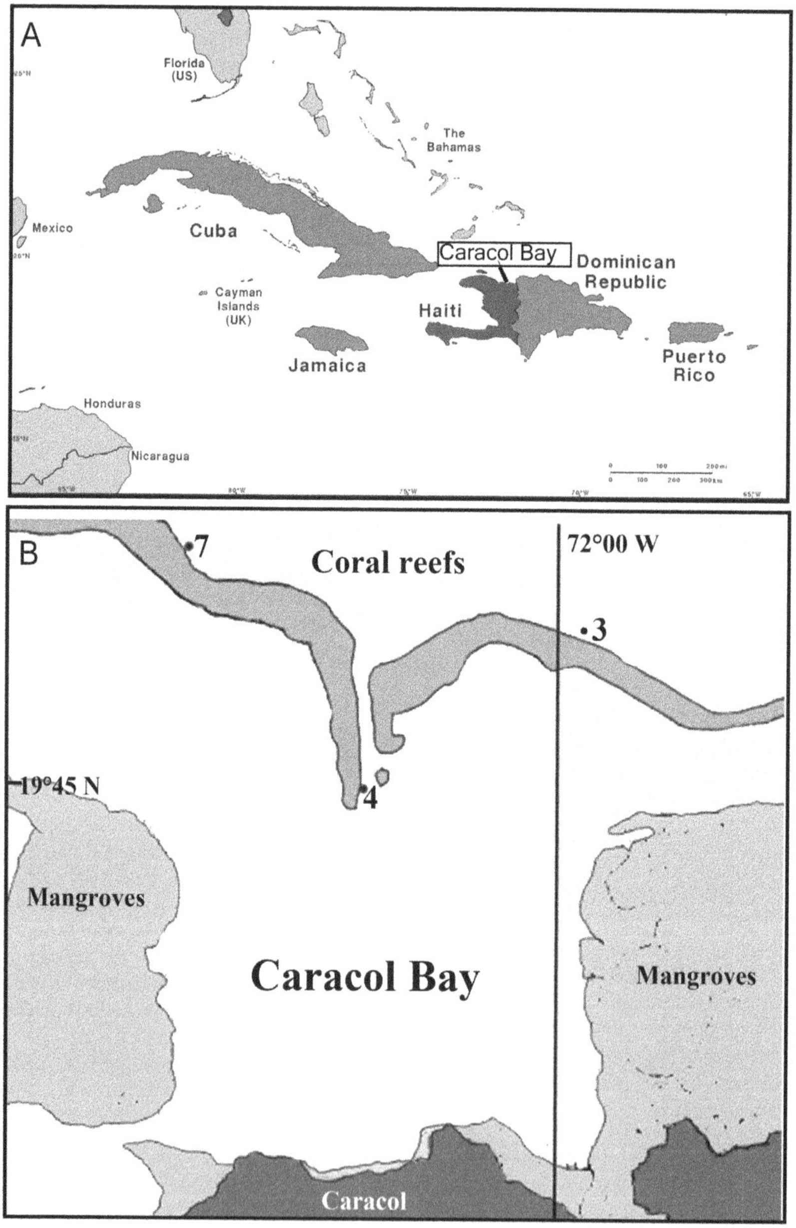

Figure 1. A. Map of the Greater Antilles showing islands, nations and the location of Caracol Bay on the north coast of Haiti. B. Map of Caracol Bay, Haiti and fringing coral reef where bryozoans were were collected 3. East forereef. Dives between 10–23 m. 4. West forereef. Dives between 3–20 m. 7. Inside channel to bay on west side of channel. Dives between 10–18 m.

and September–October. Tidal range is about 0.5 m. Circulation is mainly influenced by tides and winds, but there is outflow through the central paleo-river channel in the middle of the bay. Visibility is usually less than 15 m (Kramer et al. 2016).

Caracol Bay has the largest area of mangroves with extensive and coastal wetlands of the Three Bays region. These wetlands include large areas of seagrass, but the mangrove roots and the inner bay seagrass beds lack colonization by benthic invertebrates, perhaps because of turbidity and contamination. Caracol Bay is also the site of the largest fishing community in the region (Kramer et al. 2016) and large fishes were rare when we collected in 1981.

2.1 *Bryozoan study sites*

Dives were made at three locations (3, 4, and 7) for coral and other substrata encrusted by bryozoans (Figure 1B). Dives at other sites were made for other organisms and did not yield bryozoans.

2.1.1 *Dive location 1981-3*

This location was on the forereef. Submarine morphology showed about a 20°–30° slope from *Acropora palmata* (Lamarck 1816) reef crest, with a break in slope at about 9 m to a 45°–70° slope below, dropping off quickly to 24 m and staying nearly vertical to at least 30 m. Above the break in slope corals were mostly *Orbicella annularis* (Ellis & Solander 1786) knobs, often with skirts, plus many *Agaricia agaricites* (Linnaeus 1758) blades and plates, and *Porites astreoides* (Lamarck 1816) plates with a tendency to form buttresses. Below the break corals consisted mostly of plating *Orbicella sp.,* most likely *O. franksi* (Gregory 1895) and *Agaricia* spp., *Helioseris*, etc. Many gorgonians and sponges were also present.

Site 1981-3-1 (11 May 1981). Collection of plate corals at 10 m, with many undersurface-encrusting bryozoans including *Steginoporella magnilabris* (Busk 1854). A larger collection was made at the same place on 12 May (see below).

Site 1981-3-2 (11 May 1981). Collection of plating corals at 22–23 m. Many bryozoans, including some *Steginoporella* and *Reptadeonella*.

Site 1981-3-3 (12 May 1981). Collection of plating corals at 20 m. Many bryozoans including some *Steginoporella* and *Reptadonella* (approximately same site as 1981 3-1-2) and to 12 m (approximately same site as 3-1-1).

2.1.2 *Dive location 1981-4 (13 May 1981)*

This location was located in the channel into Caracol Bay, on the west side, as far into the bay as the reef breaks surface bordering the channel. Water was turbid and brown at the surface but clear below 4.5 to 6 m. The maximum depth of the channel at this point was about 24 to 27 m. Gorgonians were spectacularly abundant from <3–6 m, including *Gorgonia*, *Plexaura*, *Leptogorgia*, etc. at densities up to 10 colonies/m^2, many of them large. Abundant *Millepora* colonies pursuing and overgrowing gorgonians (i.e., Wahle 1980) were also present. There were also coral mounds built mostly of *Orbicella* in knobby and plating forms, often shingled with *Agaricia agaricites* and *Helioseris cucullata* (Ellis & Solander 1786), as well as large sponges, ascidians, and algae.

Site 1981-4-1. Collection of plating corals for bryozoans at 3 m. Included *Steginoporella magnilabris* and a variety of other species. Sponges and ascidians much more common here than in other Caribbean sites visited. This was true for all sites collected at Caracol Bay but especially in the channel.

Site 1981-4-2. Collections of plating corals for bryozoans at 10 m depth, including *Steginoporella magnilabris*. This part of the reef was characterized by abundant algae, sponges, and plating corals, especially *Agaricia agaricites*, *Porites astreoides*, and *Orbicella franksi*.

Site 1981-4-3. Collection of plating corals for bryozoans, including *Steginoporella magnilabris,* at 20 m depth. Here also the site was characterized by abundant algae, sponges, and platey corals (mostly various *Agaricia* spp. and *Orbicella annularis*).

2.1.3 *Location 1981-7 14 May 1981*

The last two sites sampled were on the west side of the channel into Caracol Bay. The reef is broader here, with a gentler slope with spur and groove buttress formation. Large knobby *O. annularis* to wide plating *Orbicella* sp. on buttresses. Abundant branching coral including *Porites porites* (Pallas 1766), *Madracis mirabilis* (Duchassaing & Michelotti 1860) and *Acropora cervicornis* below about 3–6 m. Frequent dense clumps of damselfish with gardens occurred.

Site 1981 7-1. Collection of *P. porites* and *M. mirabilis* for bryozoans at 10–12 m. Many Bryozoa.

Site 1981 7-2. Collections of *M. mirabilis* at 18 m. Many Bryozoa.

2.2 *Examination and analysis*

Dried substrata collected were later sorted and examined for bryozoan colonies in the laboratory. Bryozoan colonies on undersurfaces of dead coral and in a few cases attached algal substrata were marked and given preliminary identifications using a stereo microscope at 10–100 times magnification. Representative colonies of all bryozoan taxa found were also examined and documented in digital images using a scanning electron microscope (Hitachi STEM) located at the USDA Horticultural Research Laboratory in Fort Pierce. Specimens from the study are deposited in the American Museum of Natural History (AMNH) invertebrate collection.

Standard measurements of zooid and polymorph characters were made on the species described here.

Measurements are abbreviated in descriptions as follows: LZ, zooid length; WZ, zooid width; Lo, orifice length; Wo, orifice width; Do, orifice diameter; Lop, opesia length; Wop, opesia width; Lov, ooecium length; Wov, ooecium width; Lovo, ooecium orifice length; Wovo, ooecium orifice width; Lgz, gonozooid length; Wgz, gonozooid width; Lgzo, gonozooid orifice length; Wgzo, gonozooid orifice width; Lav, avicularium length; Wav, avicularium width, Llav, large avicularium length, Wlav, large avicularium width Lav2, small avicularium length, Wav2, small avicularium width, Lspir, distance from proximal edge of orifice to spiramen. N, number. N of measurements for each character = 18 unless noted in parentheses. Means for each character given in bold type, followed by standard deviation, and range.

3 RESULTS AND DISCUSSION

3.1 *Checklist of species collected*

Table 1A lists the species found at each of the Caracol Bay dive sites where bryozoans were collected. The greatest diversity in term of number of

Table 1. A. Species found at Caracol Bay, Haiti by Site. B. Haiti species shared with other Greater Antilles and Gulf of Mexico locations.

		A. Haiti. Species found by Site								B. Shared with Other Sites			
Genus	**Species**	**3-1**	**3-2**	**3-3**	**4-1**	**4-2**	**7-1**	**7-2**	**Haiti**	**Cuba**	**Jamaica**	**Puerto Rico**	**Gulf of Mexico**
Aspiscellaria	*hildae*	+							+		+		
Caberea	*marleyi*		+		+				+		+		
Calyptooecia	*Insidiosa*		+	+					+		+		
Canda	*munita*	+	+	+					+		+		
Celleporaria	***soleata***					+			+				
Cigclisula	*winstonae*	+	+						+		+		
Cradoscrupocellaria	*jamaicensis*	+							+		+		
Crassimarginatella	*tuberosa*	+							+	+	+	+	+
Crepidacantha	*longiseta*	+			+			+	+	+	+		+
Cribrilaria	*harmelini*	+	+	+			+	+	+		+		
Cribrilaria	*minervae*		+				+	+	+		+		
Cribriliaria	*smitti*	+			+		+		+		+		+
Entalophoroecia	sp.	+							+		+		
Crisia	*corallina*			+				+	+		+		
Drepanophora	*coronata*	+						+	+		+	+	
Exechonella	*panamensis*		+						+		+		
Exechonella	*rangifer*		+						+		+		
Gemelliporidra	*typica*		+						+				
Gemellitheca	*belikina*	+					+		+		+		
Gemellitheca	*multilamellosa*	+	+				+		+	+			
Gephyrophora	*rubra*	+	+	+		+	+	+	+		+		
Hippaliosina	***caracolensis***	+							+				
Hippothoa	*flagellum*				+			+	+	+			+
Klugerella	*petasus*			+					+	+	+		+
Labioporella	*octofigurata*	+						+	+		+		
Marcusadorea	*jamaicensis*			+				+	+		+		+
Margaretta	*buski*	+				+			+		+		
Microporella	*paterifera*	+	+					sp	+		+		+
Mollia	*smitti*			+					+		+		
Nevianipora	sp.	+		+					+		+		+
Palliocella	*palliolata*			+					+		+		
Paralicornia	*electilis*	+		+					+		+		
Parasmittina	*indigenella*	?		+					+		+		
Parasmittina	*eccentrica*	+							+		+	+	+
Parellisina	*latirostris*	+	+	+	+	+	+		+		+	+	+
Parellisina	*antillana*	+							+	+	+	+	+
Patinella	*purpurea*						+	+	+		+		
Petraliella	***calceata***	+	+	+			+	+	+				
Plesiocleidochasma	*pulvereum*	+		+					+		+		+
Plesiocleidochasma	*cleidostomum*	+						+	+	+	+		
Poricella	*truncatorostris*	+		+					+	+	+		
Poricella	*cayensis*	+		+					+				
Reptadeonella	***haitiensis***		+	+			+		+				
Reteporellina	sp.			+					+				
Retevirgula	*spinosissima*		+						+		+		
Rhynchozoon	*decoratum*	+		+					+		+		+
Rhynchozoon	*unispinatum*		+	+					+		+		
Rhynchozoon	*notialis*	+							+		+	+	+
Rhynchozoon	*verruculatum*	+							+		+		
Smittipora	*levinseni*		+		+			+	+	+	+		+
Schizoporella	***trimaxillata***	+		+					+				
Steginoporella	*magnilabris*	+	+	+	+	+	+	+	+	+	+	+	+
Stylopoma	***sloanei***			+		+			+		+		+
Synnotum	*aegyptiacum*				+				+	+	+		
Therenia	*porosa*	+		+					+		+		+
Thornelya	*subcorallina*	+				+			+		+		
Trematooecia	*aviculifera*		+			+			+		+		+
Trypostega	*bigeyae*	+		+				+	+		+		

species collected occurred at location 3, the three dive sites on the forereef on the east side of the channel opening into Caracol Bay where 58 species were collected at depths between 10 and 23 m. Four of the new species occurred at Location 4 on the west side of the reef fringing the channel into the Bay where 14 species were found, including one new species, *Celleporaria soleata*. Location 7 on the forereef northwest of the channel was intermediate in species diversity with 21 species, including two of the new species, *Petraliella calceata* and *Reptadeonella haitiensis*. It was dominated by finger corals, *Madracis mirabilis* and *Porites porites*, which provided less space for bryozoan settlement; lower regions of dead coral on older regions of erect branches were more often completely encrusted by leathery and calcareous algae and by *Gypsina* foraminiferans, leaving only a narrow ring between other epibionts and live coral tissue available for settlement by bryozoans.

3.2 *Comparison of species collected in Haiti with those from adjacent areas*

Table 1B compares the species found in Haiti with those recorded from other greater Caribbean areas where bryozoans have been sampled. Forty-seven of the 58 Haitian species (81%) also occur at Jamaica while 10 (17% of the Haitian fauna) including the five that are new did not. This, combined with the fact that more than half of the Jamaican species were new to science suggests that reef bryozoans are highly endemic; a fact supported by increasing realization that supposedly widespread species such as *Stylopoma "spongites"* and other common taxa are turning out to be complexes of geographically isolated cryptic species based upon genetic and quantitative morphological analyses (Herrera Cubilla et al. 2006, 2008; Jackson & Cheetham 1990, 1994). Similar problems have emerged among well studied species of corals that brood their larve. *Seriatopora hysterix*, for example, was thought to comprise a single species ranging from the Red Sea throughout the Indian Ocean to the Coral Triangle and Australia, but now we know it is a complex of dozens upon dozens of endemic species (Bongaerts et al. 2021). Furthermore, most previous bryozoan collections from the Caribbean and Gulf of Mexico have been obtained by dredging and from sites accessible from shore or from highly specific habitats so that it is difficult to make meaningful comparisons with shallow reef faunas. Meaningful biogeographic analysis will require more rigorous taxonomic analysis of extensive collections from each region examined to begin to make sense of true patterns of bryozoan diversity.

3.3 *Systematics*

Phylum Bryozoa Ehrenberg, 1831
Class Gymnolaemata Allman, 1856
Order Cheilostomata Busk, 1852
Superfamily Adeonoidea Busk, 1884

Family Adeonidae Busk, 1884
Genus *Reptadeonella* Busk, 1884

Reptadeonella haitiensis new species
Figure 2

Holotype: AMNH IZC 00382977. Caracol Bay, Haiti.

Etymology: The combination species name is geographic, based on the country it was collected: Haiti plus the Latin suffix *-ensis*.

Diagnosis: Relatively small *Reptadeonella* species, characterized by a narrow transversely oval orifice, a broadly triangular, distally directed avicularium whose tip reaches the proximal edge of the orifice, a frontal wall with a partial second group of round pores within the marginal pores, and a transverse oval to crescentic spiramen located below the midpoint of zooid frontal surfaces.

Description: Colonies encrusting coral undersurfaces (Figure 2A) Their dark purplish pigment remains when dry. Zooids have thick frontal calcification with ridges extending inward from the large round marginal pores and an overall finely beaded surface texture (Figure 2B). The frontal wall may have a second grouping of round pores inward of the evenly spaced marginal pores, The zooid orifice is narrow and transversely oval with a short, hooded peristome (Figure 2C). A broadly triangular avicularium with a rounded base is on the raised midline. Its mandible ends in a downcurved hook (Figure 2D). Its tip is usually distally directed, but in some zooids it may be slightly offset from the orifice center (Figure 2E, F). In mature zooids the avicularium rostrum tip may reach the proximal edge of the orifice. On the frontal surface above the tip of the avicularium and below the orifice there are two (usually) or three small pores. The large laterally oval to crescentic spiramen is located below the midpoint of zooid on the proximal third of the frontal surface. Gonozooids are distinguished by a narrower more elongate orifice and the presence of an extra row of pores above the distal rim of the orifice (Figure 2C).

Measurements: Lz **0.691,** 0.09, 0.540–0.918 mm; Wz **0.398**, 0.047, 0.324–0.504 mm; Lo **0.078**, 0.013, 0.054–0.090 mm; Wo **0.124**, 0.014, 0.090–0.144 mm; Lgz **0.660**, 0.068, 0.612–738 mm (3); Wgz **0.492**, 0.045, 0.450–0.540 mm (3); Lgzo **0.57**, 0.005, 0.54–0.630 mm (3); Wgzo **0.132**, 0.010, 0.126–0.144 mm (3); Lav **0.172**, 0.022, 0.126–0.216 mm; Wav **0.072**, 0.010, 0.054–0.090 mm; SpirD **0.267**, 0.039, 0.216–342 mm.

Discussion: This species belongs to the "*bipartita*" group, with its dominating distally directed avicularia, but it differs from the other species described so far in zooid morphology, particularly its smaller size, less projecting avicularian tip, and more proximal position of the spiramen on zooid

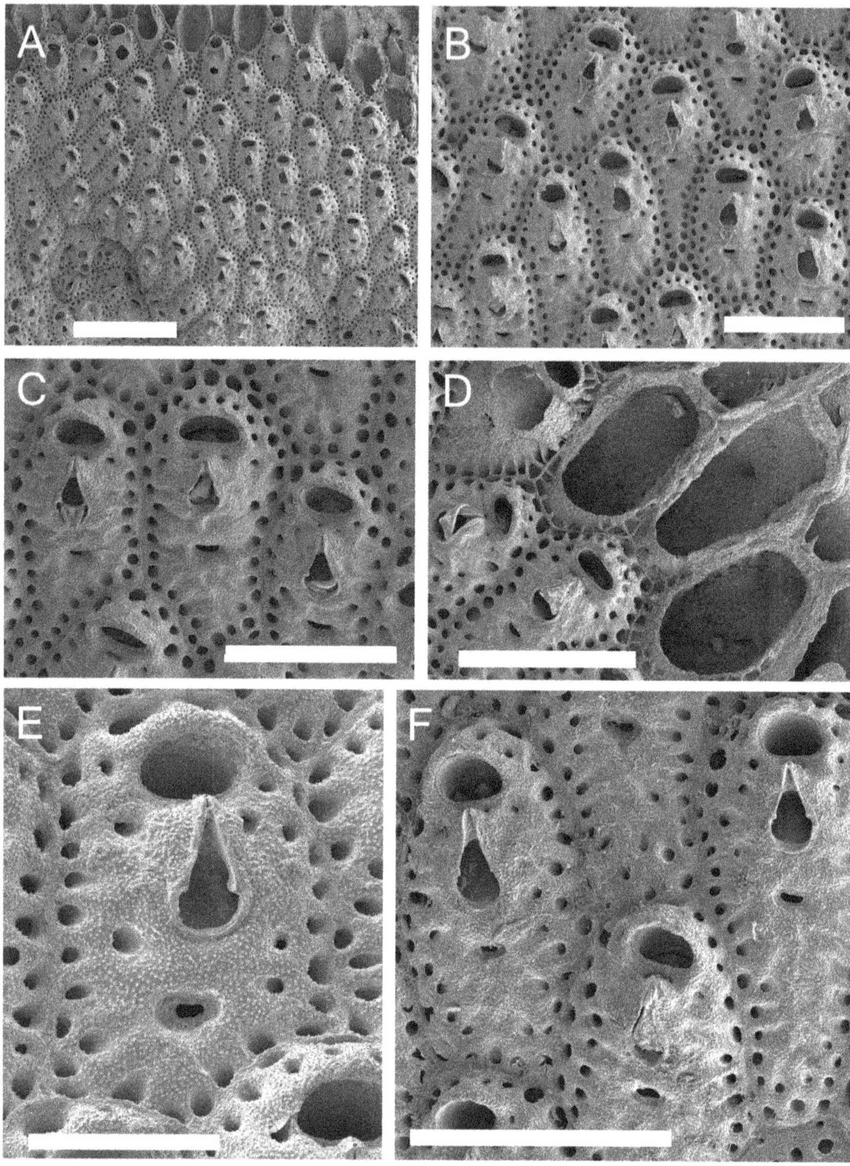

Figure 2. *Reptadeonella haitiensis* new species. A. Encrusting colony (Scale bar, 1 mm). B. Autozooids and gonozooids (Scale bar, 500 μm). C. Gonozooid between two autozooids (Scale bar, 400 μm). D. Autozooids (Scale bar, 500 μm). E. Single autozooids with very proximal spiramen (Scale bar, 200 μm). F. Growing edge of colony, note sharp down-curved mandibles on the avicularia (Scale bar, 500 μm).

frontal wall. Winston (2005) used the name *Reptadeonella bipartita* for Smitt's "*Porina violacea*", but more recent works by Cheetham et al. (2007) and Almeida et al. (2015) point out differences in Smitt's species from Florida: shorter avicularia, spiramen placed midway in the zooid frontal wall, a continuous double row of marginal pores, and a gonozooid that is distinct from autozooids. In Smitt's species also, the tips of the raised rostra project above the proximal edge of the orifice and the suborificial slit extends almost the length of the orifice of both gonozooid and autozooid (see Winston 2005, fig. 122).

Distribution: Haiti.

Superfamily Schizoporelloidea Jullien, 1883

Family Petraliidae Cook, Bock, Hayward & Gordon, 2018

Genus *Petraliella* Canu & Bassler, 1927

Petraliella calceata new species

Figure 3

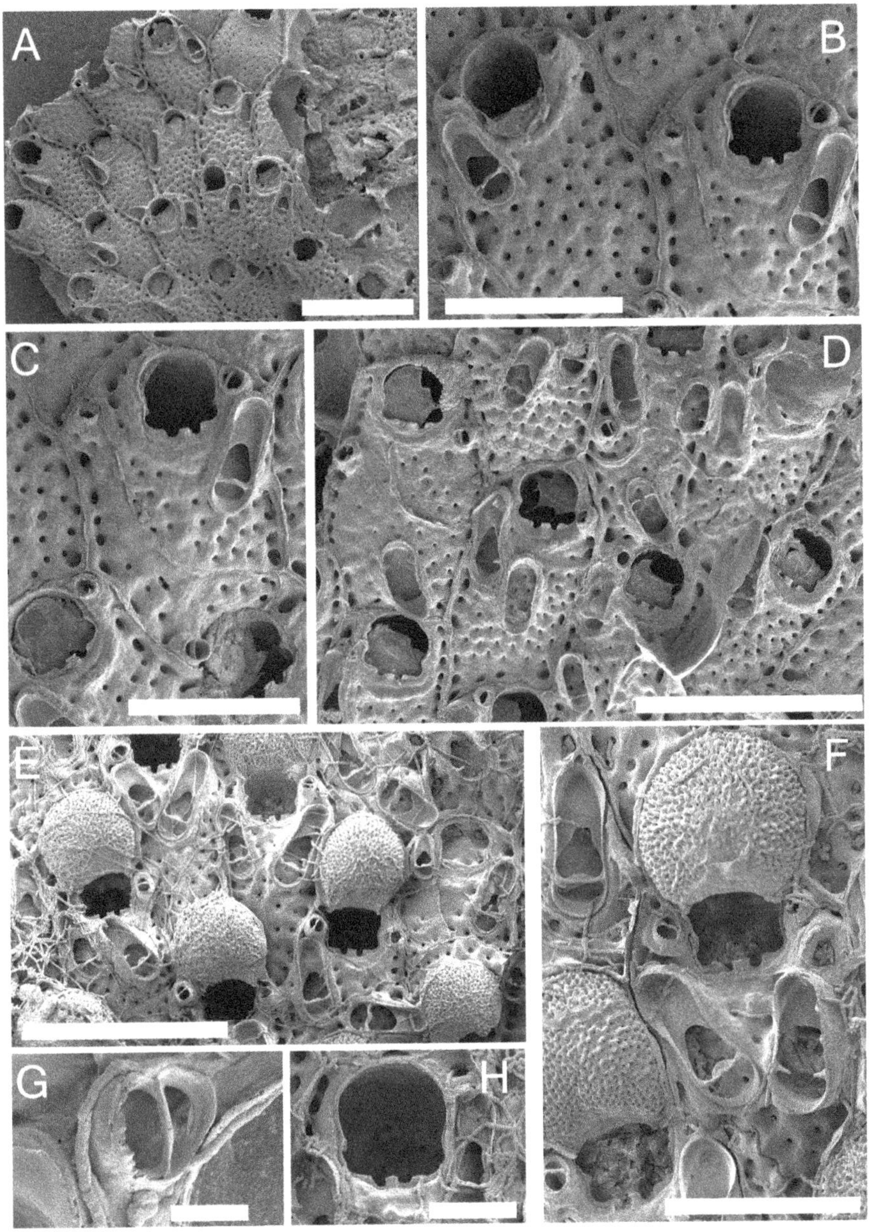

Figure 3. *Petraliella calceata* new species. A. New layer of colony overgrowing old, fouled layer (Scale bar, 1 mm). B. Two zooids at growing edge (Scale bar, 500 μm). C. Group of zooids, showing large sole-shaped avicularia (Scale bar, 1 mm). C-2. Close up of zooids to show orifices and both small and spatulate avicularia (Scale bar, 500 μm). D. Two autozooids (Scale bar, 500 μm). E. Close up of small avicularium (Scale bar, 50 μm). E. Group of zooids with large porous ovicells (Scale bar, 1 mm). E-2. Close-up of ovicelled zooids (Scale bar, 500 μm). F. Close-up of an orifice (Scale bar, 200 μm).

Holotype: AMNH IZC 00382983. Caracol Bay, Haiti.

Etymology: from the Latin *calceatus*, provided with slippers, for the outline of the large avicularia, like long narrow ballet slippers.

Diagnosis: *Petraliella* with orifice with two sinuses, finely porous ovicell frontal wall, with about 30–40 larger pores on frontal surface, in addition to marginal pores, a flat frontally multiporous helmet shaped ooecium, and two types of avicularia, small oval avicularia on one or both sides of the orifice and large oblong avicularia with straight or indented sides, rounded at both ends.

Description: Colonies encrusting, dried material brownish orange in color. Colonies not cemented to the substratum, but loosely attached by uncalcified radicles with brushy tips. The undersides of zooids show the pores where these protrude. Zooids are rectangular to elongate oval in outline, their margins distinct by means of a row of marginal pores and the raised lateral walls of adjoining zooids (Figure 3A, B). Frontal surfaces relatively flat, calcification texture rugose, with about 30–40 small sunken pores. The orifice is bell-shaped distally, the proximal edge is slightly concave with a short anvil shaped denticle between two small U– to almost O–shaped sinuses (Figure 3C–D, I). One to two or more small oval avicularia are placed beside the orifice, they have serrated rostra, and may have various orientations (Figure 3B, E). The large avicularia are oblong in outline, often narrower at midsection, with a very short body, a complete crossbar and long rostrum with serrations near the tip (Figure 3C–E). They are most commonly placed near the zooid margin at or below the level of the orifice and directed distally, but may sometimes be found in more proximal positions. The ovicell is helmet shaped, with a raised outer rim surrounding a somewhat flat frontal surface with numerous small pores, and sharply rugose texture between them (Figure 3E, F).

Measurements: Lz **1.158**, 0.099, 1.044–0.422 mm; Wz, **0.596**, 0.064, 0.468–0.702 mm; Lo **0.243**, 0.020, 0.216–0.288 mm; Wo **0.277**, 0.019, 0.252–0.324 mm; Lov **0.501**, 0.027, 0.450–0.540 mm; Wov **0.461**, 0.030, 0.378–0.504 mm; Lav **0.418**, 0.109, 0.360–0.540 mm; Wav **0.175**, 0.022, 0.144–0.216 mm.

Discussion: The species is closely related morphologically to *Petraliella bisinuata* (Smitt 1873) from Florida and the Gulf of Mexico and *Petraliella marginata* (Canu & Bassler 1928), also from the Gulf of Mexico. However, there are differences, especially in the shape of the large avicularia, which in these two Western Atlantic species are triangular, pointed at the rostrum, rather than oblong. *Petraliella bisinuata* has avicularia in two sizes, one larger than the other, both with curved triangular rostra, but on some zooids only one may occur. If there are two the smaller one is placed on one side of the orifice, the larger placed on the opposite site, its body extending proximally, its curved pointed rostrum close to the orifice and curved towards it. The orifice is similar to that of *P. calceata*, bell-shaped distally, with two sinuses in its proximal rim. In SEM illustrations of a Smitt specimen (Winston 2005: fig. 225, 226, p. 82) the middle denticle is proportionally larger and the two sinuses narrower. *Petraliella marginata* has two narrow triangular avicularia, one on each side of the orifice and oriented distally. There may also be a large avicularium similar in narrow triangular shape to the orificial pair, located below the orifice near the lateral margin and oriented distolaterally. The orifice is more semicircular than bell-shaped distally with two very short condyles separating it from a poster with a straight and finely serrate proximal border. In both species the ovicells appear similar to those of *P. calceata,* finely porous frontally with a raised outer rim.

Distribution: Caracol Bay, Haiti.

Family Schizoporellidae Jullien 1882
Genus *Schizoporella* Hincks, 1877

Schizoporella trimaxillata new species
Figure 4

Holotype: AMNH IZC 00382979. Caracol Bay, Haiti.

Etymology: The species name is a descriptive combination from Latin prefix *tri-* three, and *maxilla*, jaw, surrounded by three jaws, descriptive of the pointed avicularia adjacent to the orifice with mandibles aimed at its opening.

Diagnosis: Encrusting *Schizoporella* having zooids with porous frontal calcification with sunken pores, thick raised lateral margins, and globular porous ooecium with smaller and sparser pores. Orifice with semicircular anter and poster with V-shaped sinus. Two or three short triangular avicularia loacated beside and above the orifice.

Description: Colonies encrusting on coral undersurfaces Figure 4A–B). Zooids rectangular to irregular in outline, with scattered sunken pores covering most of the frontal surface outward to the marginal pores and the distinctly raised edge where zooid walls adjoin (Figure 4B, C). The orifice is semicircular distally with large angular condyles and a shallow V-shaped central proximal sinus (Figure 4D). It is surrounded by a low rough textured peristome in which paired lateral avicularia, and often a third distal triangular avicularium with complete calcified pivots develop Figure 5B, D, E. They have mandibles pointing toward the orifice (Figure 4F). The large globular ovicell has fewer and sparser pores and develops lateral hoodlike calcification that extends to the level of the condyles (Figure 4C, E). In ovicelled zooids only the lateral avicularia are visible (Figure 4E).

Measurements: Lz **0.749**, 0.122, 0.558–1.008 mm (12); Wz **0.549**, 0.093, 0.450–0.740 mm (12); Lo **0.163**, 0.014, 0.144–0.189 mm (12), Wo **0.139**, 0.010, 0.126–0.153 mm (12); Lov **0.326**, 0.014, 0.306–0.342 mm (9); **Wov** 0.350, 0.022, 0.324–0.378 mm (9).

Discussion: At first glance this species resembles *Gemelliporidra typica* in its flat frontal and distinct raised margins, but *G. typica* (Now placed in Bitectiporidae) has a much more calcified frontal wall, a hoof shaped orifice, and narrow more elongate avicularia whose mandibles barely reach the side of the avicularia. Like *S. trimaxillata*, *G. typica* has a sparsely porous ovicell, but it lacks the hoodlike extension found in S. *trimaxillata*. The Haitian species shares the greatest similarities in characters with species in the genus *Schizoporella*, including an evenly porous frontal wall with

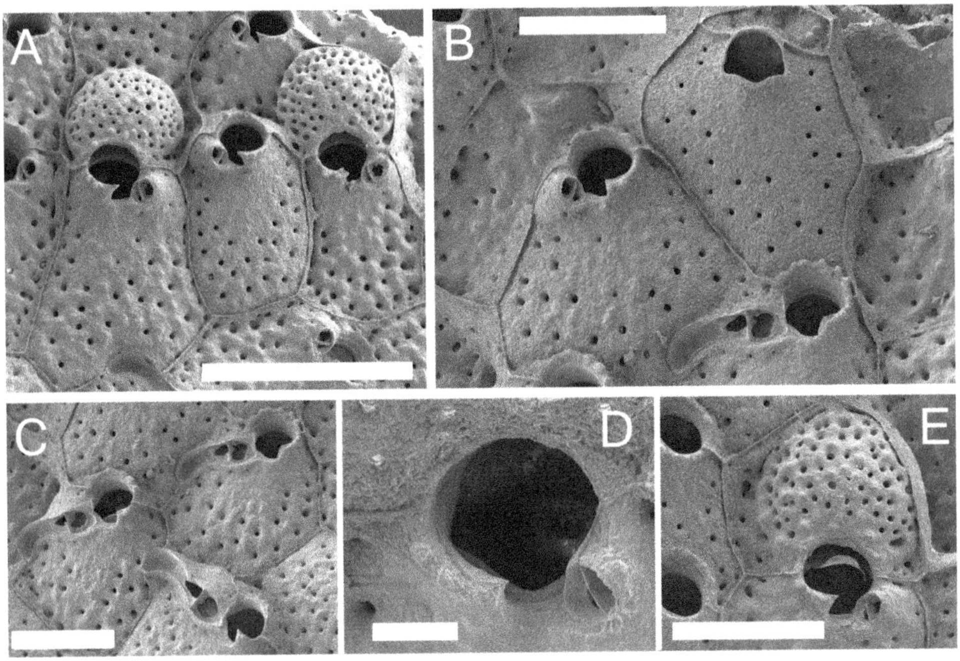

Figure 4. *Gemellitheca multilamellosa* (Canu & Bassler 1928) A. Zooids, two with ooecia (Scale bar, 1 mm). B. Three zooids, top one with just-formed orifice, (Scale bar, 500 μm). C. Three zooids with large curved avicularia (Scale bar, 500 μm). D. Close-up of fully formed orifce with peristomial sinus and oval avicularium (Scale bar, 100 μm). E. Close-up of ovicell (Scale bar, 500 μm).

sunken pores, an orifice with a rounded anter and central U to V shaped sinus, and small triangular avicularia located near the orifice. Like some other *Schizoporella* species it has a porous ovicell without obvious ridges. Morphologically it most closely resembles the cold water invasive species *Schizoporella japonica.* That species also has a rather flat, thickly calcified and porous frontal wall, with distinct raised margins, but unlike *S. trimaxillata*, *S. japonica* has a shallow U-shaped central sinus, small condyles, and a single triangular avicularium on one side of the orifice and extending proximally. In addition, although its ovicell also has some pores, it also has a central ridge in development and with maturity the number and depth of the ridges increases.

Distribution: Caracol Bay, Haiti.

Infraorder Lepraliomorpha Gordon, 1989
Superfamily Smittinoidea Levinsen, 1909
Family Bitectiporidae MacGillivray, 1895
Genus *Gemellitheca* Winston & Jackson, 2021
Gemellitheca multilamellosa (Canu & Bassler 1928) n. combination
Figure 5

Cyclicopora multilamellosa Canu & Bassler, 1923: 138, pl. 46, figs. 3–6.

Gemelliporidra multilamellosa Canu & Bassler, 1928a: 104.

Gemelliporidra multilamellosa Powell, 1971: 771.

Material examined: AMNH IZC 00382989, 00382992. Caracol Bay, Haiti.

Description: Colonies encrusting, forming multilayered encrustations that may cover up to several cm of area and have rims raised in a saucer shape. Zooids are oval to irregular in outline, with a raised line of calcification visible where lateral walls adjoin in primary layers (Figure 5A, B). The frontal shield is rough textured with an area with large sunken pores visible proximally and laterally, then imperforate around the the distal end with its peristomial avicularia (Figure 5C). The primary orifice is semicircular distally with a very shallow U to V-shaped curve to its proximal edge. It can only be seen on zooids in which frontal calcification has just developed (Figure 5B). As a zooid calcifies further a low peristomial collar with an irregular distal rim, and a sharp edged proximal sinus develops, obscuring the primary orifice except distally (Figure 5A, C). A small oval, diagonally raised avicularium with a serrated rostrum is

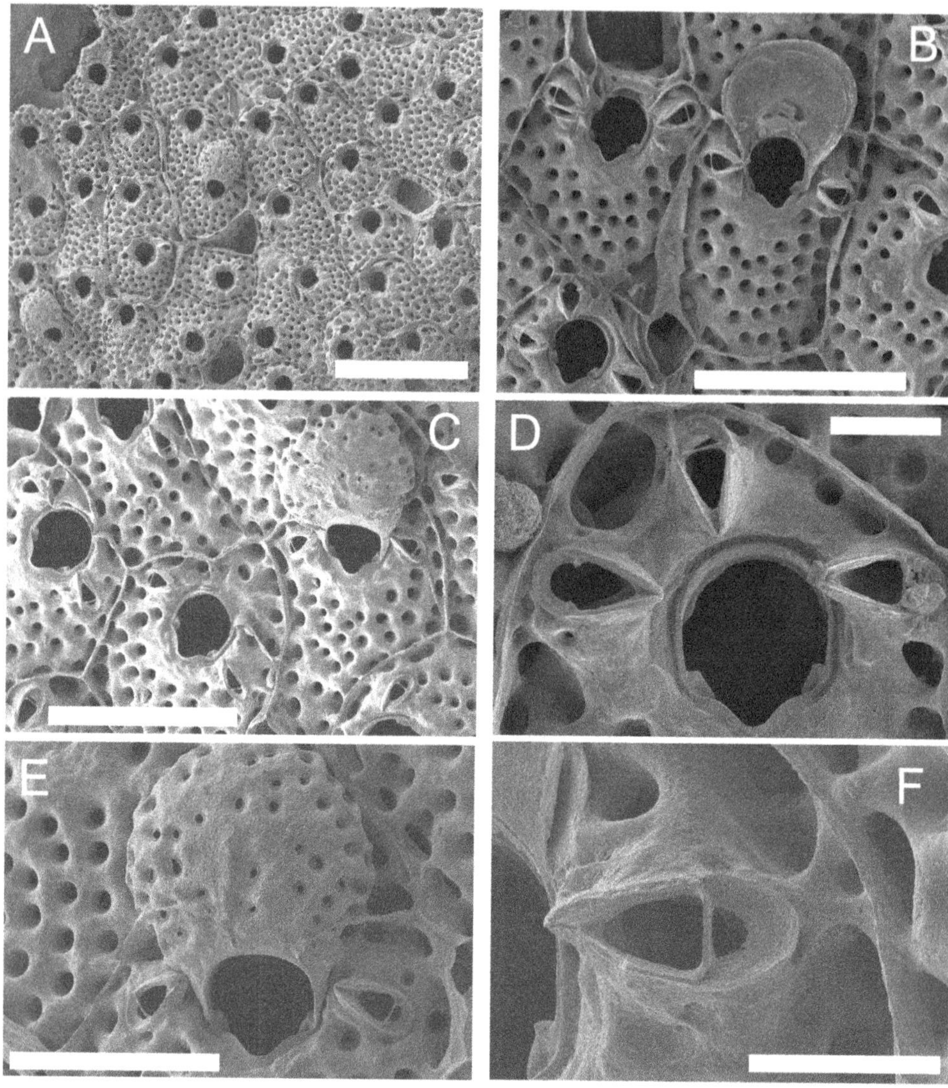

Figure 5. *Schizoporella trimaxillata* new species. A. low magnification view of part of colony (Scale bar, 1 mm). B. Group of zooids with ovicells developing and two avicularia (Scale bar, 500 μm). C. Autozooids and an ovicelled zooid with two avicularia (Scale bar, 500 μm). D. Close-up of an orifice with three avicularia (Scale bar, 100 μm). E. Distal end of an ovicelled zooid, showing orifice and avicularia (Scale bar, 300 μm). F. Close-up of an orificial avicularium (Scale bar, 100 μm).

nested on one side of the peristomial sinus, the other side of the sinus rim comes to a sharp point (Figure 5D). Large avicularia with curved mandibles and rostra occur on some zooids, replacing (originating in the same position as) the small oval avicularia (Figure 6C). The ooecium is round with an opening separate from the orifice, its surface texture rough and bumpy, with more and smaller sunken pores than are present on zooids (Figure 6A, E).

Measurements: Lz **1.087**, 0.176, 0.882–1.476 mm; Wz **0.757**, 0.126, 0.594–0.1062 mm; Lo **0.196**, 0.213, 0.162–0.234 mm; Wo **0.209**, 0.0235, 0.180–0.288 mm; Lov **0.551**, 0.0457, 0.504–j0.630 mm (12); Wov **0.588**, 0.0462, 0.540–0.666 mm (12); Lav1 **0.496**, 0.0728, 0.342–0.666 mm; Wav1 **0.127**, 0.0190, 0.108–0.180 mm; Lav2 **0.148**, 0.0169, 0.126–0.180; Wav2 **0.101**, 00150, 0.090–0.126 mm.

Discussion: This species was described from fossil material (Pleistocene), but it is still extant, as was first noted by Powell (1971), who noted its large greenish colonies, but did not illustrate or describe them in his checklist. It has been mentioned in at least two other publications (Jackson & Herrera Cubilla 2000; Simpson et al. 2017), but not described or illlustrated.

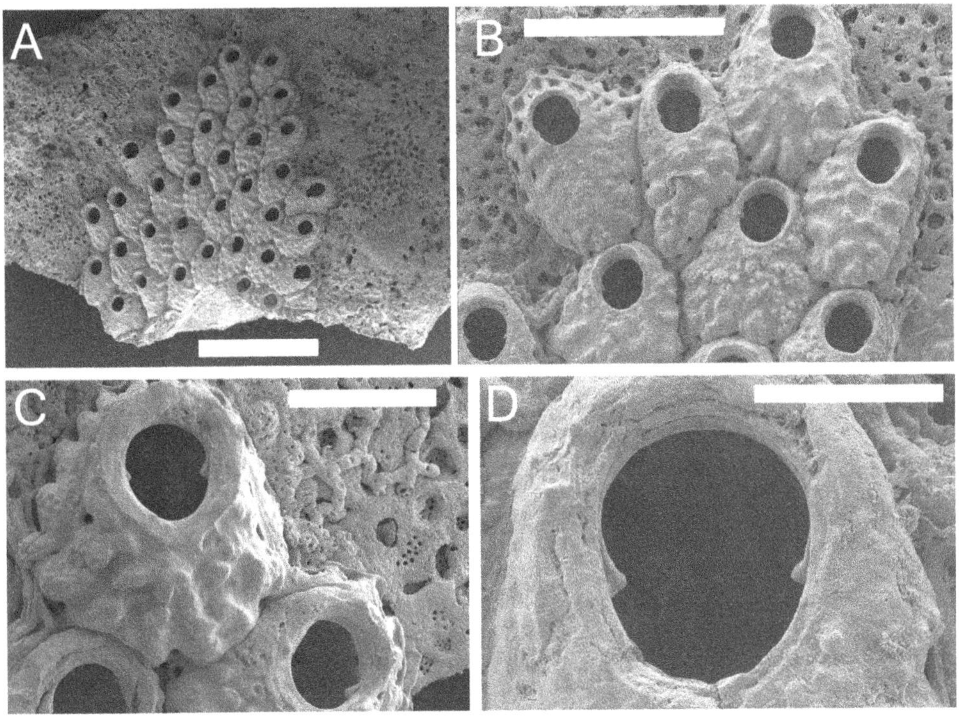

Figure 6. A–D. *Hippaliosina caracolensis* new species. A. Colony on empty snail shell (Scale bar, 1 mm). C. Close-up of autozooid (Scale bar, 200 μm). D. Close-up of an orifice (Scale bar, 100 μm).

At Galeta reef in Panama colonies were attached like bracket mushrooms to the sides of ledges and under large coral heads, their concave undersurfaces showed regularly spaced skeletally reflected excurrent mounds and incurrent channels. Their tentacle crowns were large, about 850 μm in diameter with 24 tentacles (Winston 1978). Coloration of living colonies ranged from dark purple to brownish orange to orange white to greenish. Tentacles were white, introverts a dark brownish orange (Winston, unpublished notes). Powell (1971) found living colonies at Galeta reef at low tide, but wrote only a sentence about their large greenish colonies.The colonies found in Haiti were not as large as those seen in Panama, but were developing similar concave multilaminar colonies.

Distribution: Pleistocene–Recent: Panama, Haiti, probably elsewhere in the Caribbean.

Family Hippaliosinidae Winston, 2005
Genus *Hippaliosina* Canu, 1919

Hippaliosina caracolensis new species
Figure 6 A–D

Holotype: AMNH IZC 00382980. Caracol Bay, Haiti.

Etymology: The species name is geographic, in recognition of the place the specimen was collected: Caracol plus the Latin place suffix *-ensis*.

Diagnosis: Snail shell encrusting *Hippaliosina* species with small zooids, thick rough textured calcification with raised ridges and thick raised peristome.

Description: The single colony found was encrusting a dead snail shell that was already coated with calcarous algae (Figure 6A). Zooids are small and oval with a thick rough textured and pustulose calcification, with raised ridges extending from the marginal pores Figure 6B). The surface is imperforate except for marginal pores The orifice is oval and takes up the distal quarter of the frontal surface, the condyles divide it into two almost equal parts, rounded anteriorly, with a convex poster almost as broad as the anter (Figure 6C–D). The smooth rim of the primary orifice is surrounded by a thick raised peristome that is higher and thicker distally (Figure 6C). No avicularia or apparent reproductive zooids present in this colony.

Measurements: Lz **0.456**, 0.037, 0.396–0.504 mm (6); Wz 0.**291**, 0.033, 0.252–0.324 mm (6); Lo **0.129**, 0.007, 0.126–0.144 mm (6); Wo **0.153**, 0.010, 0.144–0.162 mm (6).

Discussion: Despite its lack of avicularia the new species is best placed in *Hippaliosina* on the basis of its hoof-shaped orifice with its very deep and broad poster and its frontal shield with its rugose calcification, imperforate except for marginal pores. The ooecium in *Hippaliosina* is endozooidal, not visible on the zooid surface. Instead fertile zooids are larger in size and have a broader orifice than autozooids. No fertile zooids were found in this small colony. The genus has more than a dozen living and fossil species, mostly tropical (Cook et al. 2018). Two other species are known from the Western Atlantic. *Hippaliosina rostrigera* occurs from Florida to the Gulf of Mexico and Caribbean. Like *H. caracolensis*, *H. rostrigera* has rough-textured frontal calcification, although it is more regularly developed in the form of large beads. It has single or paired triangular avicularia originating on the side of the orifice and pointed distomedially. *Hippaliosina imperfecta* (Canu & Bassler 1928) from Brazil (Canu & Bassler 1928b, Vieira et al. 2008) is very similar to *H. rostrigera*. Hastings (1930) and Osburn (1952) recorded *H. rostrigera* from the eastern Pacific. Osburn also described two new Eastern Pacific species under *Hippaliosina*: *H. inarmata*, and *H. costifera*, but the presence of surficial ovicells, as well as much smaller posters on zooid orifices in those species makes that generic assignment questionable.

Distribution: Caracol Bay, Haiti.

Superfamily Celleporoidea Johnston, 1838
Family Celleporidae Johnston, 1838
Genus *Celleporaria* Lamouroux, 1821

Celleporaria soleata new species
Figure 7

Holotype: AMNH IZC 00382981. Caracol Bay, Haiti.

Etymology: The species is named *soleata*, from the Latin, *soleatus*, wearing sandals, descriptive of the sandal-shaped outline of its many large interzooecial avicularia.

Diagnosis: *Celleporaria* species having an orifice with 2–3 distal spines, a single central proximal sinus, large, flat sandal-shaped avicularia, and an elongate flask-shaped ovicell with a narrow lipped orifice.

Description: Colony encrusting, with frontal budding. Zooids oval in primary layer (Figure 7 A), more irregular in secondary layers. Orifice semicircular to elongated with a single central sinus and 2–3 orificial spines (Figure 7 A–C). It is surrounded by a thick rim from whose proximal edge a cone- to pillar-shaped projection bearing an small oval suboral avicularium, is raised vertically and oriented proximally. Sometimes the suboral avicularian tubercle is tilted to project over the proximal edge of the orifice (Figure 7B). Frontal calcification is imperforate except for marginal pores, and a pore at the base of the suboral avicularium tubercle. Surface texture pustulose. Interzooecial avicularia have marginal pores and similar calcification to zooids (Figure 7B). Their surface may be raised, but the rim of the avicularian openings are low and sandal-shaped, a complete cross bar separating the body from elongate and wider rostrum (Figure 7B–D). The ovicells in this species are strikingly different from the cap-shaped ovicells of most *Celleporaria* species. They are elongate, globular and flask-like, with flared lips that cover much of the orifice (Figure 7C, D, F). They are also imperforate, with a pustular surface texture and small marginal pores.

Measurements: Lz **0.752**, 0.087, 0.630–0.900 mm; Wz **0.543**, 0.068, 0.432–0.666 mm; Lo **0.148**, 0012, 0.126–0.162 mm; Wo **0.168**, 0.013, 0.144–0.180 mm; Lov **0.257**, 0.016, 0.360–0.414 mm (12); Wov **0.368**, 0.029, 0.324–0.432 mm (12); Lav **0.532**, 0.077, 0.360–0.666 mm; Wav **0.213**, 0.037, 0.180–0.342 mm.

Discussion: This species might be the species Canu & Bassler (1928a) called *Celleporaria vagans*, but it is not the *Celleporaria vagans* of Busk (1881), a species originally described from the subantarctic southern Indian Ocean's Crozet Islands. *Celleporaria vagans* (Busk) has also been reported from warmer water in Indo-Pacific localities (Harmer 1957; Ryland & Hayward 1992; Tilbrook et al. 2001). Doubts were expressed by the authors of both publications that these tropical specimens are conspecific with Busk's subantarctic material. The specimens they described from warm water are similar in zooidal morphology to *C. vagans* in having a pustulose frontal wall with small marginal pores, and sandal shaped vicarious avicularia, but differ in other zooidal characters. The proximal edge of the orifice of *C. vagans* forms a shallow concavity with three thin denticles projecting. The suboral avicularium tubercle in *C. vagans* is a short cone. As the zooid develops it projects distally and laterally, meeting a projection from the lateral rim of the peristome to fuse into an oval opening. The vicarious avicularia have raised tilted bodies and the rim of their sandal-shaped rostrum is raised and distally cupped in contrast to the flat-rimmed walls of those of *C. soleata*. The ooecia are more than hemispherical in shape, but do not develop into the elongate globose flasks with flanged rims that enclose much of the orifice of *C. soleata*.

The genus was formerly placed in the Lepraliellidae, but recent work by Di Martino (2022) on *Lepraliella* has changed ideas about placement of species formerly included in that family grouping, and molecular studies support a close relationship of *Celleporaria* with Celleporidae.

Distribution: Caracol Bay, Haiti. Cuba (?)

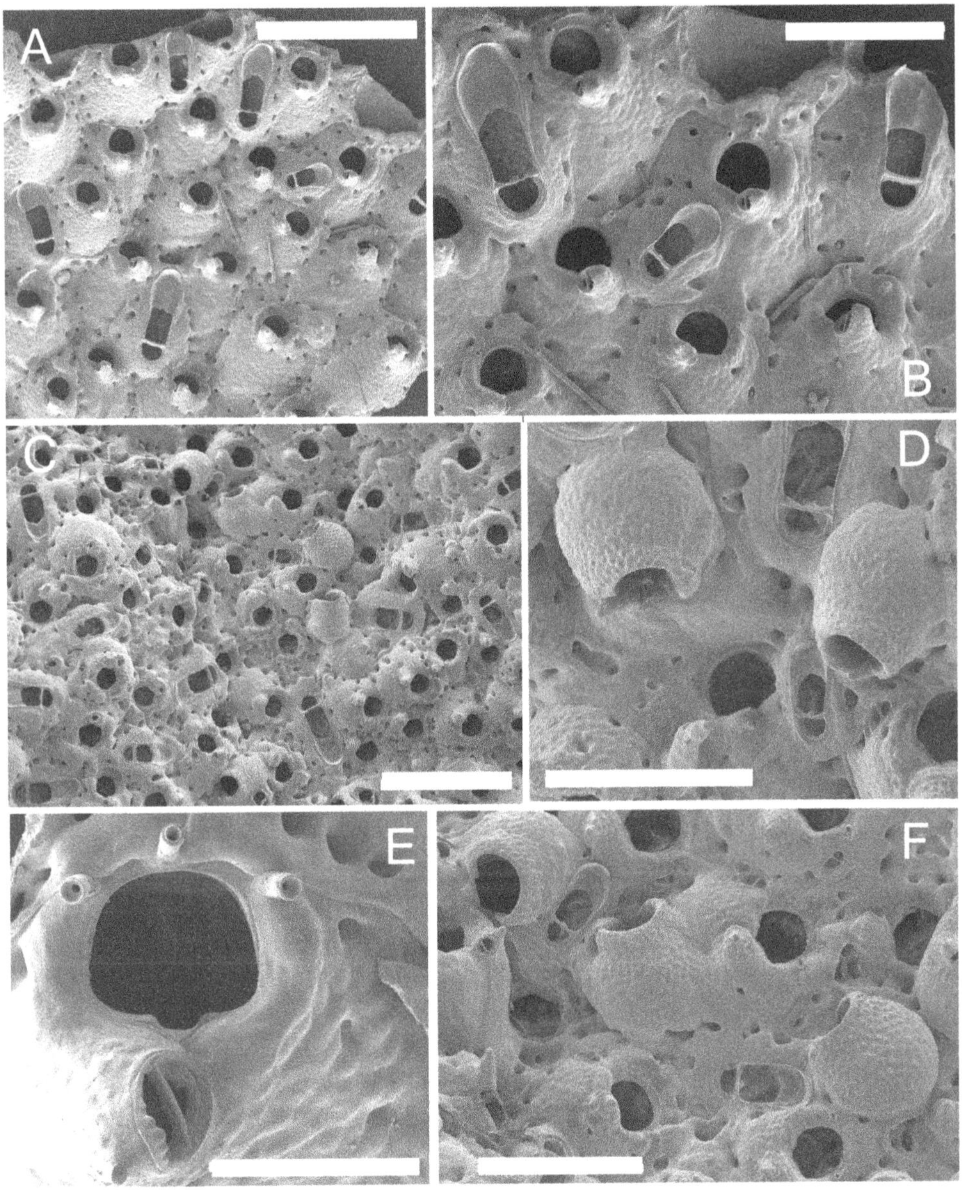

Figure 7. *Celleporaria soleata* new species. A. Autozooids and large shoe-shaped avicularia near growing edge of colony (Scale bar, 1 mm). B. Area of frontal budding, showing autozooids, ooecia, and large and small avicularia (Scale bar, 1 mm). C. Autozooids showing suboral avicularia and large shoe-shaped avicularia (Scale bar, 500 µm). D. Close-up of orifice with central sinus, three distal spines and oval suboral avicularium with serrate rostrum (Scale bar, 200 µm). E. Zooids with flask shaped avicularia (Scale bar, 500 um). F. Two flask shaped ooecia. (Scale bar, 500 µm).

ACKNOWLEDGMENTS

Walter Adey generously provided space on his cruise that made these collections possible. Valerie Paul and the staff at the Smithsonian Marine Station, Fort Pierce, Florida provided facilities and support and Wayne Hunter and Maria Gonzalez for help with use of the STEM at the USDA Horticultural laboratory in Fort Pierce, Florida where the SEM images were taken. Lily Berniker at AMNH assisted with final cataloging and deposition of specimens. Nancy Knowlton acted as a dive buddy for all the SCUBA collections. This is contribution no. 1187 from the Smithsonian Marine Station, Fort Pierce, Florida.

REFERENCES

Allman, G.J. 1856. *A Monograph of the Freshwater Polyzoa, Including All the Known Species, Both British and Foreign*. London: The Ray Society.

Almeida, A.C.S., Souza, F.B.C., Sanner, J. & Vieira, L. M. 2015. Taxonomy of Recent Adeonidae (Bryozoa, Cheilostomata) from Brazil, with the Description of Four New Species. *Zootaxa* 4013 (3): 348–368.

Bongaerts, P., Cooke, I.R., Ying, H., Wels, D., den Haan, S., Hernandez-Agreda, A., Brunner, C.A., Dove, S., Englebert, N., Eyal, G., Foret, S., Grinblat, M., Hay, K.B., Harii, S., Hayward, D.C., Lin, Y., Mihaljevic, M., Moya, A., Muir, P., Sinniger, F., Smallhorn-West, P., Torda, G., Ragan, M.A., van Oppen, M.J.H. & Hoegh-Guldberg, O. 2021. Morphological Stasis Masks Ecologically Divergent Coral Species on Tropical Reefs. *Current Biology* 31: 2286–2298.

Busk, G. 1852. *Catalogue of Marine Polyzoa in the Collection of the British Museum. I. Cheilostomata*. London: Trustees of the British Museum.

Busk, G. 1881. Descriptive Catalogue of the Species of *Cellepora* Collected on the 'Challenger' Expedition. *Journal of the Linnean Society, London, Zoology* 15: 341–356.

Busk, G. 1884. Report on the Polyzoa Collected by H.M. S. Challenger During the Years 1873–1876. Part 1. The Cheilostomata. *Report on the Scientific Results of the Voyage of the H.M.S. "Challenger", Zoology* 10: 1–216.

Canu, F. 1919. *Hippaliosina*, un Nouveau Genre de Bryozoaires. *Bulletin de la Societé géologique de France* (4) 18: 88–94.

Canu, F. & Bassler, R.S. 1923. North American later Tertiary and Quaternary Bryozoa. *United States National Museum Bulletin* 125: 1–302.

Canu, F. & Bassler, R.S. 1927. Classification of the Cheilostomatous Bryozoa. *Proceedings of the United States National Museum* 69: 1–42.

Canu, F. & Bassler, R.S. 1928a. Fossil and Recent Bryozoa of the Gulf of Mexico. *Proceedings of the United States National Museum* 72: 1–199.

Canu, F. & Bassler, R.S. 1928b. Bryozoaires du Brésil. *Bulletin de la Sociéte des Sciences de Seine-et-Oise* 9: 58–110.

Cheetham, A.H., Sanner, J. & Jackson, J.B.C. 2007. *Metrarabdotos* and Related Genera (Bryozoa: Cheilostomata) in the Late Paleogene and Neogene of tropical America. *Journal of Paleontology* 81 (Supplement 1): 1–91.

Cook, P.L., Bock, P.E., Hayward, P.J. & Gordon, D.P. 2018. Class Gymnolaemata, order Cheilostomata. In P.L. Cook, P.E. Bock, D.P. Gordon & H. Weaver (eds), *Australian Bryozoa. Vol. 2. Taxonomy of Australian families*: 61–280. Melbourne: CSIRO Publishing.

Di Martino, E. 2022. Revision of the Type Species of Some Cheilostome Bryozoan Genera in the Collection of the Swedish Museum of Natural History. *Zootaxa* 5125.2 (2022): 157–181

Ehrenberg, C.G. 1831. s.n. In P.C. Hemprich & C.G. Ehreberg (eds), *Symbolae Physicae, seu Icones et Descriptiones Corporum Naturalium novorum aut Minus Cognitorum, Quae ex Itineribus per Libyam, Aegyptum, Nubiam, Dongalam, Syriam, Arabiam et Habessiniam...Studio Annis 1820–25 redirerunt*, 1–162. Ex Officina Academica, venditur a Mittlero: Berolini.

Gordon, D.P. 1989. The Marine Fauna of New Zealand: Bryozoa: Gymnolaemata (Cheilostomida Ascophorina) from the Western South Island Continental Shelf and Slope. *New Zealand Oceanographic Institute Memoir* 97: 1–158.

Harmer, S.F. 1957. The Polyzoa of the Siboga Expedition, Part 4. Cheilostomata Ascophora II. *Siboga Expedition Reports* 28d: 641–1147.

Hastings, A.B. 1930. Cheilostomatous Polyzoa from the Vicinity of the Panama Canal Collected by Dr. A. Crossland on the Cruise of the S.Y. "St George". *Proceedings of the Zoological Society of London* 99: 697–740.

Herrera Cubilla, A., Dick, M.H., Sanner J. & Jackson, J. B.C. 2006. Neogene Cupuladriidae of Tropical America. I: Taxonomy of Recent *Cupuladria* from Opposite Sides of the Isthmus of Panama. *Journal of Paleontology* 80: 245–263.

Herrera Cubilla, A., Dick, M.H., Sanner J. & Jackson, J. B.C. 2008. Neogene Cupuladriidae of Tropical America. II: Taxonomy of Recent *Discoporella* from Opposite Sides of the Isthmus of Panama. *Journal of Paleontology* 82: 279–298.

Hincks, T.H. 1877. On British Polyzoa, Part 1. *Annals and Magazine of Natural History, Series 4*, 20, 212–218.

Jackson, J.B.C. 1984. Ecology of Cryptic Coral-reef Communities. III. Abundance and Aggregation of Encrusting Organisms with Particular Reference to Cheilostome Bryozoa. *Journal of Experimental Marine Biology and Ecology* 75: 36–57.

Jackson, J.B.C. & Cheetham, A.H. 1990. Evolutionary Significance of Morphospecies: A Test with Cheilostome Bryozoa. *Science* 248: 579–583.

Jackson, J.B.C. & Cheetham, A.H. 1994. Phylogeny Reconstruction and the Tempo of Speciation in Cheilostome Bryozoa. *Paleobiology* 20: 407–423.

Jackson, J.B.C. & Herrera Cubilla, A. 2000. Adaptation and Constraint as Determinants of Zooid and Ovicell Size Among Encrusting Ascophoran Cheilostome Bryozoa from Opposite Sides of the Isthmus of Panama. In A. Herrera Cubilla & J.B.C. Jackson (eds), *Proceedings of the 11th International Bryozoology Association Conference*: 249–258. Balboa: Smithsonian Tropical Research Institute.

Jackson, J.B.C. & Winston, J.E. 1982. Ecology of Cryptic Coral Reef Communities. I. Distribution and Abundance of Major Groups of Encrusting Organisms. *Journal of Experimental Marine Biology and Ecology* 57, 135–147.

Jullien, J. 1882. Note sur une Nouvelle Division des Bryozoaires Cheilostomiens. *Bulletin de la Sociéte Zoologique de France* 6: 271–285.

Jullien, J., 1883. Dragages du 'Travailleur'. Bryozoaires, Espèces draguées dans l'Océan Atlantique en 1881. *Bulletin de la Société Zoologique de France* 7: 497–529.

Kramer, P., Atis, M., Schill, S., Williams, S.M., Freid, E., Moore, G., Martinez-Sanchez, J.C., Benjamin, F., Cyprien, L.S., Alexis, J.R., Grizzle, R., Ward, K., Marks, K. & Grenda, D. 2016. Baseline Ecological Inventory for Three Bays National Park, Haiti. *The Nature Conservancy: Report to the Inter-American Development Bank*, 1–180

Lamouroux,J.V.F. 1821. *Exposition Méthodique des Genres de l'ordre des Polypiers, avec Leur Description et Celles des Principales Espèces Figurées dans 84 Planches; les 63 Premiers Appartenant a L'histoire Naturelle des Zoophytes d'Ellis et Solander*. Paris: V. Agasse.

Levinsen, G.M.R. 1909. *Morphological and Systematic Studies on the Cheilostomatous Bryozoa*. Copenhagen: Nationale forffatteres Forlag.

MacGillivray, P.H. 1895. A monograph of the Tertiary Polyzoa of Victoria. *Transactions of the Royal Society of Victoria* 4: 1–166.

Osburn, R.C. 1940. Bryozoa of Porto Rico with a Resume of West Indian Bryozoan Fauna. *Scientific Survey of Porto Rico and the Virgin Islands* 16: 321–486.

Osburn, R.C. 1952. Bryozoa of the Pacific coast of America, part 2, Cheilostomata–Ascophora. *Report of the Allan Hancock Pacific Expeditions* 14: 271–611.

Powell, N.A. 1971. The marine Bryozoa near the Panama Canal. *Bulletin of Marine Science* 21: 766–778.

Ryland, J.S. & Hayward, P.J. 1992. Bryozoa from Heron Island, Great Barrier Reef. *Memoirs of the Queensland Museum* 32: 223–301.

Simpson, C., Jackson, J.B.C. & Herrera Cubilla, A. 2017. Evolutionary Determinants of Morphological Polymorphism in Colonial Animals. *The American Naturalist*, 190 (1), 17–28.

Tilbrook, K.J., Hayward, P.J. & Gordon, D.P. 2001. Cheilostomatous Bryozoa from Vanuatu. *Zoological Journal of the Linnean Society*, 131 (1): 35–109.

Vieira, L.M., Migotto, A.E. & Winston, J.E. 2008. Synopsis and Annotated Checklist of Recent Marine Bryozoa from Brazil. *Zootaxa* 1810 (1): 1–39.

Wahle, C.M. 1980. Detection, Pursuit, and Overgrowth of Tropical Gorgonians by Milleporid Hydrocorals: Perseus and Medusa revisited. *Science* 209 (4457): 689–691.

Williams, S.M., Chollett, I., Roff, G., Cortés, J., Dryden, C.S. & Mumby, P.J. 2015. Hierarchical Spatial Patterns in Caribbean Reef Benthic Assemblages. *Journal of Biogeography*. doi:10.1111/jbi.12509

Winston, J.E. 1978. Polypide Morphology and Feeding in Marine Ectoprocts. *Bulletin of Marine Science* 28: 1–31.

Winston, J.E. 2005. Redescription and Revision of Smitt's "Floridan Bryozoa in the Collection of the Museum of Comparative Zoology, Harvard University. *Virginia Museum of Natural History Memoir* 7: 1–150.

Winston, J.E. & Jackson, J.B.C. 2021. Coral Reef-associated Bryozoans of Jamaica. *Zootaxa*, 4988 (1): 1–281.

Bryozoan Studies 2022 – Key, Porter & Wyse Jackson (eds)

Skeletonisation in the stenolaemate bryozoan orders Cryptostomata (Suborder Rhabdomesina) and Trepostomata: the role of the Bryozoan Skeletal Index (BSI) as a taxonomic character

Patrick N. Wyse Jackson*
Department of Geology, Trinity College, Dublin, Ireland
ORCID ID: 0000-0001-5605-0670

Marcus M. Key, Jr.
Department of Geosciences, Dickinson College, Carlisle, PA, USA
ORCID ID: 0000-0003-4097-0143

Catherine M. Reid
School of Earth and Environment, University of Canterbury, Christchurch, New Zealand
ORCID ID: 0000-0002-7204-2037

ABSTRACT: The Bryozoan Skeletal Index (BSI) is a measure of the relative proportion of skeleton to open space in stenolaemate bryozoans. Here the measure is calculated for the Families of the Suborder Rhabdomesina of the Order Cryptostomata and in the Order Trepostomata. Patterns show that in the former, the members of the Family Arthrostylidae yield lower BSI values than members of other cryptostome families, while in general the BSI values in the trepostomes are higher than those of the cryptostomes. A number of taxa are evaluated with respect to BSI as their ordinal or familial placement has varied according to different authors; these include *Coelotubulipora* and *Tebitopora*. For both, BSI values are low, suggesting that placement in the Cryptostomata is correct. However, other morphological characters indicate that the former is a trepostome and the latter a cryptostome. The genera *Nematotrypa* and *Cyclophaenopora* that had been placed in either the Arthrostylidae or the Nematotrypidae are considered and shown to likely belong in the latter family rather than the former. BSI is a measure that can be utilized with some value in systematic placement but only alongside salient and characteristic morphological data.

1 INTRODUCTION

The suprageneric arrangement of Paleozoic stenolaemate bryozoans now consists of a number of orders of which the Trepostomata erected by Ulrich (1882) and Cryptostomata (Vine 1884) both contain taxa that may develop erect growth forms that appear superficially similar. For a period in the mid-twentieth century, use of the Order Cryptostomata fell into abeyance and was only reestablished by Blake (1975) during revision of the *Treatise on Invertebrate Paleontology* which documented two cryptostome suborders, the Suborder Rhabdomesina and Suborder Ptilodictyina and their constituent genera (Blake 1983a, 1983b,1983c; Karklins 1983a, 1983b).

The Bryozoan Skeletal Index (BSI) was devised as a measure of the relative proportion of skeleton to open space in stenolaemate bryozoans (Wyse Jackson *et al.* 2020), and it is computed for bryozoans of the Order Trepostomata and Cryptostomata (Suborder Rhabdomesina). In this study the Suborder Ptilodictyina is not included as colonies comprise bifoliate branches for which BSI is difficult to compute as they grossly differ from ramose cylindrical colonies common in the Trepostomata and Rhabdomesinids. Earlier assessments have documented the pattern of BSI across the Calcium-Aragonite transition in the Mississippian (Key *et al.* 2022) and in the latitudinal distribution of these bryozoans through the Paleozoic (Reid *et al.* 2022).

The purpose of this paper is two-fold. Firstly, it ascertains whether the BSI values in trepostome and some rhabdomesine cryptostome families reveal trends within and between families and in the orders. Secondly, the paper examines some specific genera in which BSI may be helpful in providing taxonomic insights. The placement of a number of

*Corresponding Author: wysjcknp@tcd.ie

DOI: 10.1201/9781003454915-18

genera has remained problematic to the present day due to them possessing morphologies that are at least in part characteristic of either trepostomes or cryptostomes, and the emphasis placed on these morphologies for suprageneric assignment. This issue occurs in the Ordovician as well as in the Late Paleozoic and in some Triassic stenolaemates. Spjeldnaes (1984) erected the rhabdomesine Family Nematotrypidae to accommodate the new genus *Cyclophaenopora*. It was not clear as to which other genera he considered to belong to this family, that he noted showed both cryptostome and trepostome affinities. Gorjunova (1985) reassessed the family and included *Nematotrypa*, *Moyerella*, *Kielanopora* and *Clausotrypa*, although she excluded *Cyclophaenopora* which Bock (2022) lists in the Family Arthrostylidae. For the purpose of this exercise, we have assessed *Nematotrypa, Cyclophaenopora* and *Moyerella* and use BSI to assess whether all taxa might be better placed in the Arthrostylidae on this measure.

Gorjunova (2011) placed her new Family Coelotubuliporidae in the Trepostomata although it appeared that the constituent genera *Coelotubulipora* Yang, Hu & Xia, 1988 and *Dunaevella* Gorjunova, 2011 possessed morphological characteristics that parallel rhabdomesonid cryptostomes, an opinion advocated earlier by Morozova *et al.* (2006) when they described the species *C. rara* from the Devonian of Poland. The Triassic genus *Tebitopora* Hu, 1984 has been enigmatic and was originally assigned to the Trepostomata (Hu 1984) and retained there by Schäfer and Grant-Mackie (1998) but subsequently reassigned to the Cryptostomata (Ma *et al.* 2020; Powers & Pachut 2008, p. 364). What are the parallels in skeletal characters that have led to these taxonomic conundrums and can the computation of BSI in these taxa aid their taxonomic placement?

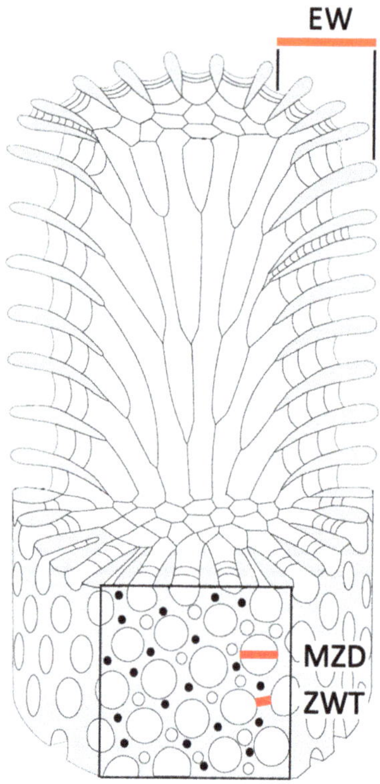

Figure 1. Morphological characters used to compute the Bryozoan Skeletal Index (BSI). From Wyse Jackson *et al.* (2020, fig. 2). Abbreviations: EW: exozone width; MZD: autozooecial aperture diameter; ZWT: autozooecial wall thickness.

2 METHODS

The Bryozoan Skeletal Index (BSI) is a measure of the relative proportion of skeleton to open space in the outer exozonal portion of the colony (Wyse Jackson *et al.* 2020). It is derived from measures of three characters easily obtained from colonies: (1) maximum autozooecial apertural diameter at the zoarial surface or in shallow tangential section [MZD], (2) thickness of the zooecial wall between adjacent autozooecial apertures [ZWT], and (3) the thickness of the exozone [EW] (Figure 1) in the formula:

$$BSI = ((EW * ZWT)/MZD) * 100$$

The accuracy of BSI values computed from stenolaemate bryozoans may be affected by a number of morphological and taphonomic features. These are discussed in Wyse Jackson *et al.* (2020, p. 198) and include the development and positioning of maculae, inclusion of intrazooecial features, and whether ramose colonies are crushed taphonomically. In addition, where exozonal walls are moniliform and irregular in thickness as in *Stenopora* (Reid 2021), the values of MZD and ZWT may vary depending on the depth of the tangential section at which it was measured. This variation can be reduced through taking a number of measurements at different depths through the exozone, although preparation of multiple sections of peels can be time consuming, and these may not be available where literature sources are utilised. In order to provide an accurate BSI value in some globular and dome-shaped trepostome genera such as *Diplotrypa* Nicholson, 1879, recognition of the exozone-endozone boundary is required given all zooecial walls are very thin-walled unlike in most taxa in the Order.

For 83 genera of rhabdomesinid cryptostomes and 155 genera of trepostomes, the BSI was calculated from published data in type species descriptions, or from measurement of photographs of tangential and longitudinal sections except when data were published in tabular form. The age of the type species defined as the midpoint of the type locality stratigraphic age ranges from Shanan Peters' Macrostrat.org (Key *et al.* 2022), and all numerical ages were based on the latest version of the International Chronostratigraphic Chart (Cohen *et al.* 2020). Genus stratigraphic ranges were derived from the descriptions by Boardman and Buttler (unpublished data) and augmented with Philip Bock's webpage Bryozoa.net.

BSI values were tabulated within geological periods for the type species of genera and sorted in a series of bins from low (BSI <10) to high (>200) for three groups: the arthrostylid crypostomes, all other rhabdomesinid cryptostomes, and the trepostomes (Table 1). To investigate ordinal and family level BSI, values were listed for the genera assigned to each Family; these familial designations follow Blake (1983c), Boardman and Buttler (unpublished data) and Taylor (1993). These BSI values were then tabulated, grouped into the same bins as in Table 1, and the total number of genera per Family is listed as are the minimum, maximum and mean BSI values and the age range (Table 2). Families are listed in order of their first appearance at the level of stratigraphic stages from the Tremadocian (Lower Ordovician) to the Rhaetian (Late Triassic) as given in the International Chronostratigraphic Chart (Cohen *et al.* 2020). This has allowed for differences and similarities in BSI between families to be easily picked out and for trends through time to be demonstrated.

In addition to the type species used to compute the BSI for genera, additional species contained within specific taxa such as *Tebitopora* and those placed in the Family Coelotubuliporidae (*Coelotubulipora* Yang, Hu and Xia, 1988 and *Dunaevella* Gorjunova, 2011) are examined in more detail to assess whether BSI can aid taxonomic placement. BSI for the Coelotubuliporidae was computed on the basis of descriptions in original publications in which the species were first published (see listing in Table 2) and for Tibetopora on illustrations of the type species in Ma *et al.* (2020).

Table 1. BSI values in Cryptostome and Trepostome families. Number of genera in bins of BSI values from low (<10) to high (>200). NG = Number of Genera; X = Mean BSI; Min = Minimum BSI; Max = Maximum BSI; Age range of genera type species given in Ma. Families are listed according to their earliest appearance.

BSI value range:	<10	10–49.9	50–99.9	100–149.9	150–199.9	>200	NG	X	Min	Max	Age
Cryptostomata							**84**				
Arthrostylidae	14	5	–	–	–	–	19	13.6	0.5	24	462–267
Nematotrypidae	–	–	1	1	–	–	2	90.4	70.9	110	457–452
Rhabdomesidae	4	17	6	–	1	–	28	39.6	3.6	179.4	425–286
Hyphasmoporidae	3	4	1	1	–	–	9	35.3	4	129.2	403–265
Rhomboporidae	–	8	3	–	–	–	11	34.8	10.1	66.1	399–244
Nikiforovellidae	4	3	–	–	–	–	7	11.3	6.3	25.8	398–265
Bactroporidae	–	1	–	–	–	–	1	14	–	–	387
Lenaporidae	–	2	–	–	–	–	2	37.3	30.7	44	385–377
Pseudoascoporidae	1	–	1	–	–	–	2	31.3	6	56.6	365–286
Nudymiellidae	–	1	–	–	–	–	1	12.3	–	–	338
Goldfussitrypidae	–	1	–	–	–	–	1	20	–	–	334
Medvedkellidae	–	–	1	–	–	–	1	60.8	–	–	305
Trepostomata							**155**				
Aisenvergiidae	–	4	3	–	–	–	7	62.7	39	98	480–353
Dittoporidae	–	4	1	2	–	–	7	62	18	148	471–452
Heterotrypidae	1	4	2	–	–	–	7	35.3	6	77	465–243
Trematoporidae	2	4	2	2	–	1	11	65.4	4	200	464–385
Ralfimartitidae	2	3	–	–	–	–	5	16	7	35	463–450
Bimuroporidae	1	2	–	–	–	–	3	19	8	34	461–451
Diplotrypidae	–	–	–	1	–	–	1	140	–	–	461
Atactotoechidae	4	5	4	–	–	–	13	36.8	4	97	458–366
Monticuloporidae	–	3	–	–	–	–	3	16	12	18	456–447
Mesotrypidae	4	1	1	1	–	–	7	30.3	4	108	456–449
Halloporidae	–	4	–	–	–	–	4	26.5	16	35	454–428
Amplexoporidae	–	7	1	–	–	–	8	32.6	12	78	454–339
Batostomellidae	1	–	1	–	–	–	2	38.5	6	71	450–449
Leioclemidae	–	4	1	–	–	–	5	32	14	60	439–256

(*continued*)

Table 1. Continued

BSI value range:	<10	10–49.9	50–99.9	100–149.9	150–199.9	>200	NG	X	Min	Max	Age
Stenoporidae	1	6	6	1	–	1	15	57.3	7	211	432–225
Hemieridotrypidae	–	2	1	–	–	–	3	45.7	23	82	425–421
Eridotrypellidae	1	7	5	–	–	–	13	44.2	7	94	425–255
Anisotrypidae	–	2	3	–	–	–	5	62.4	28	86	419–256
Stereotoechidae	–	1	–	–	–	–	1	48	–	–	384
Ulrichotrypellidae	–	2	1	–	–	–	3	37.3	23	59	375–256
Dyscritellidae	–	5	1	1	–	–	7	59	11	131	366–232
Crustoporidae	–	3	1	–	1	–	5	51.6	15	160	359–305
Coelotubuliporidae	1	1	–	–	–	–	2	48.8	10.6	87	352–326
Cycloporidae	–	2	–	–	–	–	2	21.5	14	29	346
Nipponstenoporidae	–	1	–	–	–	–	1	29	–	–	335
Astralochomidae	–	1	–	–	–	–	1	25	–	–	331
Stenoporellidae	–	2	–	–	–	–	2	30	12	48	328–255
Mishulgellidae	1	–	–	–	–	–	1	3	–	–	308
Maychellinidae	–	2	1	–	–	–	3	27.3	13	56	274–329
Araxoporidae	–	3	–	–	–	–	3	29	18	35	260–215
Zozariellidae	–	–	1	1	–	–	2	94.5	83	106	245–232

Table 2. BSI values in species of the genera *Coelotubulipora* and *Dunaevella* in the Family Coelotubuliporidae.

	BSI	Geological horizon
Coelotubulipora irregularis (Nekhoroshev 1956)	1.3	Devonian
Coelotubulipora maculata (Nekhoroshev 1956)	9.4	Mississippian
Coelotubulipora euspinusa Yang *et al.*, 1988	10.6	Mississippian
Coelotubulipora euspinusa ferodiaphragma Yang *et al.*, 1988	3.3	Mississippian
Coelotubulipora euspinusa claustubulosa Yang *et al.*, 1988	13.1	Mississippian
Coelotubulipora obliqua Yang *et al.*, 1988	7.6	Mississippian
Coelotubulipora superhemisepta Yang *et al.*, 1988	3.3	Mississippian
Coelotubulipora superhemisepta claustubulosa Yang *et al.*, 1988	2.8	Mississippian
Coelotubulipora superhemisepta torquicuticula Yang *et al.*, 1988	12.0	Mississippian
Coelotubulipora interminuta Yang *et al.*, 1988	1.3	Mississippian
Coelotubulipora bicurva Yang *et al.*, 1988	6	Mississippian
Coelotubulipora irinae Gorjunova, 2011	5	Mississippian
Coelotubulipora intervacara (Trizna 1958)	15	Mississippian
Coelotubulipora varionodata (Trizna 1958)	5	Mississippian
Dunaevella shishovae (Schulga-Nesterenko 1955)	87	Mississippian
Dunaevella peristomata (Schulga-Nesterenko 1955)	24.8	Mississippian

3 RESULTS

BSI values calculated for 84 cryptostome and 155 trepostome genera are presented in Table 1 where values are compiled into six value bins between <10 and >200. Data by families (12 in the Order Cryptostomata and 31 in the Order Trepostomata) are provided in this table. Six trepostome genera that have not been assigned to a Family yielded BSI values while some other trepostome families contain genera for which BSI values could not be computed either due to taphonomic preservational issues or because not all parameters were measureable, i.e., the Pseudocampylidae, Helenoporidae, Dianulitidae, Revalotrypidae and the Girtyporidae. Table 2 gives BSI values of species contained within those genera in the Family Coelotubiporidae which have been placed at various times in either of the orders.

4 DISCUSSION

4.1 *General BSI trends in cryptostome and trepostome families*

The smallest most delicate cryptostomes are the arthrostylids that first appeared in the early Ordovician, but which rapidly diversified so that during this period they reached their greatest diversity. From the Devonian onwards their diversity declined, and only one genus *Permoheloclema* Ozhgibesov, 1983 persisted into the Permian. Structurally the Arthrostylidae are varied with simple narrow erect or branched zoaria, sometimes articulated, generally with thin exozones and small stylets that may only extend marginally beyond the zoarial surface. In many taxa, autozooecial apertures are large with narrow interapertural walls. The relatively large autozooecial apertures and narrow interapertural walls lead to low BSI values within the Arthrostylidae with a mean of 13.6 (Table 1, Figure 2).

By contrast, the members of the other families in the Suborder Rhabdomesina exhibit a greater range of BSI values from 3.6 in the genus *Promediapora* Gorjunova, 1992 to 179.4 in *Muromipora* Gorjunova, 2011 that reflect their greater structural complexity (Figure 3A-B). They largely comprise erect cylindrical zoaria with a well-defined axis from which are budded short cylindrical zooecial chambers in which hemisepta or other structural elements may be developed. Apertures tend to be oval to rhombic in shape, frequently surrounded by styles of a variety of sizes and structure, whereas interapertural areas may be skeletal or marked by the presence of space-filling metapores, especially in the Family Hyphasmoporidae. *Muromipora* has a very wide exozone comprised of thickened interapertural walls with narrow rhombic apertures.

The three most diverse Rhabdomesina families, the Rhabdomesidae, Rhomboporidae and the Hyphasmoporidae all show congruence in their mean BSI values that for their constituent genera cluster around the mid-30s (Table 1, Figure 2A). This moderately high BSI value is mainly related to increased exozonal width when compared to the Arthrostylids.

Genera within the Order Trepostomata exhibit a wide variety of zoarial forms from robust ramose erect branching forms, to hemispherical forms, and thin encrusting unilaminar or multilaminar sheets. The bryozoan skeleton may be subdivided into zooidal and extrazooidal skeleton with the latter contained within zooidal boundaries and the latter outside (Boardman & Buttler 2005). Trepostome zooids may be polymorphic with feeding autozooids most frequent, and in some families, exilazooecia and mesozooecia may be common. Within these, internal partitions such as diaphragms, hemiphragms or cystiphragms may be present that subdivide the zooidal chamber or slightly reduce its living chamber volume. Extrazooidal features include wall structures of several constructions and skeletal styles. It is difficult to distinguish the boundary between zooecial and extrazooecial skeleton (Boardman & Buttler 2005,

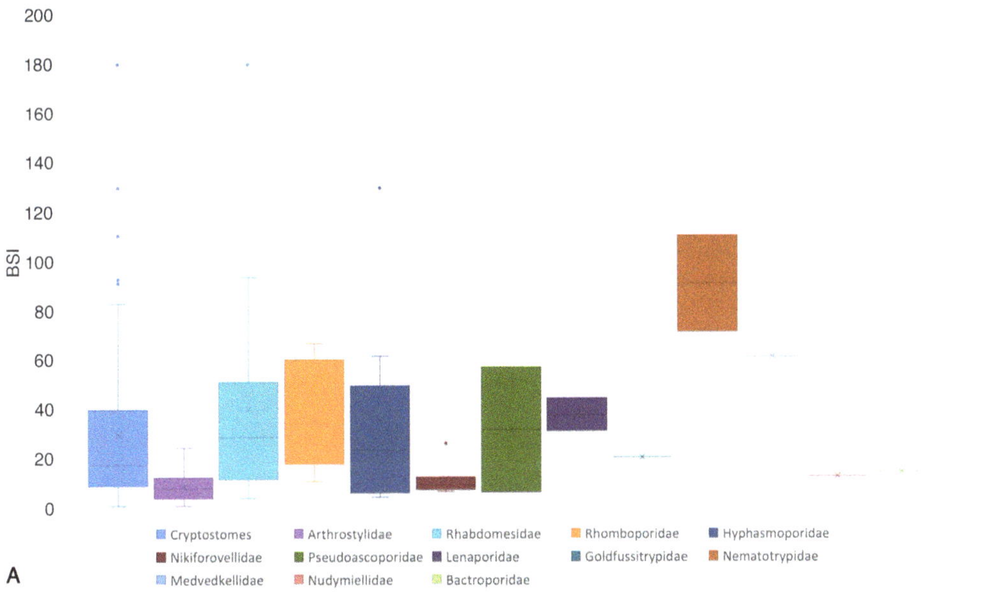

Figure 2. BSI in the orders (A) Cryptostomata and (B) Trepostomata and selected families and genera. In these box and whisker plots the Minimum and Maximum values are given by the position of the lower and upper whiskers respectively, the Quartile 1 and 3 values by the lower and upper edges of the box, while the Median is plotted with the dashed line and the Mean by an X.

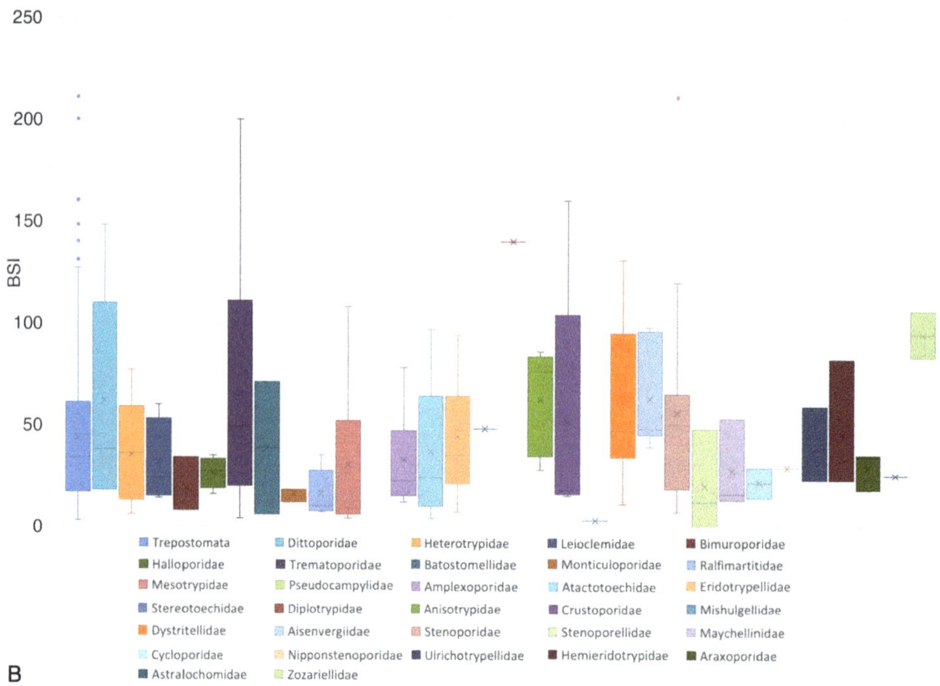

Figure 2. (Continued)

p. 1095) and some of the exozone skeleton is probably extrazooidal. Autozooecia in ramose colonies are budded from a well or ill-defined axial region, have long chambers in the endozone, and vestibules orientated at a high angle to the zoaria surface in the exozone. Diaphragms and other structural elements in chambers are frequently developed as are polymorphs such as mesozooecia and exilazooecia that are situated within the exozonal tissue. The exozone is usually composed of thickened skeletal walls in which acanthostyles may be developed. In some taxa, such as *Diplotrypa*, the exozone walls remain thin.

In the Order Trepostomata, BSI ranges from 3 in *Mishulgella* Gorjunova, 2001, an encrusting form from the Pennsylvanian of Russia (Figure 4A) to 211 in the robust erect Permian genus *Stenodiscus* Crockford, 1945. Highest mean BSI values occur in the trepostome families Crustoporidae (51.6), Stenoporidae (57.3), Dyscritellidae (59), Dittoporidae (62), Anisotrypidae (62.4), Aisenvergiidae (62.7) and Trematoporidae (65.4)

(Figure 2B). The lowest are in the Monticuliporidae (16), Ralfimartitidae (16), and Bimuroporidae (19). However, high mean values in a family don't necessarily demonstrate that all constituent genera have high BSI values. In the Stenoporidae the range is 204 from a low of 7 in *Pycnopora* Girty, 1911 to 211 in *Stenodiscus*, resulting from the fact that the former creates thin encrusting adnate sheets and the latter is ramose with a wide exozone of thick moniliform walls. In the Trematoporidae the range in BSI is 196 (from 4 in *Acanthotrypina* Vinassa de Regny, 1921 to 200 in *Minussina* Morozova, 1960 in which the exozone thickness is pronounced (Tolokonnikova 2009)). Conversely in some families such as the Monticuliporidae the BSI range is low (12 in *Acantholaminatus* Marintsch, 1998 to 18 in both *Monticulopora* d'Orbigny, 1849 and *Gortanipora* Vinassa de Regny, 1921) reflecting the consistent narrow exozone shared by the genera in the family. Of the five genera in the Crustoporidae for which BSI was calculated, three returned low values and all have an encrusting habit as suggested they should have by the family appellation and placement, but two yielded high values including *Hunanopora* Yang, 1950 (48) and *Tabuliporella* Nikiforova, 1933 (160) with their ramose, erect colonies. Perhaps BSI might be beneficial in distinguishing potential systematic outliers in some taxa whether they be at family or below.

4.2 *Morphology, extrazooidal characters and BSI*

Which of the three defining morphological characters most influence BSI? We use Spearman's rank order correlation r_s, as the data do not have a normal distribution, and p values between the BSI characters and the BSI value to answer this question. For r_s values closer to -1 or 1, they indicate a negative or positive correlation, and closer to 0 indicate no correlation.

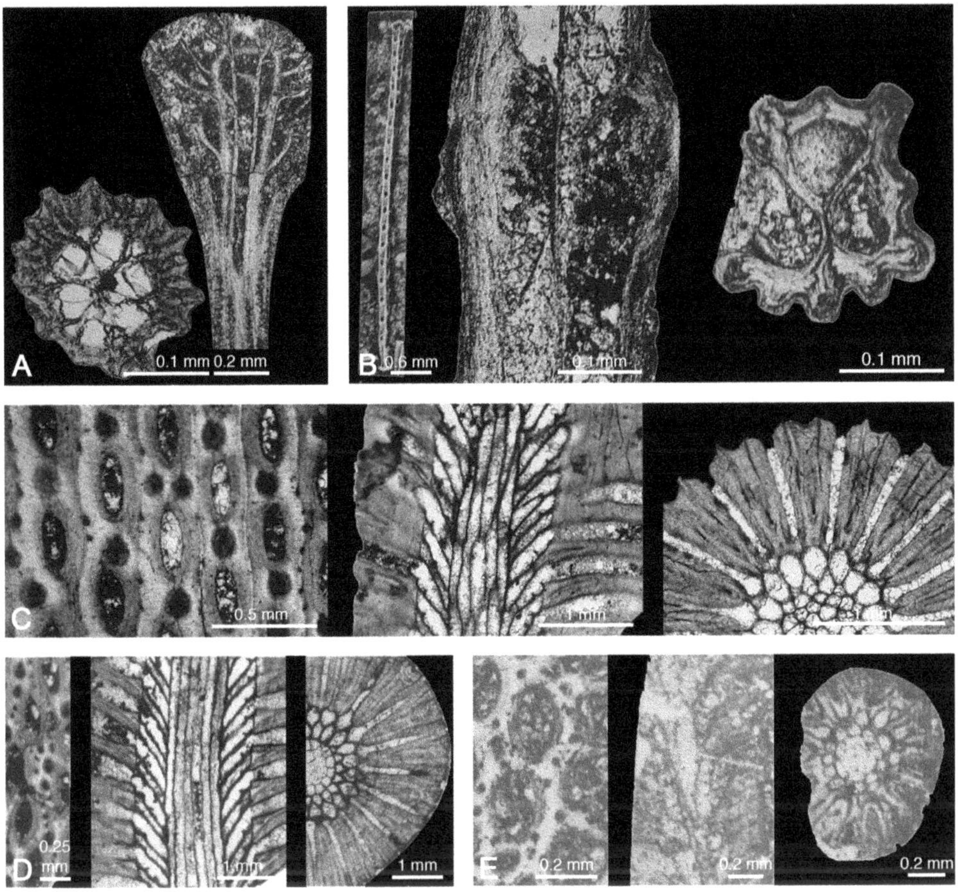

Figure 3. Examples of genera in the cryptostomata Family Arthrostylidae (A-B) and Suborder Rhabdomesina (C-E) with either high or low BSI values. (A) *Sceptropora facula* Ulrich, 1888 (BSI: 24); (B) *Arthrostylus tenuis* (James 1878) (BSI: 0.5); (C) *Muromipora smirnovi* Gorjunova, 2011 (BSI: 179.4); (D) *Ascopora florata* Schulga-Nesterenko, 1955 (BSI: 90.5); (E) *Promediapora transcaucasica* Gorjunova, 1992 (BSI: 3.6). [A-B from Utgaard 1983, C from Gorjunova 2011a, D from Gorjunova 2001, E from Gorjunova 1996].

Values for p above 0.05 indicate no significant correlation, but p values below 0.05 indicate more influence one of the three variables (i.e., exozone width (EW), autozooecial aperture diameter (MZD), and zooecial wall thickness (ZWT) has on predicting BSI. In both the trepostomes and the cryptostomes, r_s values are closer to 0 and p values are > 0.05 for MZD – there is no significant control on BSI by MZD. For EW and ZWT, r_s values are higher and p values all well below 0.05 – EW and ZWT have significant influence on BSI (Figure 5A-F), but the latter may show variation depending on the depth of section.

4.3 *Insights from specific taxa or families*

4.3.1 *Family Nematotrypidae*

If taken as a member of the Arthrostylidae, *Cyclophaenopora* Spjeldnaes, 1984 is an outlier with BSI values of 110.0 which is five times higher than the next lower genus in the family, *Sceptropora* Ulrich, 1888. This suggests that *Cyclophaenopora* may not be correctly assigned to this family. *Nematotrypa* Bassler, 1911 has a BSI of 70.9 which is closer to *Cyclophaenopora* to which it may be closely allied; this supports the placement of these two taxa in the Family Nematotrypidae (Nekhorosheva 2007). The taxonomic relationship of these two taxa is currently under investigation in a separate study by Ernst, Tolokonikova and Wyse Jackson. In contrast, *Moyerella*, which Blake (1983c) included as an arthrostylid and Gorjunova (1985) and Ernst & Carrera (2008) as a nematotrypid, has a lower BSI of 8.5. If one takes a low BSI value as a familial defining character, then the genus would be best placed in the Arthrostylidae as per Blake (1983c) but a detailed morphological assessment of a species of *Moyerella* from the Ordovician of Argentina (Ernst & Carrera 2008 suggests that it belongs to the Family Nematotrypidae as Gorjunova (1985) originally wrote.

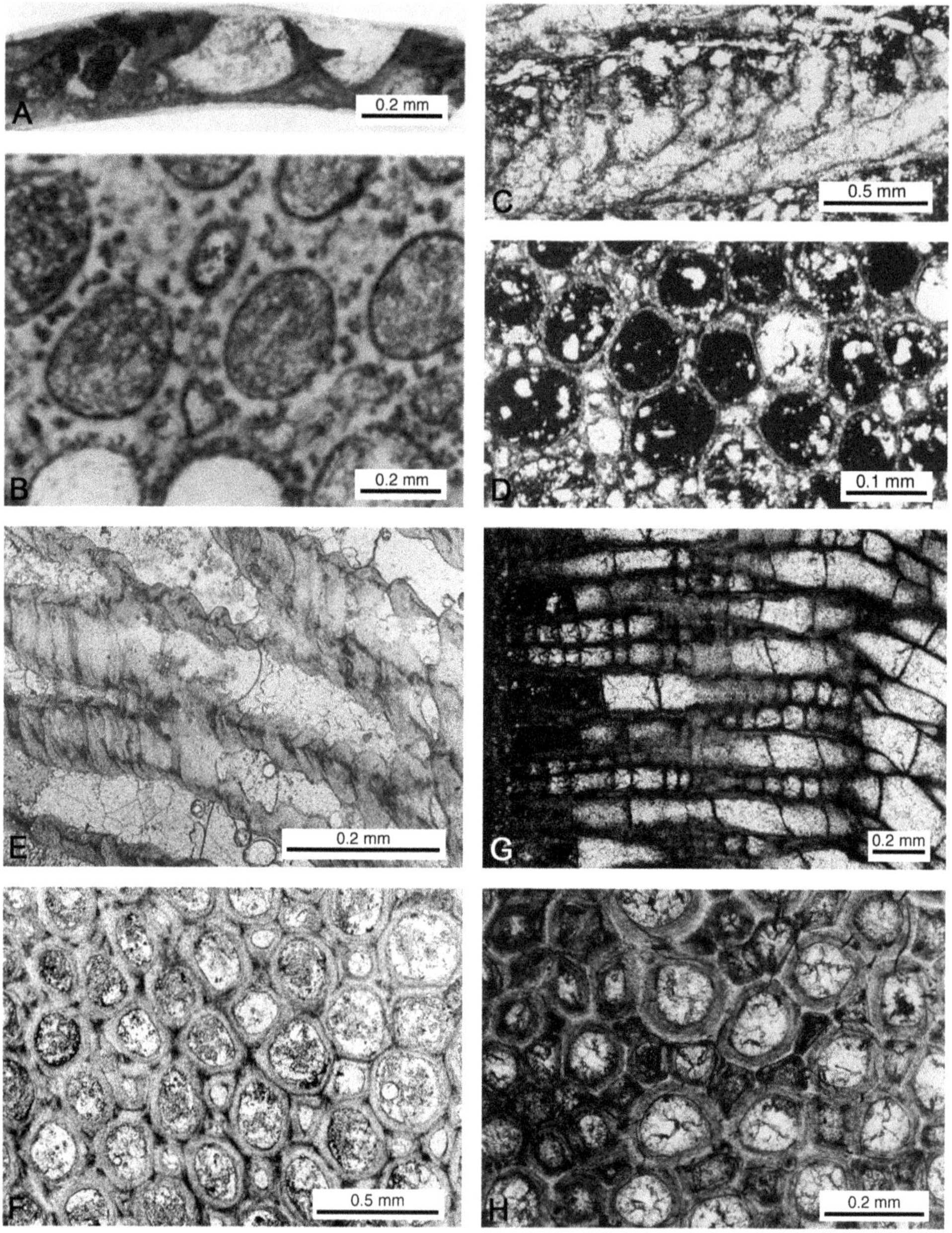

Figure 4. Examples of genera in the trepostomata with either low or high BSI values. (A-B) *Mishulgella stellata* Gorjunova, 2001 (BSI: 3); (C-D) *Pycnopora regularis* Girty, 1911 (BSI: 7); (E-F) *Stenodiscus moniliformis* Crockford, 1945 (BSI: 211); (G-H) *Minussina maculosa* Morozova, 1960 (BSI: 200) [A-B from Gorjunova 2001, C-H from Boardman and Buttler unpublished].

4.3.2 *Family Coelotubuliporidae*

In the genera assigned to the Family Coelotubuliporidae by Gorjunova (2011b), BSI values range from 1.3 in *Coelotubulipora irregularis* (Nekohorshev 1956) from the Devonian of Siberia and *C. interminuta* Yang *et al.* 1988 from the Mississippian of Hunan, China to 13.1 in *C. euspinusa claustubulosa* Yang *et al.*, 1988 from the same horizon and location in China. The mean BSI for the genus *Coelotubulipora* is 5.1 which is in contrast with the other constituent genus *Dunaevella* in which BSI is higher in both of its species: 24.8 and 87 (Table 2).

Species in *Coelotubulipora* are characterized by possessing a central axial tube that resembles closely that which is the diagnostic generic feature of

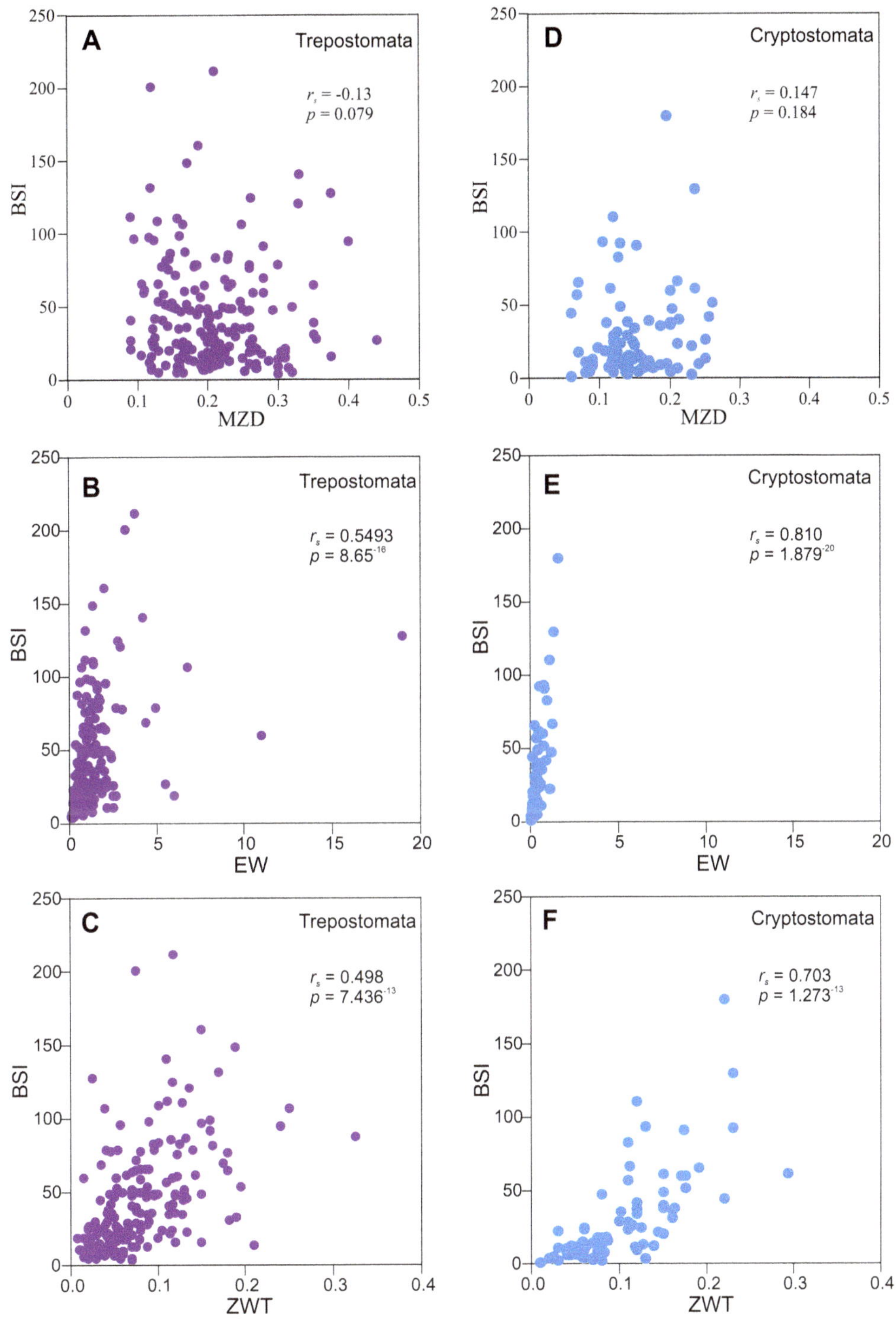

Figure 5. (A-C) BSI vs. morphological characters in trepostome genera; (D-F) BSI v. morphological characters in cryptostome genera. Measurements of morphological characters in mm.

the cryptostome genus *Rhabdomeson* Young & Young, 1874 (Wyse Jackson & Bancroft 1985) (Figure 6). It is therefore not surprising that Morozova *et al.* (2006) transferred the genus from the trepostomes, where it was initially placed by Yang *et al.* (1988), to the cryptostomes. However, Gorjunova (2011b) argued that *C. rana* described by Morozova *et al.* (2006) belongs to

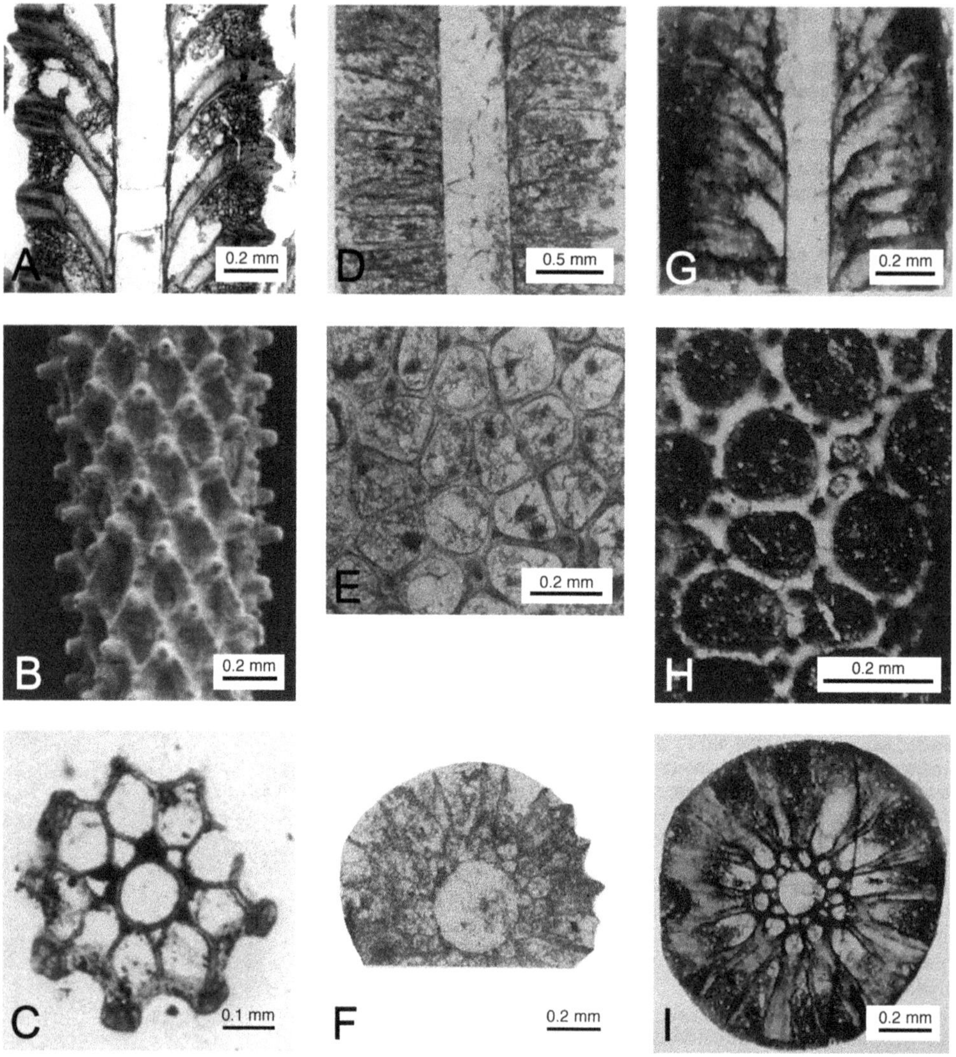

Figure 6. *Rhabdomeson* and taxa in the Family Coelotubuliporidae showing gross similarities in central axis but differences in autozooecial apertural arrangement and acanthostyle distribution; (A-C) *Rhabdomeson progracile* Wyse Jackson & Bancroft, 1985 (BSI: 3.9); (D-F) *Coelotubulipora euspinusa* Yang, Hu & Xia, 1988 (BSI: 10.6); (G-I) *Dunaevella shishovae* (Schulga-Nesterenko 1955) (BSI: 87) [A-C from Wyse Jackson & Bancroft 1985, D-F from Yang, Hu & Xia 1988, G-I from Gorjunova 2011b].

Rhabdomeson and not *Coelotubulipora,* and that species placement in the latter is based not only on the central axial tube but other additional morphological characteristics. In *Coelotubulipora*, the autozooecial apertural arrangement is not regular in longitudinal rows, the budding pattern from the axis is more irregular as is the distribution of stylets on interapertural walls, and exozonal zooecial walls are very thin, unlike those developed in *Rhabdomeson.* Although the low mean BSI is similar to *Rhabdomeson* and suggests that *Coelotubulipora* is closely allied to it using this solitary measure, the other morphological characteristics counter this argument and Gorjunova's (2011b) placement of *Coelotubulipora* in the Trepostomata is supported by BSI.

4.3.3 *Tebitopora*

Identification of the complexities of skeletal material and the budding patterns of autozooecia, together with the development or absence of polymorphs allow for assignment of taxa to stenolaemate families and orders. The genus *Tebitopora* described from the Triassic of Tibet by Hu (1984) was originally considered to contain two species, *T. orientalis*, the type species, and *T. depressa* but is now known to be monospecific (Ma *et al.* 2020). BSI may well be a crude measure for supra-generic taxonomic determinations, but in the case of *Tebitopora* which has been placed in two orders since its initial description, its BSI value is 22 (Table 2) which is in keeping with Late Paleozoic cryptostomes and is lower than all but one Triassic genera that have been assigned to the Trepostomata: *Arcticopora* (131), *Buria* (64), *Dyscritellopsis* (47), *Metastenodiscus* (78), *Phragmotrypa* (36), *Vysokella* (83) and *Zozariella* (106); the only exception being *Reptonoditrypa* (13).

5 CONCLUSIONS

In the orders Cryptostomata and Trepostomata, BSI is largely controlled by exozone width and interapertural wall thickness while autozooecial apertural diameter is not as significant.

The placement of a number of Upper Paleozoic stenolaemate genera remains problematic on account of the presence of structural and zooecial morphologies considered by researchers to be characteristic of both orders. In rhabdomesonid cryptostomes such as *Muromipora* (BSI: 179.4) exozone thickness increases to constitute over two-thirds of branch diameter so that taxa are more robust like their trepostome counterparts, or as in *Ogbinopora* Shishova, 1965 (BSI: 129.2) exozone width is significantly thicker than in most cryptostomes thus contributing a high degree of skeletal material.

In the Suborder Rhabdomesina, BSI mean values increase from the Silurian to the Pennsylvanian but decreases in later geological periods. Over the same geological timespan, BSI mean values in the Order Trepostomata are more consistent. The BSI value of *Tebitopora* is low and differs from that in unequivocal Triassic trepostomes and so BSI supports its placement in the Order Cryptostomata.

Moyerella which has a low BSI akin to the arthrostylids may be more correctly placed in the Family Nematotrypidae in the Trepostomata on structural and morphological grounds as demonstrated by Ernst and Carrera (2008). The same conclusion and reasoning should be applied in retaining the Family Coelotubuliporidae in the Trepostomata rather than in the Suborder Rhabdomesina even though BSI values of *Coelotubulipora* species are low like other rhabdomesinids.

BSI values calculated for genera within families may not be consistently similar. This study demonstrates that utilization of BSI as a taxonomic distinguishing characteristic is not reliable, but simply an indicator to be utilized alongside all available morphological data on which a greater emphasis should be placed. It should be viewed as a similar derived character such as Boardman's axial ratio (Boardman 1960) which is commonly calculated in systematic descriptions. BSI should be computed for new taxa or those undergoing revision, so that generic BSI values may be considered against that from other genera in the same family (as might be done for members in the trepostome Family Crustoporidae), and for new or earlier described species so that comparisons can be made. Similarly, any suprageneric revisions may find value in computing BSI values of candidate genera for placement in families. Outliers in values would merit further investigation and carefully considered subsequent taxonomic placement of taxa.

ACKNOWLEDGEMENTS

We are most grateful to both Caroline Buttler for providing unpublished images and descriptions of trepostome genera and a listing of trepostome families that were assembled by herself and Richard Boardman, and to Andrej Ernst for providing published images of some Palaeozoic bryozoans. The reviews by Caroline Buttler and Juan Suaréz Andrés helped improve this paper.

REFERENCES

Astrova, G.G. 1978. Historical Development, Systematics and Phylogeny of Bryozoa. Order Trepostomata. *Trudy Paleontologischeskogo Instituta Akademiya Nauk SSSR* 169: 1–240 [In Russian].

Bassler, R.S. 1911. The Early Paleozoic Bryozoa of the Baltic Provinces. *United States National Museum Bulletin* 77: 1–382.

Blake, D.B. 1975. The Order Cryptostomata Resurrected. In S. Pouyet (ed.), Bryozoa 1974. *Documents de Laboratoires de Géologie Faculté de sciences de Lyon*, hors-serie 3 (1): 211–223.

Blake, D.B. 1983a. The Order Cryptostomata. In R.A. Robison (ed.), *Treatise on Invertebrate Paleontology, Part G, Revised, Bryozoa*: 440–452. Boulder & Lawrence: Geological Society of America & University of Kansas Press.

Blake, D.B. 1983b. Introduction to the Suborder Rhabdomesina. In R.A. Robison (ed.), *Treatise on Invertebrate Paleontology, Part G, Revised, Bryozoa*: 530–549. Boulder & Lawrence: Geological Society of America & University of Kansas Press.

Blake, D.B. 1983c. Systematic Descriptions of the Suborder Rhabdomesina. In R.A. Robison (ed.), *Treatise on Invertebrate Paleontology, Part G, Revised, Bryozoa*: 550–592. Boulder & Lawrence: Geological Society of America & University of Kansas Press.

Boardman, R.S. 1960. Trepostomatous Bryozoa of the Hamilton Group of New York State. *United States Geological Survey Professional Paper* 340: 1–87.

Boardman, R.S. & Buttler, C.J. 2005. Zooids and Extrazooidal Skeleton in the Order Trepostomata (Bryozoa). *Journal of Paleontology* 79: 1088–1104.

Bock, P. *Bryozoa Home Page; Family Arthrostylidae.* http://bryozoa.net/cryptostomida/arthrostylidae/index.html (accessed 3 July 2022)

Cohen, K.M., Harper, D.A.T., Gibbard, P.L. & Fan, J.-X. 2020. *The ICS International Chronostratigraphic Chart, version 2020/03*. International Commission on Stratigraphy. https://stratigraphy.org/ICSchart/ChronostratChart2020-03.pdf

Crockford, J.M. 1945. Stenoporids from the Permian of New South Wales and Tasmania. *Proceedings of the Linnean Society of New South Wales* 70: 9–24.

d'Orbigny, A. 1849. *Prodrome de Paléontologie Stratigraphique Universelle des Animaux Mollusques et Rayonnés faisant suite au Cours élémentaire de Paléontologie et géologie Stratigraphiques* Vol. 1, 1–394. Paris: Masson.

Ernst, A. & Carrera, M. 2008. Cryptostomid Bryozoans from the Sassito Formation, Upper Ordovician Cool-water Carbonates of the Argentinean Precordillera. *Palaeontology* 51: 1117–1127.

Girty, G.H. 1911. New Genera and Species of Carboniferous Fossils from the Fayetteville Shale of Arkansas. *Annals of the New York Academy of Science* 20: 189–238.

Gorjunova, R.V. 1985. Morphology, System and Phylogeny of Bryozoa (Order Rhabdomesida). *Trudy Paleontologicheskogo Instituta Akademii Nauk SSSR* 208: 1–152 [In Russian].

Gorjunova, R.V. 1992. Morphology and System of the Paleozoic bryozoans. *Trudy Paleontologischeskogo Instituta Akademiya Nauk SSSR* 251: 1–168 [In Russian].

Gorjunova, R.V. 1996. Phylogeny of the Paleozoic bryozoans. *Trudy Paleontologischeskogo Instituta* 267: 1–165 [In Russian].

Gorjunova, R.V. 2001. Bryozoa. In A.S. Alekseev & S. M. Shik (eds), *Middle Carboniferous of Moscow Syneclise (southern part). Volume 2. Biostratigraphy*: 80–91. Moscow: Scientific World [In Russian].

Gorjunova, R.V. 2011a. Carboniferous Bryozoans of the Order Rhabdomesida of the East European Platform. *Paleontological Journal* 45: 1–106.

Gorjunova, R.V. 2011b. Family Coelotubuliporidae Fam. Nov. and Morphological Parallelisms in the Evolution of Bryozoans. *Paleontological Journal* 45: 510–524.

Hu, Z. 1984. Triassic Bryozoa from Xizan (Tibet) with Reference to Their Biogeographical Provincialism in the World. *Acta Palaeontologica Sinica* 23: 568–577.

James, U.P. 1878. Descriptions of Newly Discovered Species of Fossils from the Lower Silurian Formation, Cinncinnati group. *The Palaeontologist* 1, 1–8.

Karklins, O.L. 1983a. Introduction to the Suborder Ptilodictyina. In R.A. Robison (ed.), *Treatise on Invertebrate Paleontology, Part G, Revised, Bryozoa*: 453–488. Boulder & Lawrence: Geological Society of America & University of Kansas Press.

Karklins, O.L. 1983b. Systematic Descriptions of the Suborder Ptilodictyina. In R.A. Robison (ed.), *Treatise on Invertebrate Paleontology, Part G, Revised, Bryozoa*: 489–529. Boulder & Lawrence: Geological Society of America & University of Kansas Press.

Key, M.M., Jr., Wyse Jackson, P.N. & Reid, C.M. 2022. Trepostome Bryozoans Buck the Trend and Ignore Calcite-aragonite Seas. *Palaeobiodiversity and Palaeoenvironments* 102: 253–263.

Ma, J., Buttler, C.J. & Taylor, P.D. 2020. The Last Known Cryptostome Bryozoan? *Tebitopora* from the Tibetian Triassic. In P.N. Wyse Jackson & K. Zagorsek (eds), *Bryozoan Studies 2019*: 91–98. Prague: Czech Geological Survey.

Marintsch, E.J. 1998. Systematic Paleontology, Biostratigraphy and Paleoecology of the Middle Ordovician Bryozoan (Trepostomata) from the Hermitage Formation of East-central Tennessee. *Bulletins of American Paleontology* 112: 1–121.

Morozova, I.P. 1960. Devonian Bryozoa of the Minusinsk and Kuznetsk Basins. *Trudy Paleontologischeskogo Instituta Akademiya Nauk SSSR* 86, 1–207 [In Russian].

Morozova, I.P., Weiss, O.B. & Racki, G. 2006. New Devonian and Carboniferous Bryozoans of the Holy Cross Mountains (Central Poland). *Paleontological Journal* 40: 529–540.

Nekhoroshev, V.P. 1956. Lower Carboniferous Bryozoa of the Altai and Siberia. *Trudy Vsesoyuznyy Nauchno-Issledovatelskiy Geologicheskiy Institut (VSEGEI)*, n. s., 13: 1–419 [In Russian].

Nekhorosheva, L.V. 2007. Results of Paratype Research on the Type Species *Cyclophaenopora* Spjeldnaes, 1984 (Bryozoa, Family Nematotrypidae). *Nature and Economy of Kuzbass. Regional proceedings*. Kuzbass Pedagogical Academy), Novokuznetsk: KuzGPA Publishing House, 11: 35–37 [In Russian].

Nicholson, H.A. 1879. *On the Structure and Affinities of the "Tabulate Corals" of the Paleozoic Period, with Critical Descriptions of Illustrative Species*. Edinburgh: William Blackwood and Sons.

Nikiforova, A.I. 1933. The Carboniferous Deposits of Central Asia: Contributions to the Knowledge of the Lower Carboniferous Bryozoa of Turkestan. *Trudy Vsesoyuznogo Geologorazvedochnogo Ob'edineniya* 207, 1–76.

Ozhgibesov, V.P. 1983. *Permoheloclema* – a New Bryozoan Genus of the Order Rhabdomesonida. *Paleontologicheskii Zhurnal* 1983 (4): 96–98.

Powers, C.M. & Pachut, J.F. 2008. Diversity and Distribution of Triassic Bryozoans in the Aftermath of the End-Permian Mass Extinction. *Journal of Paleontology* 82: 362–371.

Reid, C.M. 2021. A redescription of *Stenopora* and the Type Species *Stenopora tasmaniensis* Lonsdale, 1844 (Trepostomata, Bryozoa). *Journal of Paleontology* 95: 1147–1157.

Reid, C.M., Wyse Jackson, P.N. & Key, M.M., Jr. 2022. Latitudinal Influences on Bryozoan Calcification Through the Paleozoic. *Paleobiology*. doi: 10.1017/pab.2022.31

Schäfer, P. & Grant-Mackie, J.A. 1998. Revised Systematics and Palaeobiogeography of Some Late Triassic Colonial Invertebrates from the Pacific Region. *Alcheringa* 22: 87–122.

Schulga-Nesterenko, M.I. 1955. Carboniferous Bryozoa from the Russian Platform. *Trudy Paleontologischeskogo Instituta Akademiya Nauk SSSR* 58: 1–207 [In Russian].

Shishova, N.A. 1965. The Systematic Position and Size of the Family Hyphasmoporidae. *Paleontologicheskii Zhurnal* 1965 (2): 55–62 [In Russian].

Spjeldnaes, N. 1984. Upper Ordovician Bryozoans from Ojl Myr, Gotland, Sweden. *Bulletin of the Geological Institutions of the University of Uppsala, n.s.*, 10: 1–66.

Taylor, P.D. 1993. Bryozoa. In M.J. Benton (ed.), *The Fossil Record 2*: 477–489. London: Chapman & Hall.

Tolokonnikova, Z.A. 2009. Biogeography and Evolution of Bryozoans of the Genus *Minussina* Morozova, 1961. *Paleontological Journal* 43: 391–400.

Trizna, V.B. 1958. Early Carboniferous Bryozoans of the Kuznetz Depression. *Trudy Vsesoyuznyi Nauchno-Issledovatel'skii Geologo-Razvedochnyii Neftanoi* 122: 1–436 [In Russian].
Ulrich, E.O. 1882. American Palaeozoic Bryozoa. *Journal of the Cincinnati Society of Natural History* 5: 121–175, 233–257.
Ulrich, E.O. 1888. On *Sceptropora*, a New Genus of Bryozoa, with Remarks on *Helopora*, Hall, and Other Genera of That Type. *American Geologist* 1, 228–234.
Utgaard, J. 1983. Systematic Descriptions for the Order Cystoporata. In R.A. Robison (ed.), *Treatise on Invertebrate Paleontology, Part G, Revised, Bryozoa*: 358–439. Boulder & Lawrence: Geological Society of America & University of Kansas Press.
Vinassa de Regny, P.E. 1921. Sulla Classificazione Dei Trepostomata. Atti della Società italiana di scienze naturali, e del Museo civico di storia naturale. *Milano* 59: 212–231.
Vine, G.R. 1884. Fourth Report of the Committee, Consisting of Dr. H. C. Sorby and Mr. G. R. Vine, Appointed for the Purpose of Reporting on Fossil Bryozoa. In *Report of the British Association for the Advancement of Science (Southport, 1883)*: 161–209. London: John Murray.
Wyse Jackson, P.N. & Bancroft, A.J. 1985. Generic Revision of the Cryptostome Bryozoan *Rhabdomeson* Young and Young, 1874, with Descriptions of Two Species from the Lower Carboniferous of the British Isles. *Journal of Paleontology* 69: 28–45.
Wyse Jackson, P.N., Key, M.M., Jr., & Reid, C.M. 2020. Bryozoan Skeletal Index (BSI): A Measure of the Degrees of Calcification in Stenolaemate Bryozoans. In P.N. Wyse Jackson & K. Zagorsek (eds), *Bryozoan Studies 2019*: 193–206. Prague: Czech Geological Survey.
Yang, J. 1950. Some Bryozoans from Upper Devonian and Lower Carboniferous of Hunan. *Paleontological Society China* 6: 1–16 [In Chinese].
Yang, J., Hu, Z. & Xia, F. 1988. Bryozoans from Late Devonian and Early Carboniferous of Central Hunan. *Palaeontologia Sinica* 174, Series B (23): 1–197.
Young, J. & Young, J. 1874. On a New Genus of Carboniferous Polyzoa. *Annals and Magazine of Natural History, series 4*, 13: 335–339.

Bryozoan Studies 2022 – Key, Porter & Wyse Jackson (eds)

Volgia (Bryozoa, Cystoporata): a rare occurrence from the Mississippian of Ireland

Patrick N. Wyse Jackson*
Department of Geology, Trinity College, Dublin, Ireland
ORCID ID: 0000-0001-5605-0670

Andrej Ernst
Institut für Geologie, Universität Hamburg, Hamburg, Germany
ORCID ID: 0000-0002-1299-2559

Malgorzata Shaikh-Horajska
Kerry Geo, Tralee, Co. Kerry, Ireland

John Murray
Earth & Ocean Sciences, School of Natural Sciences, University of Galway, Galway, Ireland
ORCID ID: 0000-0003-4640-4214

ABSTRACT: The Mississippian (Viséan) Oyster Hall Breccia Formation of Co. Kerry in southwestern Ireland has yielded a moderately diverse bryozoan fauna containing 14 taxa including two cryptostomes, seven fenestrates, three trepostomes and two cystoporates. The latter includes the first record of the distinctive hexagonellid genus *Volgia* Stuckenberg, 1905 from outside the Pennsylvanian type area in Russia or the Mississippian of mainland Europe. Preservation of a single colony, while not complete, is sufficient to allow a reconstruction of the overall cruciform zoarial form. It comprises four slender primary branches that emerge orthogonally to each other from a hollow basal attachment area, and which themselves further quadrifurcate producing another series of four short secondary branches.

1 INTRODUCTION

The Mississippian rocks in Ireland, formed between 359.3 and 323.4 Ma (Aretz *et al.* 2020), have been the subject of considerable study over several centuries and much is now known of the stratigraphic succession and palaeoenvironments they represent (Sevastopulo & Wyse Jackson 2009). Bryozoans comprise an important element of the marine benthic fauna from this time interval, but despite being the subject of studies since the mid-nineteenth century (e.g. M'Coy 1844; Phillips 1836) and subsequently in both the twentieth and twenty-first centuries (e.g. Bancroft & Wyse Jackson 1995; Bancroft 1985, 1986, 1988a; Hageman *et al.* 2011; Miller 1961; Olaloye 1974; Owen 1973; Tavener-Smith 1973; Wyse Jackson & Bancroft 1995; Wyse Jackson 1996; Wyse Jackson *et al.* 2006, 2017), on-going fieldwork continues to reveal new fossil taxa in Ireland, as reported herein.

*Corresponding Author: wysjcknp@tcd.ie

2 GEOLOGICAL SETTING AND BRYOZOAN FAUNA

2.1 *Geological Setting*

A unique specimen of the cystoporate stenolaemate *Volgia* Stuckenberg, 1905 described herein was recovered from upper Viséan strata outcropping at Kilfenora foreshore, which is located c. 2 km east from Fenit harbour in Co. Kerry (Figure 1A-B). The bedrock of Fenit peninsula consists of Carboniferous sediments deposited in the southern part of the Shannon Basin (Murray 2010). Carbonate sedimentation predominated in the basin during the Tournaisian and Viséan, with at least two phases of carbonate mudbank development (e.g., Hudson *et al.* 1966; Lees & Miller 1995; Sevastopulo & Wyse Jackson 2009; Thornton 1966; for Fenit specifically see Labiaux 1997). A relatively rapid (but diachronous) basin-wide switch from carbonate to siliciclastic sedimentation occurred subsequently during the Serpukhovian, marked initially by the deposition of organic-rich black shales. These are regionally

DOI: 10.1201/9781003454915-19

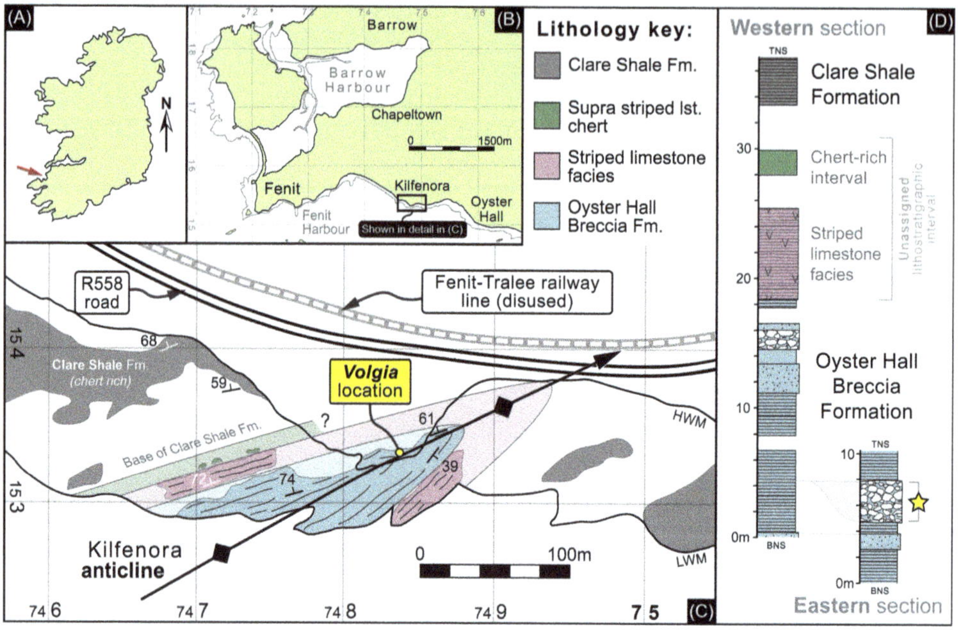

Figure 1. Geographical and geological setting of the *Volgia* bryozoan locality at Kilfenora. (A) Location of Tralee Bay in southwestern Ireland (red arrow); (B) Kilfenora location on Fenit peninsula. Note that the location of Oyster Hall further to the east along the coast – the type section for the Oyster Hall Breccia Formation is located on the foreshore there. The location of the geological map in (C) is also indicated; (C) Outcrop geological map of the Kilfenora anticline structure, which plunges towards east-northeast. The sample location for the *Volgia* specimen is shown (yellow circle); (D) Composite Mississippian stratigraphy for the sample location at Kilfenora. Both of these sections were measured on the northern limb of the Kilfenora anticline. The yellow star (eastern section) shows the large lens/channel fill breccia unit where the *Volgia* specimen was recovered. Geographic grid shown in (B) and (C) is Irish National Grid. Modified from Murray (2010).

mapped as the Clare Shale Formation. They range in age from the Serpukhovian to the Bashkirian and are generally interpreted as representing dysoxic to anoxic bottom water conditions (e.g., Fallon & Murray 2015; Hodson & Lewarne 1961; Sevastopulo 2009; Wignall & Best 2000; see also O'Sullivan *et al.* 2021).

The stratigraphic section at Kilfenora records this switch in sedimentation style within the Shannon Basin. A plunging anticline brings the upper part of the Viséan carbonate sequence to outcrop (Figure 1C), beginning with a series of interbedded dark cherty limestones and shales. These well-bedded units are deep-water in character and conspicuously incised by a lensoidal unit of dark argillaceous matrix-supported breccia (Figure 1D). The latter contains limeclasts (Figure 2) (observed on all scales), along with well-preserved silicified bioclasts, and may represent a mass-flow deposit (Murray 2010). Very similar brecciated carbonate facies outcrops c. 1.25 km further east at Oyster Hall (Figure 1B), which has also yielded a diverse silicified fauna, including bryozoans (Figure 2). Murray (2010) referred these units to the Oyster Hall Breccia Formation and noted the presence of the conodont *Gnathodus bilineatus* in the lower part of the eastern section at Kilfenora, supporting a late Viséan age (no older than mid Asbian; see Barrick *et al.* 2020; Varker & Sevastopulo 1985) for this part of the stratigraphy. The *Volgia* specimen was recovered a short distance above this level in the prominent lensoidal sedimentary breccia unit (see Figure 1D) and had thus evidently been reworked.

A distinctive carbonate laminite facies ('striped limestone') overlies the Oyster Hall Breccia Formation at Kilfenora (Figure 1C-D). This generally unfossiliferous facies is quite widely developed in the Shannon Basin; it first occurs around the level of the Asbian/Brigantian regional substage boundary and may represent a phase of basinal restriction and/or hypersalinity (e.g., MacDermot & Sevastopulo 1972; Murray 2010; Sevastopulo & Wyse Jackson 2009; Sleeman & Pracht 1999). The Clare Shale Formation outcrops a relatively short distance above this stratigraphic level on either side of the Kilfenora anticline (Figure 1C). The actual contact with the underlying carbonate-dominant sequence is not exposed. However, dark chert-rich limestones sporadically outcrop in the transition zone between the two (Figure 1C-D).

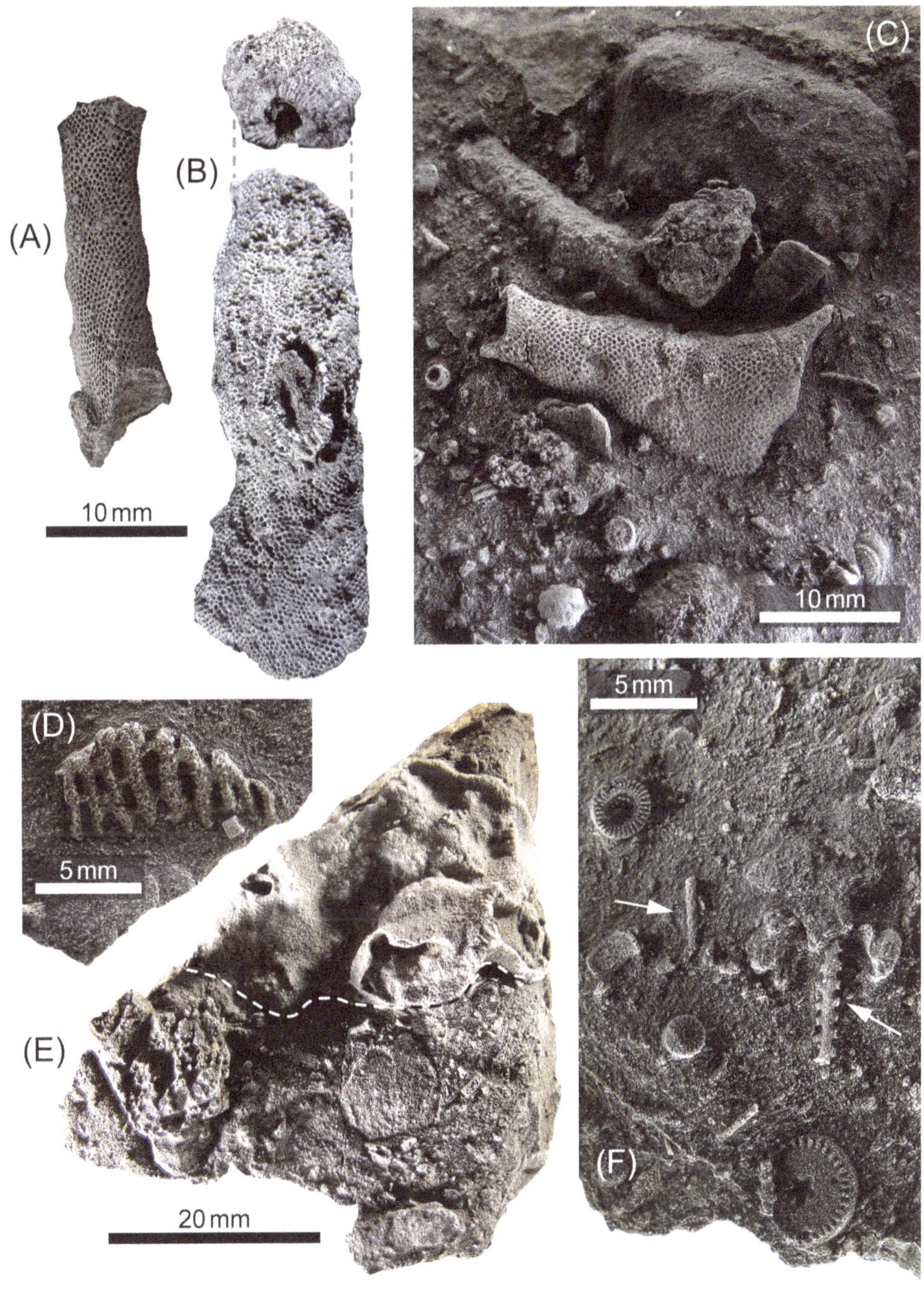

Figure 2. Silicified bryozoans and associated brecciated limestone fabric, Oyster Hall foreshore, Fenit peninsula, Co. Kerry (see Figure 1B for general location). (A) *Stenophragmidium* sp.; (B) *Tabulipora urii*, transverse view (top), longitudinal surface view (bottom); (C) *Tabulipora urii*, disarticulated crinoidal debris, and brecciated carbonate limeclast (top); (D) *Polypora dendroides*; (E) *Tabulipora howsii* (margins of colony are indicated by dashed white line); (F) *Baculopora megastoma* (left arrow), *Penniretepora flexicarinata* (right arrow) and associated crinoidal ossicles. TCD.60427 = (A), (D) and (F); TCD.60428 = (B); TCD.60429 = (C); TCD.60430 = (E). Adapted from Murray (2010).

2.2 *Bryozoan fauna*

The sedimentary breccias at Oyster Hall and Kilfenora have yielded a moderately diverse suite of 14 bryozoan taxa and an example of the enigmatic palaeocorynid-type appendage (Table 1). The majority of these taxa are common and well-known from the Viséan of Ireland and Britain (Olaloye 1974; Tavener-Smith 1973; Wyse Jackson 1996). As many of these have been documented in detail, they are not redescribed here. Palaeocorynid-type appendages are outgrowths from the surfaces of fenestellid bryozoans (Bancroft 1988b; González-Mora *et al.* 2020) that were once considered to be independent organisms associated with the bryozoans and given various generic names. These have no taxonomic status.

Table 1. Stenolaemate bryozoans recovered from the Viséan of Co. Kerry, Ireland.

Order Cryptostomata (2 taxa): *Rhabdomeson progracile* Wyse Jackson & Bancroft, 1995 and *Rhabdomeson rhombiferum* (Phillips, 1836).

Order Fenestrata (7 taxa): *Baculopora megastoma* (M'Coy, 1844) [Figure 2F], *Penniretepora flexicarinata* Young & Young, 1875 [Figure 2F], *Penniretepora tortuosa* Olaloye, 1974, *Fenestella s.l.* sp., *Rectifenestella frutex* (M'Coy, 1844), *Polypora dendroides* M'Coy, 1844 [Figure 2D], *Polypora* sp. and Palaeocorynid-type appendage.

Order Trepostomata (3 taxa): *Stenophragmidium* sp. [Figure 2A], *Tabulipora urii* (Fleming, 1828) [Figures 2B-C] and *Tabulipora howsii* (Nicholson, 1881) [Figure 2E].

Order Cystoporata (2 taxa): *Fistulipora incrustans* (Phillips, 1836) and *Volgia* sp. [Figures 3–5].

Of these, *Penniretepora tortuosa* Olaloye, 1974 is noteworthy having only previously been reported from coeval successions in Co. Fermanagh (Olaloye 1974), as is *Volgia* sp. which has not been recorded previously in Carboniferous sequences from either Ireland or Britain. As a result, the latter taxon is described in more detail below (Figures 3 and 4).

Volgia is a distinctive member of the Family Hexagonellidae Crockford, 1947 that bears cruciform branches. It has a very limited geographical and stratigraphical distribution. The type species *V. arborescens* (Stuckenberg 1895) occurs in the Pennsylvanian of Russia and only recently a second species, *V. deftera* Ernst, Wyse Jackson & Aretz, 2015, was described from younger Mississippian (Viséan) strata of Roque Rodonde, southern France and the Viséan of Germany (Ernst *et al.* 2015).

2.3 *Repository and institutional abbreviation*

Specimens examined in this study are deposited in the collections of the Geological Museum, Trinity College, Dublin (Accession numbers: TCD.60426–60432).

Figure 3. Field photograph of *Volgia* specimen (TCD.60426) *in situ* (circled) to right of hammer at Kilfenora (see Figure 1 for location). Note the brecciated texture of the enclosing host rock, with paler coloured limeclasts set within a darker fine-grained carbonate matrix.

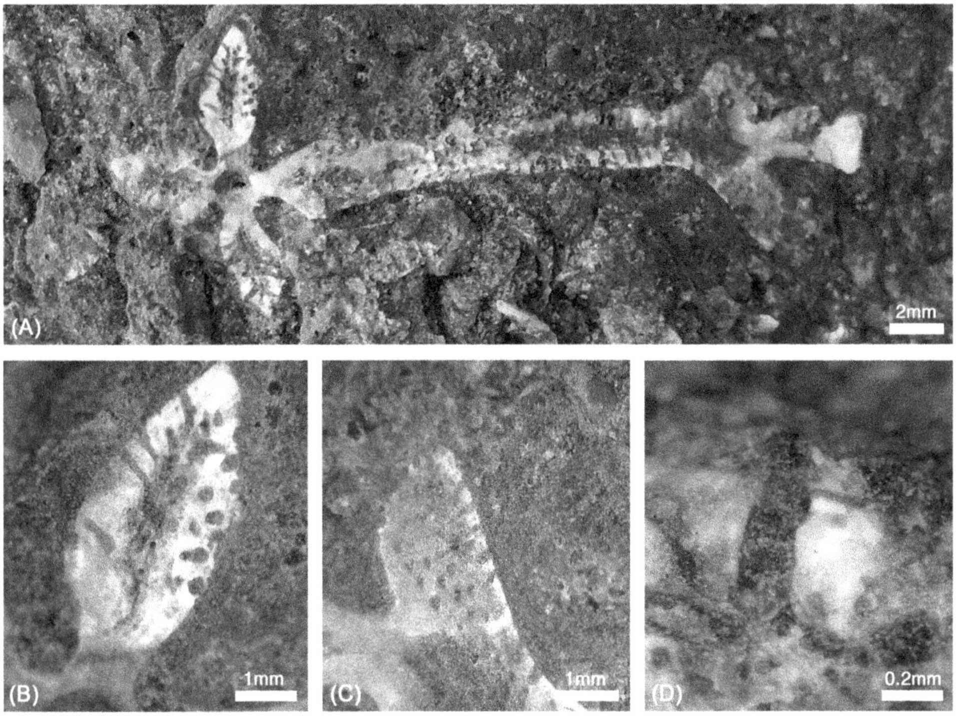

Figure 4. *Volgia* sp. TCD.60426 from Kilfenora, Co. Kerry, Ireland. (A) View of cruciform colony with central hollow base; (B) Cross-section through one branch showing sharp carina and rounded underside of branch with central mesotheca, vesicles in endozone and thickened exozone; (C) Branch surface with oval autozooecial apertures arranged in linear rows; (D) Tangential section through autozooecial chamber recumbent on mesotheca and extending through exozone.

3 SYSTEMATIC PALAEONTOLOGY

Order Cystoporata Astrova, 1964
Suborder Fistuliporina Astrova, 1964
Family Hexagonellidae Crockford, 1947
Genus *Volgia* Stuckenberg, 1905
(= *Ramiporina* Schulga-Nesterenko, 1933)

Type species. Coscinium arborescens Stuckenberg, 1895 from the Pennsylvanian of Samarskaya Luka, Samara Oblast, Russia. Original designation Stuckenberg (1905, p. 120).

Original diagnosis. "*Volgia* nov. gen. The solid skeletal parts of the bryozoans belonging to this genus have the shape of whorled stalks, which assumed a vertical position during the life of the animal. They are laterally flattened and widen towards the apex. Their cross-sections are ellipsoidal, but rather strongly thickened. Each stalk divides at the top into four new ones, which are completely similar to the crest. The new stalks are whorled and divided lengthwise by fine lamellae into two equal parts, which, as in the genus *Coscinium*, serve as a base for small tubular cells that open on both sides of the stalk into rounded calyxes. The cells are connected to each other by dense calcareous tissue. I include in this genus a species already described by me (*Coscinium arborescens*), which I had assigned to the genus *Coscinium* Keyserling. At present, however, I can add another species that I discovered in the Coal Limestone at the Zarev Kurgan." (Stuckenberg 1905, p. 120).

Diagnosis. "Colony consisting of erect bifoliate lobes, with secondary branches diverging at right angles in crucifix-shaped pattern. Mesotheca with a dark and thin middle layer and two light and thick outer layers, containing median tubuli and longitudinal ridges parallel to growth direction. Autozooecia recumbent on the mesotheca or epitheca for a long distance, then bending upwards abruptly or gently, intersecting the surface almost perpendicularly. Hemisepta absent. Thin, complete diaphragms in autozooecia, common to abundant. Autozooecial apertures circular to oval, with thick peristomes. Lunaria obscure. Autozooecia separated by vesicular skeleton in endozone and by thick stereom in exozone. Microacanthostyles (tubuli) in stereom. Maculae lacking." (Ernst *et al.* 2015, p. 162).

Discussion. Whorling of the colony branches noted by Stuckenberg (1905) is not observed in the

Kerry specimen nor by Ernst *et al.* (2015) and this morphological feature is not considered herein to be of generic value. The genus has also been reported from the Permian of Xizang (Tibet) and Beishan-Tianshan provinces in China (Yang & Lu 1981), but these occurrences remain unsubstantiated, and the upper range limit is taken to be Pennsylvanian.

Schulga-Nesterenko (1933, p. 58) placed *V. arborescens* into the new subgenus *Ramiporina* on the basis of her contention that Stuckenberg (1895) mistook vesicular tissue for autozooecia and that the species resembled material from the Pennsylvanian of the district of Pechora in northwestern Russia. Subsequently this subgenus was suppressed as it was demonstrated that *Ramiporina* had been applied to material identical to *Coscinium arborescens* Stuckenberg, 1895 which he, in 1905, had selected as the type species of *Volgia* (Nikiforova 1938). *Ramiporina* is a junior synonym of *Volgia* (Nikiforova 1938; Nakrem & Spjeldnaes 1995, p. 831; Utgaard 1983).

Geographical distribution. Russia, France, Germany and Ireland

Geological range. Mississippian (Viséan) to Pennsylvanian (probably Gzhelian).

Volgia sp.
Figures 3, 4, 5

Material. One specimen (TCD.60426) from the Oyster Hall Breccia Formation, Viséan, Mississippian; Kilfenora, Fenit Peninsula, north-side of Tralee Bay, Co. Kerry, Ireland (Figure 1).

Description. Colony comprises four bifoliate branches diverging orthogonally in a cruciform pattern from a hollow cylindrical base 1.3 mm in diameter. In this specimen, only one primary branch of any significant length (27 mm) is preserved. Secondary branches develop after 18 mm and diverge at a 90° angle to the main stem. Only two secondary branches in the plane of preservation are observed, but two other branches may have been developed perpendicular to the primary branch at this point which replicated the cruciform pattern. Branches are ovate in cross-section averaging 2.4 mm in width laterally and 4.3 mm from the rounded margin on the proximal side to the sharp carinal crest on the distal side opposite.

Autozooecial apertures arranged in four to five rows either side of branches, circular to oval in shape, 0.14 mm by 0.20 mm in diameter, with weak indistinct lunaria, spaced 0.15 mm apart. Interapertural surface smooth, lacking spines, ornamentation or heterozooecia.

Internally, branches are divided by a thin, straight mesotheca 0.1 mm in thickness. Exozone comprises dense stereom lacking styles, 0.8 mm wide on either side of branches; endozone 0.7 mm in width. Autozooecia budded from mesotheca, recumbent for 0.25 mm in initial phase becoming orientated at a c. 60° angle to the branch surface. Autozooecial chambers in exozone 0.5 mm in length, parallel sided, thin-walled and lacking diaphragms.

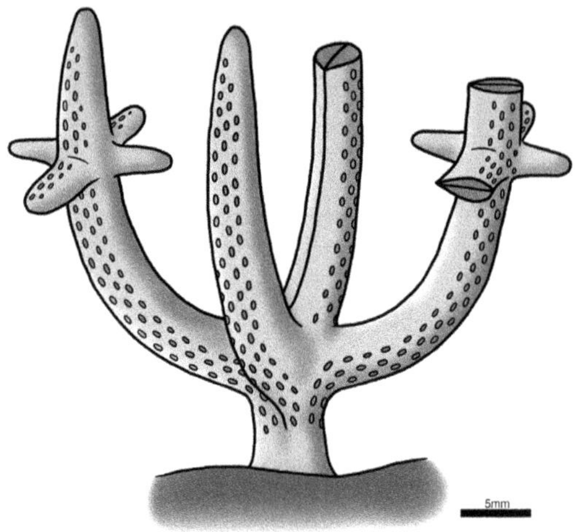

Figure 5. Reconstruction of colony form of *Volgia* based on specimen TCD.60426. For clarity, secondary branches are only shown developed on two of the primary branches rather than on all four and not all autozooecial apertures on the secondary branches are shown.

4 DISCUSSION

To date only three species of *Volgia* have been described: the type species *V. arborescens* (Stuckenberg 1895) and *V. minima* Stuckenberg, 1905 from the Pennsylvanian of Russia (Nikiforova 1938; Schulga-Nesterenko 1933; Stuckenberg 1895, 1905) and *V. deftera* Ernst *et al.*, 2015 from the Viséan of Roque Redonde, Montagne Noire in France and the Rhenish Mountains in Germany (Ernst 2005). However, *V. minima* was shown by Nikiforova (1938, p. 280) to be conspecific with *V. arborescens* and so only two species are currently accepted.

The specimen from Kilfenora, Co. Kerry (Figures 3–5) is similar to *V. arborescens* in terms of autozooecial apertural size and in lacking metazooecia but differs from *V. deftera* in which the apertures are larger and metazooecia developed. Detailed internal features cannot be obtained through utilisation of either thin sectioning or acetate peels owing to the irregular eroded surface of this

Irish specimen. As the branch is very thin it is likely that it would be lost through grinding down and polishing the surface required to produce sections or peels. As a result this taxon is left in open nomenclature.

The diversity of late Viséan bryozoans from Ireland is high, with 72 species in 30 genera reported from a number of localities, including Carrick Lough and Sillees River, Co. Fermanagh and Bundoran, Co. Donegal (Wyse Jackson & Buttler, 1994). However, the number of cystoporates recorded to date from this time interval is low, with only four taxa recognised: *Fistulipora incrustans* (Phillips, 1836), *Goniocladia cellulifera* (Etheridge, 1873) and *Sulcoretepora parallela* (Phillips, 1836) all recorded from Co. Fermanagh (Bancroft & Wyse Jackson, 1995; Wyse Jackson, 1996) and *Prismopora* sp. from Co. Armagh (see Wyse Jackson *et al.* 1999, p. 260). *Sulcoretepora* ? *ramosa* Owen, 1973 described from the Viséan of Tullaghoge, Co. Tyrone (Owen, 1973) was reassigned to the cryptostome *Clausotrypa* by Wyse Jackson (1996, p. 137).

5 CONCLUSIONS

This first record of *Volgia* sp. from the Mississippian of Co. Kerry constitutes only the fifth cystoporate bryozoan to be recorded from Viséan strata in Ireland, and it extends the geographical range of this particular genus to Ireland.

The preservation of the specimen allows for reconstruction of the unusual cruciform colony form in which from a central hollow base four primary branches are budded and which each quadrifurcate to form secondary branches. Most hexagonellid cystoporates colonies are composed of bifurcating bifoliate branches, but in some, such as *Prismopora* Hall, 1883 or *Scalaripora* Hall, 1883 branches trifurcate, while in *Evactinopora* Meek & Worthen, 1865 or *Evactinostella* Crockford, 1947 six to seven short lateral branches develop from a central point (Yancey *et al.* 2019). *Volgia* is unique in amongst cystoporates in that it exhibits a cruciform branching habit.

ACKNOWLEDGEMENT

The authors are most grateful to Susanna and Katie Wyse Jackson for their joint skillful reconstruction of *Volgia* illustrated in Figure 5.

REFERENCES

Aretz, M., Herbig, H.G. & Wang, X.D. 2020. The Carboniferous Period. In F.M. Gradstein, J.G. Ogg, M.D. Schmitz & G.M. Ogg (eds), *The Geological Time Scale 2020*, Volume 2: 811–873. Amsterdam: Elsevier.

Astrova, G.G. 1964. A New Order of Paleozoic Bryozoa. *Paleontologičeskii Zhurnal* 2: 22–31 [in Russian].

Bancroft, A.J. 1985. The Carboniferous Fenestrate Bryozoan *Ptylopora pluma* M'Coy. *Irish Journal of Earth Sciences* 7: 35–45.

Bancroft, A.J. 1986. The Carboniferous Fenestrate Bryozoan *Hemitrypa hibernica*. *Irish Journal of Earth Sciences* 7: 111–124.

Bancroft, A.J. 1988a. Ovicells in *Rectifenestella* from the Lower Carboniferous of Ireland. *Irish Journal of Earth Sciences* 9: 19–22.

Bancroft, A.J. 1988b. Palaeocorynid-type Appendages in Upper Palaeozoic Fenestellid Bryozoa. *Palaeontology* 31: 665–675.

Bancroft, A.J. & Wyse Jackson, P.N. 1995. Revision of the Carboniferous Cystoporate Bryozoan *Fistulipora incrustans* (Phillips, 1836), with Remarks on the Type Species of *Fistulipora* M'Coy, 1849. *Geological Journal* 30: 129–143.

Barrick, J.E., Alekseev, A.S., Blanco-Ferrera, S., Goreva, N.V., Hu, K., Lambert, L.L., Nemyrovska, T.I., Qi, Y., Ritter, S.M. & Sanz-López, J. 2020. Carboniferous Conodont Biostratigraphy. In S.G. Lucas, J.W. Schneider, X. Wang & S. Nikolaeva (eds), *The Carboniferous Timescale*, Geological Society, London, Special Publications 512: 695–768.

Crockford, J. 1947. Bryozoa from the Lower Carboniferous of New South Wales and Queensland. *Proceedings of the New South Wales Linnean Society* 72: 1–48.

Ernst, A. 2005. Lower Carboniferous Bryozoa from Some Localities in Sauerland, Germany. In H.I. Moyano, J.M. Cancino & P.N. Wyse Jackson (eds), *Bryozoan Studies 2004*: 49–62. Rotterdam: A.A. Balkema Publishers.

Ernst, A., Wyse Jackson, P.N. & Aretz, M. 2015. Bryozoan fauna from the Mississippian (Viséan) of Roque Redonde (Montagne Noire, Southern France). *Geodiversitas* 37: 151–213.

Etheridge, R., jr. 1873. Description of *Carinella*, a New Genus of Carboniferous Polyzoa. *Geological Magazine* 10: 433–434.

Fallon, P. & Murray, J. 2015. Conodont Biostratigraphy of the Mid-Carboniferous Boundary in Western Ireland. *Geological Magazine* 152: 1025–1042.

Fleming, J. 1828. *A History of British Animals*. Edinburgh: Bell and Bradfute, pp. xxiii+565.

Gonzalez-Mora, S., Wyse Jackson, P.N., Bancroft, A.J. & Sour-Tovar, F. 2020. Palaeocorynid-type Structures in Fenestellid Bryozoa from the Carboniferous of Oaxaca, Mexico. In P.N. Wyse Jackson & K. Zágoršek (eds), *Bryozoan Studies 2019*: 35–42. Prague: Czech Geological Survey.

Hageman, S.J., Wyse Jackson, P.N., Abernethy, A.R. & Steinthorsdottir, M. 2011. Calendar Scale, Environmental Variation Preserved in the Skeletal Phenotype of a Fossil Bryozoan (*Rhombopora blakei* n. sp.), from the Mississippian of Ireland. *Journal of Paleontology* 85: 853–870.

Hall, J. 1883. Bryozoans of the Upper Helderberg and Hamilton Groups. *Transactions of the Albany Institute* 10: 145–197.

Hodson, F. & Lewarne, G.C. 1961. A Mid-Carboniferous (Namurian) Basin in Parts of the Counties of Limerick and Clare, Ireland. *Quarterly Journal of the Geological Society, London* 117: 307–333.

Hudson, R.G.S., Clarke, M.J. & Brennand, T. 1966. The Lower Carboniferous Stratigraohy of the Castleisland Area, Co. Kerry, Ireland. *Scientific Proceedings of the Royal Dublin Society* A2: 292–317.

Labiaux, S. 1997. *Sponges in Waulsortian-type Mudmounds at Tralee Bay, Co. Kerry, Southwest Ireland. Facies* 36: 253–257.

Lees, A. & Miller, J. 1995. Waulsortian Banks. In C.L.V. Monty, D.W.J. Bosence, P.H. Bridges & B.R. Pratt (eds), *Carbonate Mud-mounds: Their Origin and Evolution*. Special Publication of the International Association of Sedimentologists, Blackwell Science, Oxford 23: 191–271.

M'Coy, F. 1844. *A Synopsis of the Characters of the Carboniferous Limestone Fossils of Ireland*. Dublin: Dublin University Press, pp. 207.

MacDermot, C.V. & Sevastopulo, G.D. 1972. Upper Devonian and Lower Carboniferous Stratigraphical Setting of Irish Mineralisation. *Geological Survey of Ireland Bulletin* 1: 267–280.

Meek, F.B. & Worthen, A.H. 1865. Descriptions of New Crinoidea, &c., from the Carboniferous Rocks of Illinois and Some of the Adjoining States. *Proceedings of the Academy of Natural Sciences of Philadelphia* 17: 155–166.

Miller, T.G. 1961. New Irish Tournaisian Fenestellids. *Geological Magazine* 98: 493–500.

Murray, J. 2010. *Mid to Upper Viséan Facies and Palaeoenvironments of the Shannon Basin, Western Ireland.* Unpublished Ph.D. thesis, University of Dublin, Trinity College. http://www.tara.tcd.ie/handle/2262/78115.

Nakrem, H.A. & Spjeldnaes, N. 1995. *Ramipora hochstetteri* Toula, 1875 (Bryozoa, Cystoporata) from the Permian of Svalbard. *Journal of Paleontology* 69: 831–838.

Nicholson, H.A. 1881. *On the Structure and Affinities of the Genus Monticulipora and its Sub-genera with Critical Descriptions of Illustrative Species.* Edinburgh and London: Blackwood, pp. 240.

Nikiforova, A. 1938. Types of Carboniferous Bryozoa of the European Part of the USSR. *Paleontology of USSR* 4(5): 1–290 [in Russian with English summary].

Olaloye, F. 1974. Some *Penniretepora* (Bryozoa) from the Viséan of County Fermanagh with a Revision of the Generic Name. *Proceedings of the Royal Irish Academy* 74B; 471–506.

O'Sullivan, G.J., Daly, J.S., Murray, J., Ó Gogáin, A., Chew, D.M., Drakou, F., Guyett, P.C., Badenszki, E. & Hoare, B.C. 2021. Uranium-lead Phosphate Chronostratigraphy: A Proof of Concept from the Mid-Carboniferous Boundary. *Sedimentary Geology* 422: 105961.

Owen, D.E. 1973. Carboniferous Bryozoa from County Tyrone. *Geological Journal* 8: 297–306.

Phillips, J. 1836. *Illustrations of the Geology of Yorkshire, Part 2. The Mountain Limestone District.* London: John Murray, pp. xx+253.

Schulga-Nesterenko, M.I. 1933. Bryozoa from the Coal-bearing and Subjacent Series of Pechora Land: *Goniocladia* Etheridge and *Ramipora* Toula, Carboniferous and Permian Representatives of the Family Cystodictyonidae. *Trudy Vsesoyuznogo Geologo-razvedochnogo Obyedinenitya NKTP USSR* 259: 1–63 [in Russian].

Sevastopulo, G.D. 2009. Carboniferous: Mississippian (Serpukhovian) and Pennsylvanian. In C.H. Holland & I.S. Sanders (eds), *The Geology of Ireland* (2nd edition): 269–294. Edinburgh: Dunedin Academic Press.

Sevastopulo, G.D. & Wyse Jackson, P.N. 2009. Carboniferous: Mississippian (Tournaisian and Viséan). In C.H. Holland & I.S. Sanders (eds), *The Geology of Ireland* (2nd edition): 215–268. Edinburgh: Dunedin Academic Press.

Sleeman, A.G. & Pracht, M. 1999. *Geology of the Shannon Estuary. A Geological Description of the Shannon Estuary Region Including Parts of Clare, Limerick and Kerry, to Accompany the Bedrock Geology 1:100,000 Scale Map Series, Sheet 17, Shannon Estuary.* Dublin: Geological Survey of Ireland, pp. vii+77.

Stuckenberg, A.A. 1895. Corals and Bryozoans from the Carboniferous Sediments in the Urals and Timan. *Trudy Geologičeskogo Komiteta* 10(3): 1–244 [in Russian].

Stuckenberg, A.A. 1905. Fauna of the Upper Carboniferous Strata of the Samara Bend. *Trudy Geologičeskogo Komiteta New Series* 23: 1–144 [in Russian with German summary].

Tavener-Smith, R. 1973. Fenestrate Bryozoa from the Viséan of County Fermanagh, Ireland. *Bulletin of the Natural History Museum, London (Geology)* 23: 389–493.

Thornton, M.S. 1966. *The Lower Carboniferous Limestones of the Tralee Bay area, Co. Kerry, Ireland.* Unpublished Ph. D. Thesis, University of Cambridge.

Utgaard, J. 1983. Systematic Descriptions for the Order Cystoporata. In R.A. Robison (ed.), *Treatise on Invertebrate Paleontology, Part G, Bryozoa, Revised*: 358–439. Boulder and Lawrence: Geological Society of America and University of Kansas.

Varker, W.J. & Sevastopulo, G.D. 1985. The Carboniferous System: Part 1—Conodonts of the Dinantian Subsystem from Great Britain and Ireland. In A.C. Higgins & R.L. Austin (eds), *A Stratigraphical Index of Conodonts*: 167–209. Chichester: Ellis Horwood.

Wignall, P.B. & Best, J.L. 2000. The Western Irish Namurian Basin Reassessed. *Basin Research* 12: 59–78.

Wyse Jackson, P.N. 1996. Bryozoa from the Lower Carboniferous (Viséan) of County Fermanagh, Ireland. *Bulletin of the Natural History Museum, London (Geology)* 52: 119–171.

Wyse Jackson, P.N. & Bancroft, A.J. 1995. Generic Revision of the Cryptostome Bryozoan *Rhabdomeson* Young and Young, 1874, with Descriptions of Two Species from the Lower Carboniferous of the British Isles. *Journal of Paleontology* 69: 28–45.

Wyse Jackson, P.N. & Buttler, C.J. 1994. Fossil Bryozoan Localities in Ireland. In P.J. Hayward, J.S. Ryland & P.D. Taylor (eds), *Biology and Palaeobiology of Bryozoans:* 219–222. Fredensborg: Olsen & Olsen.

Wyse Jackson, P.N., Ernst, A. & Suárez Andrés, J.L. 2017. Articulation in the Family Rhabdomesidae (Cryptostomata: Bryozoa) from the Mississippian of Ireland. *Irish Journal of Earth Sciences* 35: 35–44.

Wyse Jackson, P.N., McKinney, F.K. & Bancroft, A.J. 2006. Fenestrate Bryozoan Genera Based on Species from Ireland Originally Described by Frederick M'Coy in 1844. *Palaeontology* 49: 741–767.

Wyse Jackson, P.N., Taylor, P.D. & Tilsley, J.W. 1999. The 'Balladoole Coral' Re-interpreted as the Unusual Cystoporate Bryozoan *Meekoporella* Moore and Dudley, 1944 from the Lower Carboniferous of the British Isles. *Proceedings of the Yorkshire Geological Society* 52: 257–268.

Yancey, T.E., Wyse Jackson, P.N., Sutton, B. & Gottfried, R.J. 2019. Evactinoporidae, a New Family of Cystoporata (Bryozoa) from the Mississippian of North America: Growth and Functional Morphology. *Journal of Paleontology* 93(6): 1058–1074.

Yang, J-Z & Lu, L. 1981. Geographical and Stratigraphical Distributions of Permian Bryozoa in China. In G.P. Larwood & C. Nielsen (eds), *Recent and Fossil Bryozoa*: 305–308. Fredensborg: Olsen & Olsen.

Young, J. & Young, J. 1875. New Species of *Glauconome* from Carboniferous Limestone Strata of the West of Scotland. *Proceedings of the Natural History Society of Glasgow* 2: 325 335.

Bryozoan Studies 2022 – Key, Porter & Wyse Jackson (eds)
© 2023 The Editors, ISBN: 978-1-032-59487-3

Author index

www.ingramcontent.com/pod-product-compliance
Ingram Content Group UK Ltd.
Pitfield, Milton Keynes, MK11 3LW, UK
UKHW052210190926
475702UK00006B/952
* 9 7 8 1 0 3 2 5 9 4 8 7 3 *